Chromosomal Proteins and Gene Expression

NATO ASI Series

Advanced Science Institutes Series

A series presenting the results of activities sponsored by the NATO Science Committee, which aims at the dissemination of advanced scientific and technological knowledge, with a view to strengthening links between scientific communities.

The series is published by an international board of publishers in conjunction with the NATO Scientific Affairs Division

A	**Life Sciences**	Plenum Publishing Corporation
B	**Physics**	New York and London
C	**Mathematical and Physical Sciences**	D. Reidel Publishing Company Dordrecht, Boston, and Lancaster
D	**Behavioral and Social Sciences**	Martinus Nijhoff Publishers
E	**Engineering and Materials Sciences**	The Hague, Boston, and Lancaster
F	**Computer and Systems Sciences**	Springer-Verlag
G	**Ecological Sciences**	Berlin, Heidelberg, New York, and Tokyo

Recent Volumes in this Series

Series A: Life Sciences

Chromosomal Proteins and Gene Expression

Edited by

Gerald R. Reeck

Kansas State University
Manhattan, Kansas

Graham H. Goodwin

Royal Cancer Hospital
London, England

and

Pedro Puigdomènech

Institute of Biology
Barcelona, Spain

Plenum Press
New York and London
Published in cooperation with NATO Scientific Affairs Divison

Proceedings of a NATO Advanced Study Institute on
Chromosomal Proteins and Gene Expression,
held September 17–26, 1984,
in Sitges, Spain

Library of Congress Cataloging in Publication Data

NATO Advanced Study Institute on Chromosomal Proteins and Gene Expression
(1984: Sitges, Spain)
  Chromosomal proteins and gene expression.

  (NATO ASI series. Series A, Life sciences; vol. 101)
  "Proceedings of a NATO Advanced Study Institute on Chromosomal Proteins
and Gene Expression, held September 17–26, 1984, in Sitges, Spain"—Verso of
t.p.
  Bibliography: p.
  Includes index.
  1. Gene expression—Congresses. 2. Chromosomal proteins—Congresses. I.
Reeck, Gerald R. II. Goodwin, Graham H. III. Puigdomenech, Pedro. IV. North
Atlantic Treaty Organization. Scientific Affairs Division. V. Title. VI. Series: NATO
advanced scientific institutes series. Series A, Life sciences; v. 101.
QH450.N37   1984   574.87'322   85-28237
ISBN 978-1-4684-7617-0   ISBN 978-1-4684-7615-6 (eBook)
DOI 10.1007/978-1-4684-7615-6

© 1985 Plenum Press, New York
Softcover reprint of the hardcover 1st edition 1985
A Division of Plenum Publishing Corporation
233 Spring Street, New York, N.Y. 10013

PREFACE

    This book stems from an Advanced Study Institute on Chromo-
somal Proteins and Gene Expression that was held in Sitges,
Spain, on September 17-26, 1984. It would be misleading to call
this volume a conference proceedings, however. The ASI was not
a conference, but a course with diverse activities, only one of
which was a set of major presentations by the lecturers.
Indeed, the concept of lecturer was intentionally obscured as we
all learned from each other through shorter presentations by
other participants and through seminars, poster sessions, and
small group discussions. Furthermore, many participants found
that exchanging ideas outside organized sessions was among the
most rewarding aspects of the course. Some even claimed to have
profitably probed the intricacies of nucleosome structure and
transcriptional regulation while basking in the sun on the
beach! Obviously, it is difficult to catch the flavor of such
varied proceedings in a book. (I cannot confirm the incident on
the beach, never having found time to set foot there. Such is
the fate of the director of a meeting.)

    The ASI was judged a success -- and enthusiastically so --
by most participants. Not only did we deepen our understanding
of our scientific field, we made new friends and learned about
scientific and nonscientific aspects of life in other countries
and about issues that transcend international boundaries in our
complex world. We hope that this volume will be as successful
as the course was. From the outset, the organizing committee
set a goal of producing a book that would be of true value to
the broader scientific community and not simply a vehicle to
help the participants recall what went on at the meeting. We
settled on the idea of the contributions taking the form of
reviews and, with one exception, that is what the reader will
find in this volume. In many cases, the reviews relate nicely
to each other, and the typical reader will find several in the
book that are of interest. We believe that, in aggregate, the
reviews provide a most valuable and timely overview of chromosome structure, chromosomal proteins, and biological processes
that involve chromosomes. The time lag between the meeting and
the submission of manuscripts to the publisher has allowed the
incorporation of some material into the book that had not been
published at the time of the ASI. Besides contributions from
lecturers, several chapters were prepared by others who partici-
pated in the course. Their participation enriched the ASI and
their chapters add greatly to the value of this book.

It is a pleasure to acknowledge the many important contributions of my co-organizers, Graham Goodwin and Pere Puigdomenech. One of the most rewarding aspects of my involvement with the ASI was getting to know each of them well. Whatever success our overall endeavor ultimately attains would not have been possible without their efforts.

The principal source of support for the course was the Scientific Affairs Division of NATO. I want to thank Dr. Craig Sinclair, the Director of the ASI program, for his advice, encouragement, and support, which have extended from a time when the idea of this ASI had not yet been formulated in detail, through the course itself, and until the present, as we finalize the financial aspects of the meeting.

We also thank the following agencies in Spain for their contributions and support:

Comisión Asesora de Investigación Científica y Técnica (CAICYT), Ministerio de Educación y Ciencia.

Comissió Interdepartamental de Recerca i Innovació Tecnològica (CIRIT), Generalitat de Catalunya.

Departament d'Ensenyament, Generalitat de Catalunya.

I would like to acknowledge the support of NIH for my research efforts on chromosomal proteins in the form of research grants CA-17782 and GM-29203. Without that support the notion of my directing a meeting such as the Sitges ASI would have been out of the question.

Finally, I want to thank Maureen Rider for her dedicated and competent secretarial help.

Gerald R. Reeck

12 July 1985
Manhattan

CONTENTS

HISTONE, NUCLEOSOME, AND CHROMOSOME STRUCTURES

TRANSCRIPTIONAL REGULATION AND CHROMATIN STRUCTURE

NUCLEOSOME STRUCTURE

Gerald R. Reeck

Department of Biochemistry
Kansas State University
Manhattan, Kansas 66506

INTRODUCTION

With the curious exception of dinoflagellates (1), the fundamental architectural unit in eukaryotic chromosomes is the nucleosome (2). Because nearly all of the DNA in a nucleus occurs in nucleosomes, one cannot think intelligently about any structural aspect of chromatin or chromosomes or about any biological process that occurs on chromosomes without taking into account the structure of the nucleosome. Athough the physical entity itself has been existence for several hundred million years, the concept of the nucleosome arose only about 12 years ago. We are thus in the early stages of unraveling the nucleosome's structural details and their physiological significance. Nonetheless, a great deal of information has already been obtained, especially in structural terms.

The makeup of the nucleosome corresponds closely to that formulated initially by R.D. Kornberg (3) for the "repeating unit of chromosome structure" in a remarkably prescient article in 1974. A nucleosome consists of from about 160 to 250 base pairs of DNA, 1 molecule of histone H1 (or one of its homologs) and 2 molecules of each of the other histones (H2A, H2B, H3, and H4). The amount of DNA in a nucleosome varies among species and among cell types within a species (4). Furthermore, there apparently is variation in nucleosome DNA length within a given cell type (5). Another type of heterogeneity in nucleosome preparations, and one that is possibly related to the variation in nucleosome DNA length, stems from the fact that the H1 family is heterogeneous (see chapter by Crane-Robinson, this volume). Much more highly conserved than the nucleosome itself is the nucleosome core -- 1.8 turns of duplex DNA wrapped around an octamer of histones. In this chapter I will deal with the structure of the nucleosome core and its constituents to draw attention to some recent major advances and to point out some important issues that remain open.

STRUCTURE OF THE NUCLEOSOME CORE PARTICLE

Despite a rather strict overall constancy in composition (145 base pairs of DNA and 2 molecules of each of the core histones -- H2A, H2B, H3, and H4), core particle preparations are not homogeneous. There is heterogeneity in both their protein and DNA components. There is more than one gene in a given organism for some histones, and the histones are subject to post-translational modifications (6). Then there is, of course, an enormous sequence heterogeneity in the DNA of core particle preparations (other than those reconstituted from cloned DNAs or synthetic polynucleotides).

Despite such sources of heterogeneity, workers in Klug's laboratory have obtained large, highly ordered crystals of rabbit kidney core particles (7). These crystals diffract to a a resolution of 5 Å, which begins to approach atomic resolution. Richmond et al. (8) have now produced an electron density map at 7 Å resolution and proposed a structural model of the core particle based on that map. Because of the large unit cell, this structure determination was demanding just from the standpoint of data collection and calculation. The work of Richmond et al. (8) was, moreover, innovative from a technical standpoint. To generate useful isomorphous replacements, they used a rather new type of reagent that contains clusters of heavy atoms and hence gives especially large contributions to the diffraction pattern. Perhaps even more important was an intentional variation in hexanediol concentration to obtain uniform unit cell dimensions from crystal to crystal by controlled dehydration.

In discussing the structure of the core particle, it is helpful to have a set of reference positions. The particle (actually, only most of it) has a 2-fold axis of rotational symmetry, and the site on the DNA on the 2-fold axis is a particularly useful landmark. Klug et al. (9) have labeled this position 0. At each 10 bp along the DNA, moving up axis of the left-handed superhelix, the reference positions are incremented by +1. Conversely, at each 10 bp along the DNA, moving down the axis of the superhelix, the reference positions are incremented by -1. Thus, integral reference positions along the core particle's 145 bp of DNA run from -7 to 7. The integral positions correspond approximately to sites of DNase I attack on DNA of the nucleosome core (10, 11).

Perhaps the most interesting aspect revealed in the work at 7 Å resolution is the path of the DNA. At this level of resolution, considerable detail can be discerned in double-stranded DNA, including identification of major and minor grooves. It was previously known that the DNA in the core particle forms 1.8 superhelical turns around the histone octamer. Now it is clear that the path of the DNA is not a smooth curve. Rather, at several places (most prominently at positions -4, -1, +1, and +4) the DNA is kinked. It is most interesting that these positions correspond to the DNase I sites in core particle DNA of lowest susceptibility to the enzyme (11). That relative resistance appears to not be due to direct blocking by histones of the enzyme's access to DNA but to a locally altered DNA conformation caused by what are presumably especially strong interactions with histones H3 and H4.

At 7 Å resolution, the information that can be obtained about protein structure is still rather limited. One is quite far, for instance, from being able to successfully follow the entire path of a polypeptide chain. Assignments of regions of electron density to individual histone molecules have been proposed, relying heavily on the cross-linking studies of Mirzabekov (12). No detailed insight into histone/histone interactions has been obtained. Long rods of electron density are seen in the protein portion of the electron density map, and these are almost certainly helices, presumably α-helices. (Circular dichroism measurements, as in reference 13, have indicated clearly that histones contain large amounts of helix.) The helical segments are notable for their length. For instance, the authors attribute two rods of length 40 Å to histone H4. Thus, this rather small protein molecule, in which about only 80 residues are in the folded domain (see below), must be at least 40 Å in its longest dimension. Necessarily, then, it must be quite asymmetric.

The generality that DNA is outside the protein in a core particle (14) is upheld in detail in the recent results (8). All of the histone/DNA contacts appear to occur on the inner surface of the DNA double helix. No electron density that can be assigned to protein occurs outside the DNA. There is no "cross-linking," by an individual histone molecule, of DNA at positions that are separated by one superhelical turn (e.g., DNA at positions -6 and +1.6) and that are thus brought into proximity by the formation of the superhelix. The lack of such crosslinks is presumably important to the dynamic behavior of nucleosomes in biological processes and to nucleosome assembly and disassembly.

STRUCTURE OF THE HISTONE OCTAMER

Another major contribution to our understanding of the structure of the nucleosome has recently been made by a group under the leadership of E. N. Moudrianakis. Burlingame et al. (15) have calculated an electron density map based on X-ray diffraction from crystals of the chicken erythrocyte histone octamer (no DNA present). Crystallization was in 69% saturated ammonium sulfate and a pH of 6.5 (16). The map was calculated at a resolution of 3.3 Å.

In several respects, the structural model advanced by Burlingame et al. (15) for the histone octamer strongly resembles the histone portion of the model of the nucleosome core particle proposed by Richmond et al. (8). Long helices are apparent within the histones in the octamer model. These are, of course, much more clearly identified at this higher level of resolution than they were in the core particle study. The high helix content in both models is consistent with solution studies of the proteins' conformations in chromatin and when free of DNA and at elevated ionic strength (e.g., see reference 13).

The spatial disposition of the individual polypeptides is similar in the two models. Thus, in each case there is a two-fold axis of rotational symmetry and the Cys 110 residues of H3 lie near that axis and on the exterior of the protein mass. The H3 and H4 polypeptides are closely associated in each model and two H2A-H2B dimers lie on the surface of the H3-H4 tetramer, on either side of the two-fold axis. In each model, the H2A molecules lie on that part of the H3 molecules that approach or touch each other on the 2-fold axis.

Despite these similarities, there is a large and most interesting difference between the models in overall dimensions and shape. If we approximate the shape of the protein portion of the core particle as a disk, its diameter would be 66 Å on the average and the disk would be about 60 Å thick (8). On the other hand, the model of Burlingame et al. (15) for the octamer has the shape of a rugby ball (in the authors' terms). The end-to-end length of the rugby ball, 110 Å, corresponds topologically to the thickness (only 60 Å) of the disk. The diameter through the midsection of the rugby—ball—octamer is 65-70 Å (15). This correponds topologically and is comparable in magnitude to the diameter of the histone portion of the core particle. See Figure 1 for a schematic depiction of the relationship between the two models.

Despite this major difference in shape and dimension, there is not irreconcilable disagreement between the two models if we recognize that they are not of the same entity: the histone portion of the core particle is tightly bound to 145 base pairs of DNA. We need only suppose that the histone octamer is more compact when bound to DNA than when free of DNA. In assessing the basis for differences between the models, I will assume that each model is correct at the level of detail at which it has been advanced.

From a mechanical standpoint, a substantial compaction in the octamer structure proposed by Burlingame et al. (15) is feasible. Because of the large amount of solvent within the

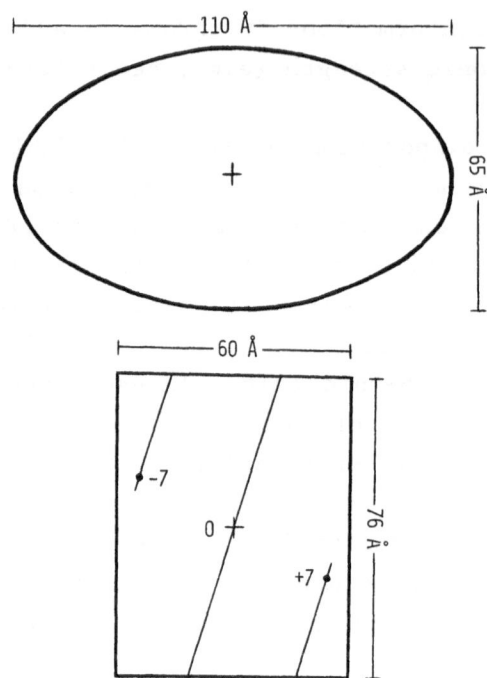

Figure 1. Schematic depiction of the relationship between the
models of Burlingame et al. (15) and Richmond et al.
(8) for the histone octamer (top) and the histone
portion of the nucleosome core particle (bottom).
The crosses indicate the 2-fold axes, which are
perpendicular to the plane of the paper. In the
bottom portion, the approximate path of the DNA on
the core particle is shown as a line. Positions -7,
0, and +7 on the DNA are indicated. A fitting of
double-helical DNA to grooves on the surface of the
octamer also gives a left-handed superhelix (15).

rugby ball shape -- a point that the authors stress -- the
octamer would appear to be eminently compressible. A reconcil-
iation of the two models would be provided by compression along
the long axis of the rugby ball to form a compact disk with a
thickness of 60 Å.

There is evidence that is consistent with the notion that
the DNA of the nucleosome holds the polypeptide chains of the
octamer in positions that, in the absence of DNA, are thermo-
dynamically unstable. In solution, in the absence of DNA,
formation of a disulfide bond between H3 molecules destabilizes
the H3-H4 tetramer (17). That is, disulfide-linked H3 molecules
will not participate in tetramer formation. They can, on the
other hand, participate in core particle formation (18) and

thereby be forced into the tetramer configuration. Another relevant piece of evidence comes from the work of Walker et al. (See their chapter, this volume.) Their results indicate that rather small peptides, derived from histones by combined tryptic and chymotryptic attack, are held in position in a nucleosome-like structure by DNA whereas those peptides would otherwise not assume that arrangement.

What might be the biological relevance of the proposed compression? If we are to look to one study or the other for a representation of the state of histones and DNA in a nucleosome, the core particle structure would seem to be the logical choice. Admittedly, a histone octamer bound to 145 bp of DNA is not the same as what exists in a nucleosome. On the other hand, in its composition it is a good deal closer than is a histone octamer with no DNA bound at all. Thus, I would postulate that in nucleosome assembly a major compaction occurs within the histone octamer upon binding to DNA. (This assumes that the dimensions of the octamer in the solvent from which it was crystallized are the same as the dimensions of the octamer at physiological ionic strength. This point is perhaps impossible to investigate, given the thermodynamic instability of the octamer at physio-logical ionic strength (19).) There is a large amount of energy released when the many ionic interactions are set up between a histone octamer and DNA (20). A sizable part of that energy could be used to impose conformational constraints on the DNA and the histone octamer and still have enough energy released to produce a very stable particle.

SEQUENCE PREFERENCES IN NUCLEOSOMES

Reconstitution studies have demonstrated that there can be strong sequence preference in the positioning of a nucleosome core on a DNA molecule (21, 22). This could conceivably have either a kinetic or thermodynamic basis, but I shall assume in the subsequent discussion that the basis is in fact thermo-dynamic -- that is, that nucleosomes (or core particles) con-structed with certain 145 base pair stretches of DNA are more stable than are core particles constructed on other 145 bp stretches.

Trifonov has presented a particularly cogent and stimulating analysis of a possible basis for such differences in stability (23, 24). His point is that nucleosomes formed on stretches of DNA that are more easily deformed into the conformation of the superhelical turns around the histone octamer will be more stable than are nucleosomes formed with DNA stretches less readily deformed into those superhelical turns. There is an inherent sequence-dependent tendency of double-stranded DNA molecules to bend (25). DNA sequences that have tendencies to bend as required in the superhelical turns of a nucleosome will require less energy for their deformation.

The direct connection between deformation energy of DNA and nucleosome stability can be seen from a simple thermodynamic analysis of core particle formation (see Figure 2). The scheme ignores changes in protein conformation or in quaternary structure that might accompany nucleosome formation, but those would be common to formation of all nucleosomes, assuming that the octamers involved were constructed from the same histone gene products and had the same level of post-translational modifications. ΔG^{o} for core particle formation (reaction III) is simply the sum of the ΔG^{o}'s for reactions I and II. It seems reasonable to assume that ΔG^{o} will be invariant for reaction II

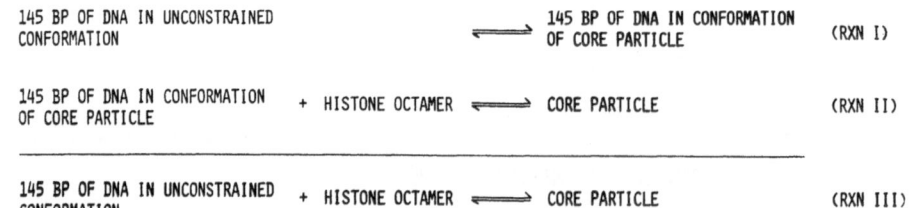

Figure 2. A formal thermodynamic analysis of formation of nucleosomal core particle from a histone octamer and 145 bp piece of DNA that is not constrained in its conformation. This simple scheme is not intended to portray the mechanism of core particle formation, but instead to show a theoretical two-step path for the process that emphasizes the direct transmission of differences in free energy for constraining 145 bp of DNA in the nucleosomal superhelical conformation into differences in stabilities of the nucleosomal core particles.

regardless of the sequence of the DNA involved. That is, a histone octamer may well bind equally strongly to any DNA sequence that is in the correct superhelical conformation. On the other hand it seems quite clear that DNA sequences will differ significantly in ΔG^O for reaction I -- the free energy required to constrain them in the conformation of the superhelical turns of DNA in the nucleosome. ΔG^O for reaction I is expected to be large and positive for any DNA sequence, but could be substantially smaller for some sequences than for others. Any difference in ΔG^O for reaction I between two DNA sequences will be transmitted directly into the values of ΔG^O for reaction III involving the two sequences -- i.e., into the stabilities of the nucleosomes formed with these two sequences. A 10-fold difference in nucleosome stability would be caused by a difference of 1.4 kcal/mole for ΔG^O for reaction I. A 100-fold difference in nucleosome stability would be caused by a difference of 2.8 kcal/mole for ΔG^O in reaction I. At this time reliable estimates of ΔG^O for that reaction are not available, nor are reliable estimates of differences in ΔG^O for that reaction for different DNA sequences. Thus it is not possible to make quantitative predictions of relative nucleosome stability, although Trifonov (23, 24) has attempted to determine the order of preference for nucleosome formation among different sequences.

Differences in nucleosome stability based on differences in the deformation energy for different DNA sequences could have a physiologically significant role in determining the positioning of nucleosomes, as has been emphasized by Trifonov (23, 24) and others (e.g., see reference 8). One could also imagine that competition for histones among potential nucleosome-forming sites on a DNA lattice could determine where nucleosomes are not found at all (24). Given the fact that ΔG^O for reaction II will be large and negative, most naturally occurring DNA sequences would be expected to have a large negative value of ΔG^O for reaction III (i.e., they will participate in the formation of stable nucleosomes). Thus differential stabilities of nucleosomes would be expected to explain nucleosome exclusion only if the amount of histones is not saturating with respect to DNA (and that equilibrium or near-equilibrium conditions hold in a nucleus).

THE CONCEPT OF MODULAR CONSTRUCTION OF PROTEINS AS APPLIED TO HISTONES

In an entirely different and very specific context I have recently used the term "modular" to describe the construction of certain proteins in the course of evolution (26). I want to now consider the structure and evolution of histones from that viewpoint.

The notion that nature has genetic mechanisms for stitching together a protein molecule by combining DNA segments that code for individual stretches of polypeptide has been in circulation for some time. Gilbert, for instance, has discussed the idea in detail, with particular emphasis on the possible role of introns in the evolution of proteins (e.g., see reference 27). The emphasis of my ideas in this area is that quite different evolutionary constraints operate on fundamentally different sorts of polypeptide units. For reasons that will become apparent, I propose to call the individual units "modules."

The type of module that we are most familiar with is the domain, which I take to be an independent folding unit with a globular conformation. (Operationally, a domain is frequently identified as a polypeptide fragment that is soluble under non-denaturing conditions and possesses a folded conformation.) Modules other than domains (i.e., sections of a polypeptide that are not folded into globular units) differ dramatically from domains not only in structural organization but in the constraints to which they are subject in the course of evolution. Those differences can be emphasized by using a term other than "domain" to describe regions that are not globular.

In the evolution of a domain, there is strong selective pressure on function (that is, on maintaining key residues directly involved in the biological activity of the domain) and strong pressure on the structural support for that function (that is, on maintaining the required spatial distribution of critical residues and the stability of the globular unit). In the evolution of a nonfolded module, such as a histone tail (28), there is evidently little constraint on the folding of the polypeptide chain, and constraints for the purpose of maintaining the function are much more relaxed than those acting on a domain. In nonfolded modules there is frequently such large variation in sequence that there is no recognizable sequence similarity between functionally analogous modules. It appears that the evolutionary constraint is at the level of composition rather than sequence. Thus, it is in domains where we generally see the greatest evolutionary constraints, and by a wide margin.

The clearest case of modular construction in histones is in the H1 family. The C terminal module in these proteins (the portion after the highly conserved globular domain) is similar in composition among different members of the family (e.g., H1, H1^o, and H5) but is variable in length and not recognizably similar in sequence (Reeck and Teller, unpublished observations). (See the chapter by Crane-Robinson in this volume for details on the H1 family.) It appears that the function of this module requires a particular composition (rich in Lys, Ala, and Pro) but that a large number of sequences having that composition may be able to serve the required function.

We have presented evidence previously that the H2A, H2B, H3 and H4 constitute a family of homologs (29), i.e., that they all diverged from a common ancestral protein. That conclusion was limited, however, to the C terminal portions of the molecules. We specifically refrained from claiming sequence similarity in the N terminal regions of the histones, in which the polypeptides have peculiar compositions that are rich in Lys and/or Arg as well as in certain other amino acids. In the current context and in light of a good deal of work on the structural organization of the core histones (for example, see the chapter by Walker et al. in this volume), I would advance the following interpretation. Each core histone has a 2-module structure: the majority of the polypeptide in each case is a single domain (a globular, folded module) and the N terminal portion (roughly a quarter of each sequence) is a nonfolded module (a tail). Introducing the term "module" may seem to be simply a minor change in nomenclature, but I think it is in fact more substantial than that. By not calling a histone's tail a domain, we emphasize the differences in evolutionary pressures and in structural constraints in this region as contrasted to the those in domains. We also call attention to the possible evolutionary origin of the histones. Our analysis of sequence similarities among the core histones is consistent with a co-ancestral origin for the domains of the 4 proteins (29). The N-terminal modules (i.e., the tails) may well not be co-ancestral but instead may be independent stretches of polypeptide of independent origin and similar compositions. To put this another way, if one says that two domains are related, one is postulating a co-ancestral relationship (i.e., homology). To say that two modules are similar, on the other hand, is merely an assertion of a compositional similarity that need not be due to co-ancestry.

We have previously pointed out that members of the HMG-14/17 family of proteins (see chapter by Goodwin et al., this volume) are similar in composition and, over limited regions, in sequence to the tails of core histones and histone H1 (30). The architecture of members of the HMG-14/17 family is that of a single, nonfolded module. That module is apparently functionally similar in at least a crude sense to the histone tails: they all bind relatively weakly to DNA and achieve a charge neutralization in the process. Recent results of Walker (20) emphasize the relatively weak binding of the histone tails to DNA in chromatin. Those tails dissociate from DNA over an ionic strength range comparable to that required for dissociation of HMG-14 or 17. The functional characteristics of the HMG-14/17 module or a histone tail module apparently requires a amino acid composition that is rich in Lys and/or Arg, amino acids with small side chains, and (usually) Pro. The weakness of binding of these modules to DNA is interesting, in light of their very high contents of Lys and/or Arg. We have argued that the weakness of the interaction is due to the enormous configurational entropy available to these polypeptide modules in their unbound states (30). (These modules appear to be unstructured in a 3-dimensional sense when not bound to DNA). We have further postulated that the relatively high Pro content in such modules serves to limit the configurational freedom of the polypeptide in the dissociated state and thus lessens to a certain extent the entropy factor favoring dissocation. We would predict that substitution of Gly or Ala for each Pro in such a module would significantly weaken the interaction with DNA because of greater configurational freedom in the dissociated state.

The precise roles of the histones tails are still not known. They are not needed for nucleosome formation (31). There is evidence that they are required for formation of higher order chromatin structure (32). It is clear that in chromatin they are not mobile. They are almost certainly bound to DNA, from which they can be dissociated at modest ionic strengths (20). It is likely that in core particles and in the histone octamer the histone tails are not in fixed conformations. They would therefore not contribute to the diffraction pattern and would not be represented in the electron density maps. We would, therefore, learn nothing about the conformations or interactions of the tails from the current structure determinations, even when carried to higher resolution.

CONCLUSIONS

If the fundamental architectural unit in chromatin can itself be said to have a key structural feature it would be the nucleosome core -- 1.8 turns of superhelical DNA wrapped around an octamer of histones. Elucidation of the structure of that core will be a major advance in understanding nucleosome and chromatin structure. We are approaching the point of knowing the structure of the nucleosome core at atomic resolution through crystallographic studies on the core particle and on the histone octamer. One might have expected that, being on the verge of having such detailed information, nucleosome structure would be fading as an area of research interest. Instead, the field is becoming more interesting. As information emerges, new questions arise and old questions are formulated more precisely. Thus, even if we restrict ourselves to the core itself and ignore the unanswered questions about histone H1, linker DNA, and higher order structure, there are many interesting issues to be investigated. What changes occur in the histone octamer upon formation of a nucleosome core? What sort of alternate configurations (presumably at somewhat higher energies) exist for the nucleosome core? Restated, what sort of conformational flexibility is there in the nucleosome core and how is that flexibility related to biological function? (Note that this is very difficult to approach through crystallography, which gives a static picture of the conformation of lowest energy.) To what do the histone tails bind? What is the basis for sequence preference for nucleosome stability? Can one experimentally measure relative stabilities? Can one predict relative stabilities based on DNA sequence? Answers to many of these questions are likely to come through studies of reconstituted particles in which both the DNA and histone components are homogeneous.

We have learned much about the nucleosome and its core over the last 12 years, but we are far from understanding the many subtleties that have been built into these structures over hundreds of millions of years of evolution. The investigation of nucleosome structure is thus a fresh and challenging area that offers many rewards to the scientist who can approach it with an infusion of new ideas and approaches.

ACKNOWLEDGMENTS

I thank D.J. Cox of this department for a critical reading of this manuscript and several valuable suggestions. Beyond that, I would like to acknowledge his influence on my thinking that has resulted from a continuous and rewarding dialogue over the last 11 years. Use of the term "co-ancestral" as a synonym of "homologous" in its strictly defined sense is a suggestion of A.D. McLachlan. Work in my laboratory on chromosomal proteins is supported by NIH grants CA-17782 and GM-029203. This is publication number 86-16-B of the Kansas Agricultural Experiment Station.

REFERENCES

1. HERZOG, M. and SOYER, M.O. (1983). The native structure of dinoflagellate chromosomes and their stabilization by Ca^{++} and Mg^{++} cations. Eur. J. Cell Biol. 30, 33-41.

2. McGHEE, J. and FELSENFELD, G. (1980). Nucleosome structure. Ann. Rev. Biochem. 49, 1115-1156.

3. KORNBERG, R.D. (1974). Chromatin structure: a repeating unit of histones and DNA. Science 184, 868-871.

4. KORNBERG, R.D. (1977). Structure of chromatin. Ann. Rev. Biochem. 46, 931 954.

5. LOHR, D., CORDEN, J., TATCHELL, K., KOVACIC, R.T., and VAN HOLDE, K.E. (1977). Comparative subunit structure of HeLa, yeast, and chicken erythrocyte chromatin. Proc. Natl. Acad. Sci. USA 74, 7983.

6. ISENBERG, I. (1979). Histones. Ann. Rev. Biochem. 48, 159-191.

7. FINCH, J.T., BROWN, R.S., RHODES, D., RICHMOND, T., RUSTON, B., LUTTER, L.C., and KLUG, A. (1981). X-ray diffraction study of a new crystal form of the nucleosome core showing higher resolution. J. Mol. Biol. 145, 757-769.

8. RICHMOND, T.J., FINCH, J.T., RUSTON, B., RHODES, D., and KLUG, A. (1984). Structure of the nucleosome core at 7 Å resolution. Nature 311, 532-537.

9. KLUG, A., RHODES, D., SMITH, J., FINCH, J.T., and THOMAS, J.O. (1980). A low resolution structure for the histone core of the nucleosome. Nature 287, 509-516.

10. SIMPSON, R.T. and WHITLOCK, J.P., Jr. (1976). Mapping DNAase I-susceptible sites in nucleosomes labeled at the 5' ends. Cell 9, 347-353.

11. LUTTER, L.C. (1978). Kinetic analysis of deoxyribonuclease I cleavages in the nucleosome core: evidence for a DNA superhelix. J. Mol. Biol. 124, 391-420.

12. MIRZABEKOV, A.D. (1981). Nucleosome structure. Trends Biochem. Sci. 6, 23-25.

13. BIDNEY, D.L. and REECK, G.R. (1978). Association products and conformations of salt-dissociated and acid-extracted histones. A two-phase procedure for isolating salt-dissociated histones. Biochemisty 16, 1844-1849.

14. HJELM, R.P., KNEALE, G.G., SUAU, P., BALDWIN, J.P., and BRADBURY, E.M. (1977). Small angle neutron scattering studies of chromatin subunits in solution. Cell 10, 139-151.

15. BURLINGAME, R.W., LOVE, W.E., WANG, B.-C., HAMLIN, R., XUONG, N.-H., and MOUDRIANAKIS, E.V. (1985). Crystallographic structure of the octameric histone core of the nucleosome at a resolution of 3.3 Å. Science 228, 546-553.

16. BURLINGAME, R.W., LOVE, W.E., and MOUDRIANAKIS, E.V. (1984). Crystals of the octameric histone core of the nucleosome. Science 223, 413-414.

17. LEWIS, R.O., COX, D.J., and REECK, G.R. (1980). Chromatographic change in histone H3-H4 preparations during short term storage and its reversal by bisulfite. Int. J. Pept. Prot. Res. 16, 219-224.

18. CAMERINI-OTERO, R.D. and FELSENFELD, G. (1977). Histone H3 disulfide dimers and nucleosome structure. Proc. Natl. Acad. Sci. USA 74, 5519-5523.

19. EICKBUSH, T.H. and MOUDRIANAKIS, E.N. (1978). The histone core complex: an octamer assembled by two sets of protein-proteins interactions. Biochemistry 17, 4955-4964.

20. WALKER, I.O. (1984). Differential dissociation of histone tails from core chromatin. Biochemistry 23, 5622-5628.

21. CHAO, M.V., GRALLA, J., and MARTINSON, H.G. (1979). DNA sequence directs placement of histone cores on restriction fragments during nucleosome formation. Biochemistry 18, 1068-1074.

22. SIMPSON, R.T. and STAFFORD, D.W. (1983). Structural features of a phased nucleosome core particle. Proc. Natl. Acad. Sci. USA 80, 51-55.

23. TRIFONOV, E.N. (1983). Sequence dependent variations of B-DNA structure and protein-DNA recognition. Cold Spring Harbor Symp. Quant. Biol. 47, 271-278.

24. MENGERITSKY, G. and TRIFONOV, E.N. (1983). Nucleotide sequence- directed mapping of the nucleosomes. Nucleic Acids Res. 11, 3833-3851.

25. DICKERSON, R.E. and DREW, H.R. (1981). Structure of a B-DNA dodecamer. II. Influence of base sequence on helix structure. J. Mol. Biol. 149, 761-786.

26. REECK, G.R. and HEDGCOTH, C. (1985). Amino acid sequence alignment of cereal storage proteins. FEBS Lett. 180, 291-294.

27. GILBERT, W. (1985). Genes-in-pieces revisited. Science 228, 823-824.

28. BOHM, L. and CRANE-ROBINSON, C. (1984). Proteases as structural probes for chromatin: the domain structure of histones. Bioscience Rep. 4, 365-386.

29. REECK, G.R., SWANSON, E., and TELLER, D.C. (1978). The evolution of histones. J. Mol. Evol. 10, 309-317.

30. REECK, G.R. and TELLER, D.C. (1984). High mobility group proteins: purification, properties, and amino acid sequence comparisons. In: "Progress in nonhistone protein research," Vol. II (I. Bekhor, ed.) pp. 1-22. CRC Press, Boca Raton, Florida.

31. WHITLOCK, J.P., Jr. and STEIN, A. (1978). Folding of DNA by histones which lack their NH_2-terminal regions. J. Biol. Chem. 253, 3857-3861.

32. ALLAN, J., HARBORNE, N., RAU, D.C., and GOULD, H. (1982). Participation of core histone "tails" in the stabilization of the chromatin solenoid. J. Cell Biol. 93, 285-297.

RESYNTHESIS OF HISTONE PEPTIDE BONDS ON A DNA MATRIX[*]

I.O. Walker[1], S.G. Davies[2], and P.N. Schofield[3]

[1]Department of Biochemistry
University of Oxford
Oxford, OX1 3QU   England

[2]Dyson Perrins Laboratory
South Parks Road
Oxford, OX1 3QU   England

[3]Imperial Cancer Research Fund
Burtonhole Lane
Mill Hill
London, NW7 1AD   England

INTRODUCTION

Digestion of core chromatin with trypsin leads to degradation of the basic amino and carboxyl terminal ends and the production of five large polypeptide fragments which are protected from further tryptic attack by their interactions with DNA. The small basic oligopeptides produced by the trypsin dissociate from the DNA whereas the large, protected fragments remain bound (1). Subsequent digestion of these bound, trypsin resistant polypeptides with chymotrypsin produces further degradation to smaller peptides which still remain bound to the DNA. Double digestion of core chromatin with trypsin and chymotrypsin does not cause any change in the secondary structure of the bound peptides, as monitored by ellipticity at 220 nm (reference 1, and unpublished data of I.O. Walker).

Here we describe the surprising effect of adding phenyl methylsulphonyl fluoride (PMSF), a serine protease inhibitor, to

_____

[*]This paper is dedicated to the memory of Ian Walker.

17

the doubly digested core chromatin. Under appropriate conditions, the PMSF promotes the resynthesis of the peptide bonds hydrolysed by chymotrypsin.

EXPERIMENTAL

Core chromatin was prepared as described previously (1) by repeated extraction of chick erythrocyte chromatin with 0.7 M NaCl, in order to strip away H1 and H5.

Digestions with tosylchlorophenyl ketone (TCPK)-trypsin were carried out for 18 h in 1 mM Tris-HCl, pH 7.4, at 20°C at a trypsin:histone ratio of 1:30. The reaction was stopped by adding soybean trypsin inhibitor (2 moles inhibitor:1 mole trypsin). Chymotrypsin (at an enzyme:histone weight ratio of 1:10) was then added and incubation continued for 1 h at 20°C. The reaction was stopped by adding TCPK. PMSF was added at room temperature, as required. Samples were removed for SDS-polyacrylamide gel analysis as described previously (1). Samples for electrophoresis on polyacrylamide-urea gels were incubated with 0.5 N HCl, dialysed against 0.9 N acetic acid, 20% sucrose, and electrophoresed on acid-urea gels prepared according to (2).

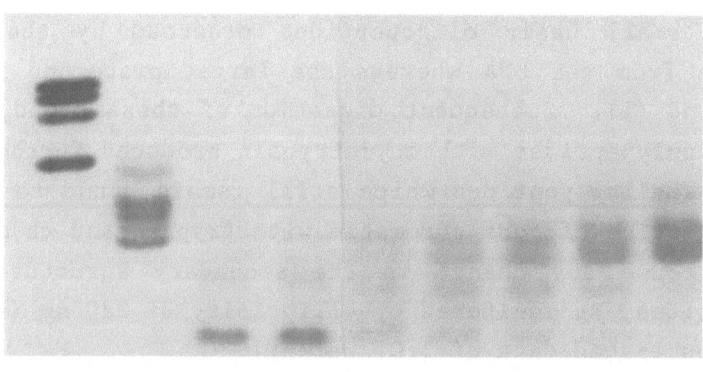

Figure 1. SDS-polyacrylamide gel electrophoresis of core histones. (a) native core histones; (b) core histones after digestion of core chromatin with trypsin for 18 h at 20°C; (c) same as (b) and then followed by chymotrypsin digestion for 1 h; (d) same as (c) but after 5 h digestion; (e-i) core histone after double digestion and addition of 0.1 mM PMSF for 0', 18', 38', 60', and 240', respectively.

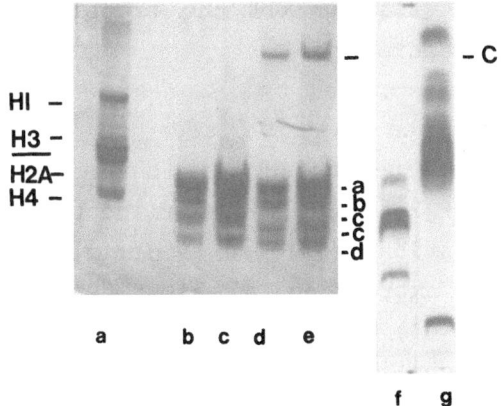

Figure 2. Polyacrylamide-urea gel electrophoresis of core histones. (a) native calf thymus histones; (b) core
histones after digestion of core chromatin with
trypsin and chymotrypsin and then followed by the
addition of 0.1 mM PMSF for 4 h; (e) same as (d) but
double loading of protein; (f) calf thymus histones;
(g) core histones extracted from core chromatin after
digestion with trypsin and chymotrypsin.

RESULTS

Trypsin/Chymotrypsin Double Digestions

The digestion of core chromatin with trypsin produces the
five large limit peptides shown in Figure 1 and described
previously (1). Subsequent digestion with chymotrypsin leads to
further degradation that produces four minor bands and one
major, heterogeneous, low molecular weight band which runs at
the electrophoretic front (Figure 1). In a more highly cross-
linked polyacrylamide gel, the major band is resolved into 7
peptides with apparent molecular weights ranging from 2500 to
6000. These bands remained stable for at least 5 hours after
stopping the reaction by adding TCPK.

In Figure 1 is shown the effect of adding PMSF to a double
digest in which the trypsin has been inhibited with soybean
trypsin inhibitor and the chymotrypsin with TCPK. With increasing time after the addition of PMSF, low molecular weight
bands disappear and bands of higher molecular weight appear. By
4 h the band pattern is identical to the pattern seen after

trypsin digestion alone. All the trypsin limit peptides have apparently been reformed.

This effect can also be seen on acid-urea gels. Acid extraction of trypsin-treated core chromatin followed by electrophoresis at pH 2.8 resolves four peptides (Figure 2, a-d). After double digestion, followed by incubation with 0.1 mM PMSF for 4 h and acid extraction, the peptide band pattern (with the exception of band C') has reverted to the pattern seen on trypsin digestion alone. Band C', in the sample which has been treated with PMSF, has a greater electrophoretic mobility than does band C in the original digest. Judging from the difference in mobilities of the two forms of native histone H4 (Figure 2, Track 2), which arise from a difference in net charge of +1 due to acetylation of a lysine residue in about half the population of H4, the difference in mobility of the C bands arises from a net difference of +1 in charge. This charge difference must be due to the gain of a positive charge rather than the loss of a negative charge because the electrophoresis takes place at pH 2.8 where all the side chain carboxyl groups would be expected to be protonated and uncharged. One possibility is that an acetylated lysine in the native histone became deacetylated on chymotrypsin treatment.

These observations show that in the presence of PMSF, the chymotrypsin peptides are covalently joined to give the original peptide fragments produced by trypsin treatment.

Chymotrypsin Digestion of Previously Undigested Chromatin

The effect of direct chymotrypsin digestion on core chromatin is shown in Figure 3. In the limit, SDS-polyacrylamide gel electrophoresis resolves 7 peptides. On adding PMSF the low molecular weight fragments disappear, to be replaced by a set of larger polypeptides, at least two of which attain the size of the original intact histones. As in the case of the double digest, small peptide fragments have clearly been covalently linked together after adding PMSF. The cleavage reaction cannot be completely reversed back to the native histones because chymotrypsin makes several cuts in the basic amino and carboxyl

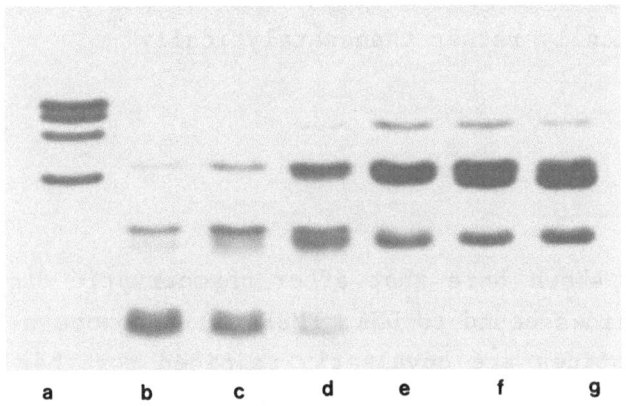

Figure 3. Chymotrypsin digest and PMSF treatment of core chromatin. (a) core histones; (b) core histones after digestion for 18 h with chymotrypsin; (c-g) core histones after adding PMSF (0.1 mM) for 11 min, 1 h, 7 h, 24 h, 101 h, respectively.

tails of the native histones. When this occurs it is expected, by analogy with the action of trypsin, which cleaves only in the basic tails and releases the small basic fragments from DNA (1), that these small chymotrypsin-produced fragments will dissociate. Once dissociation occurs, it is clear the peptides cannot be covalently linked together to form complete histones.

Stoichiometry Calculations

The extent of reversal in the double digestion experiments is dependent on the concentration of PMSF. Only partial reversal was observed at concentrations of 25, 50, and 75 μm, whereas complete reversal was obtained at 100 μm.

The following calculation shows that PMSF is acting stoichiometrically in the peptide bond resynthesis reaction. In our experiments, the concentration of each core histone is 1.7×10^{-5} M. Eleven peptides are produced by chymotrypsin digestion of the five tryptic limit peptides which themselves arise from the four core histones. Therefore, about one peptide bond is cleaved by chymotrypsin per core histone tryptic peptide. Thus, the equivalent of one peptide bond must be reformed per core histone. In total, $4 \times 1.7 \times 10^{-5} = 6.8 \times 10^{-5}$ mole of bond is resynthesized per liter. Complete resynthesis occurs between 7 and 10 x 10^{-5} M PMSF. This implies that PMSF acts stoichiometrically rather than catalytically.[*]

DISCUSSION

We have shown here that after chymotryptic digest of histone polypeptides bound to DNA, PMSF will promote a reaction in which the peptides are covalently rejoined together. Two lines of evidence suggest that the cleaved peptides are, in terms of amino acid composition and sequence, covalently bonded to give the original polypeptides. First, the molecular weights of the reformed products are identical to the original polypeptides as shown by electrophoresis on SDS polyacrylamide gels. Second, with the exception of one peptide out of four (peptide C'), the charge-to-mass ratios of the reformed products and initial peptides are identical. The possibility of ester, rather than amide formation may be ruled out because basic hydrolysis, which would be expected to cleave esters but not amides, has no effect on the apparent molecular weight of the reformed polypeptides. It is also highly unlikely that eight peptides of different sizes could be covalently linked together to give polypeptides with the same charge and same mass as the originals unless the peptides were covalently rejoined to their original partners.

The highly specific nature of the peptide synthesis reaction strongly suggests that the covalent linkage between the peptides is a peptide bond and that this bond is in each case identical to the one cleaved by chymotrypsin. We suggest the following mechanism for the synthesis of a peptide bond involving PMSF (Figure 4).

The peptide bond (i) is cleaved by chymotrypsin to give (ii). PMSF activates the carboxyl group by formation (3) of a

_____
[*]PMSF is often used to arrest chymotrypsin digestions. The above results indicate that some caution should be exercised in the interpretation of such experiments since some reformation of specific peptide bonds may occur.

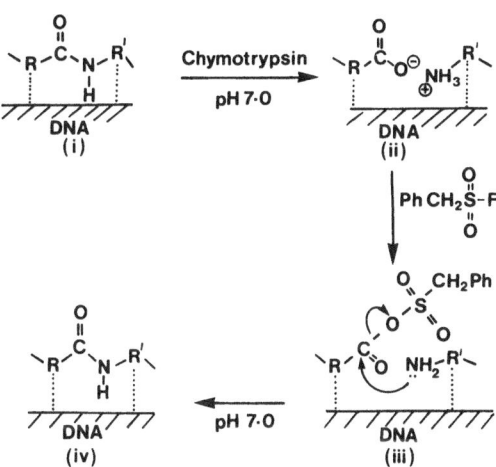

Figure 4. Proposed scheme for chymotrypsin decay and PMSF-mediated resynthesis of a peptide bond in a histone polypeptide bound to DNA.

mixed anhydride (iii), which promotes nucleophilic attack by the adjacent α-amino groups, thus reforming the original peptide bonds (iv). The energy for peptide bond synthesis is thus provided by the overall hydrolysis of PMSF. Kinetically the reaction proceeds at a measurable rate (half time of less than 2 h for the reaction conditions described in the text) because of the spatial proximity of the α-amino acid and α-carboxyl groups. These two groups must be held in a spatially favorable situation because they are at the ends of peptides which are tightly bound to DNA. (Peptide bond resynthesis did not occur when the peptides had been dissociated from the DNA in 2 M NaCl.) The specificity of these reactions is consistent with PMSF being a large hydrophobic reagent which would be expected, therefore, to favor the same sites as does chymotrypsin. (For analagous hydrophobic site preference of sulphonyl fluorides, see reference 4). This again indicates that after treatment with chymotrypsin, the small peptides remain bound in their original positions.

Our observations imply that the small polypeptide fragments produced by chymotryptic cleavage of the trypsin-resistant histone domains are held together by non-covalent forces in a highly structured complex that closely resembles that in a native nucleosome. These forces may involve protein-protein interactions, but almost certainly involve protein-DNA interactions which stabilize the secondary structure of the globular domains of the core histones.

REFERENCES

1. DIAZ, B.M., and WALKER, I.O. (1983). Trypsin digestion of core chromatin. Bioscience Rep. 3, 283-292.

2. PANYIM, S. and CHALKLEY, R. (1969). High resolution gel electrophoresis of histones. Arch. Biochem Biophys. 130, 337-346.

3. ITOH, M., NOJIMA, H., NOTANI, J., HAGIWARA, D., and TAKIA, K. (1978). Peptides VII. Some sulphonates of strongly acidic N-hydroxy compounds as novel coupling reagents. Bull. Chem. Soc. Jap. 51, 3320.

4. LIVELY, M.O. and POWERS, J.C. (1978). Specificity and reactivity of human granulocyte elastase and cathepsin G, porcine pancreatic elastase and cathepsin G, porcine elastase, bovine chymotrypsin, and trypsin, toward inhibition with sulphonyl fluorides. Biochim. Biophys. Acta 525, 171.

I.O. WALKER (1935-1984)

Ian Walker, Lecturer in Biochemistry at Oxford University, died on April 30, 1984 at the age of 48. A spirited and gifted man was thus tragically struck down in the prime of life, and we shall be deprived of his continued friendship and scientific insights.

Ian received his Ph.D. under Arthur Peacocke, with whom he applied physical techniques to the study of DNA. As an independent investigator, he turned his attention to the more complex system of chromatin.

It was in physical measurements on chromatin that Ian made his most notable contributions, although he also used his physical background to good advantage in studying enzymes (particularly phosphofructokinase) and ribosomes. Another aspect of his interests was transcriptional control in Physarum, and it was at a meeting on Physarum that he became ill with an infection (pericarditis) that was soon to claim his life. Ian had recently carried out a most interesting set of studies on histone tails and submitted a major manuscript just weeks before his death (Biochemistry 23, 5622, 1984). An intriguing set of other observations forms the basis for the chapter by Walker et al. in this volume. In addition to his research interests, Ian was a dedicated and respected tutor of undergraduates in his capacity as a Fellow at Keble College.

Devotion to his duties and success in carrying them out were trademarks of the man. One also remembers Ian's enthusiasm for life, his quick and engaging smile, his dry sense of humor, his unpretentiousness, his love of his family, and the ease with which he formed friendships. Although we shall no longer experience those traits directly, they remain indelibly imprinted in the minds of those who knew him and serve as a constant reminder that the world is a better place for his having been in it, even if for altogether too short a time.

GRR

HOW DOES H1 FUNCTION IN CHROMATIN?

Colyn Crane-Robinson

Biophysics Laboratories
Portsmouth Polytechnic
St. Michael's Building
White Swan Road
Portsmouth PO1 2DT, Hants   UK

INTRODUCTION

Histone H1 is required to form the ∼30 nm chromatin higher order structure in solution (1-3) and is therefore assumed to be the principal agent for higher order structure formation. Several recent studies of "active" chromatin suggest that H1 may be lacking from such regions (4-6) and the apparently complete absence of H1 in yeast (7) may therefore be due to a requirement for a totally uncondensed genome. The well-known mammalian H1s, such as those from calf thymus, are members of a protein family that includes molecules with considerable sequence differences. Individual members of the H1 family are often present in specialized tissues or in specific circumstances. For example, histone H5 is present in nucleated erythrocytes together with H1. Histone H1^{o} is present in mammalian tissues in addition to H1, and its level correlates with low rates of cell division (8). It seems to be a molecular hybrid between H1 and H5. Histone ϕ1 is a special form of H1 found in echinoderm sperm. It is longer than calf H1 and contains more arginine (9). In mollusks, the H1 appears to be the same size or even larger than calf H1, whereas in the sea worm P. dumerilii the H1 is 35% shorter than calf H1 (10). Certain mollusks and sea worms contain protamine-like proteins in addition to H1 (and the usual range of core histones) (11).

STRUCTURE OF HISTONE H1

H1 Sequences

Despite these differences in length and composition, there is a high degree of sequence similarity among the H1-family members and there can be no doubt that they all play a closely related role in chromatin structure. Figure 1 shows a comparison of sequences aligned for maximum sequence similarity. It is immediately apparent that there is a high degree of conservation is the central region of the molecules; that is, between the dotted lines. To the N-terminus of this is a domain of very variable length (from 4 to 39 residues) that is very basic and contains essentially no large hydrophobics and much proline. In certain cases, the very N-terminus is somewhat acidic. To the C-terminal side of the conserved domain is a long stretch of

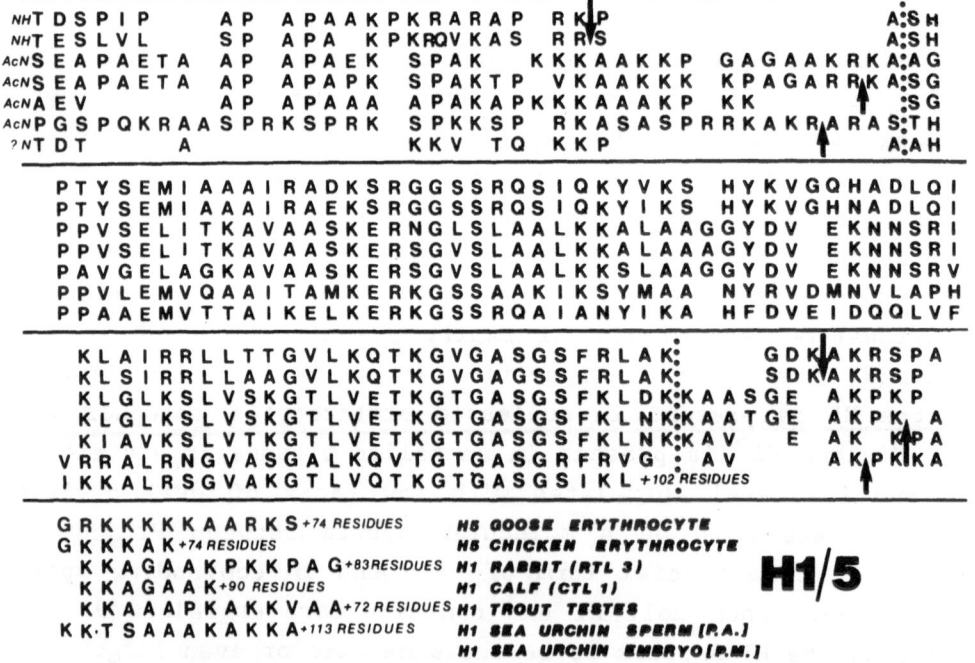

Figure 1. Sequences of H1-family molecules aligned for maximum similarity. Bold arrows indicate the limit of the 'hard-to-digest' tryptic products from calf H1, chicken H5, and sea urchin sperm φ1 (H1). Dotted lines represent the limits of the conserved sequence region. Sources of the sequences can be found in reference 14.

residues (from 60 to 200 amino acids in length, but typically about 100) that is largely lysine, proline, and alanine. In some species there are significant quantities of serine and arginine, e.g., in H5 and φ1. Neither the N- nor the C-terminal domain has a composition expected of folded protein, whereas the central domain does.

H1 Conformation

Trypsin digestion has been used to study the conformation of H1-family molecules in solution. Such experiments confirm the 3-domain structure conjectured on the basis of sequence. At the very earliest times of trypsin digestion very many cuts are made in the H1 chain and the molecule is degraded rapidly to a resistant fragment of about 80 residues (12). The bold arrows in Figure 1 show the limits of this "hard-to-digest" region for chicken H5, calf H5, and sea urchin sperm φ1 (H1). The limit fragments correspond closely to the conserved central domain. It is concluded that in free solution the central domain is a folded globule whereas the flanking domains are disordered. If nuclei (chromatin) are digested with trypsin (13, 14), the same limit product is produced from the H1 molecules and so it is assumed, as a working hypothesis, that the 3-domain structure is also to be found in chromatin. The flanking domains cannot of course be disordered in chromatin. More likely they are in an extended, unfolded conformation and thus particularly susceptible to protease attack.

Physical studies of the excised central domain indeed show it to be folded: CD and NMR indicate that the secondary and tertiary structure of intact H1 is preserved in the central fragment (12, 15, 16). Scanning calorimetry shows that the molar enthalpy of denaturation of the fragment and of intact H1 are the same, i.e., that _all_ the folded conformation is contained within the limit fragment (17). Low-angle scattering studies show that the fragment from H5 (GH5) is close to spherical and has a diameter of about 29 Å (15). Its frictional ratio is within the range found for "spherical" globular proteins whereas intact H1 has a frictional ratio of ∿2.0 (12). Figure 2 shows a diagrammatic representation of the 3-domain structure of the H1-family molecules.

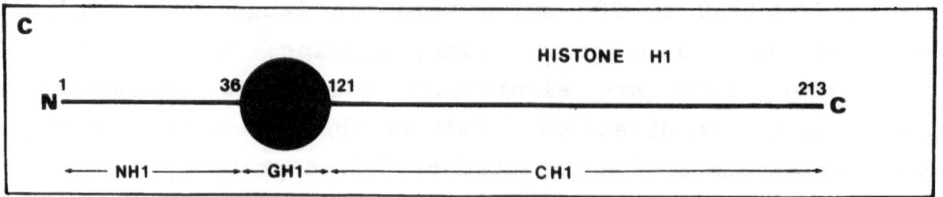

Figure 2. The 3-domain structure of H1-family molecules.

HISTONE H1 IN CHROMATIN

Stoichiometry

A recent examination of this problem using a radiolabelled lysine-specific reagent concluded that there is close to 1 molecule of H1 per nucleosome in several mammalian somatic cell types, whereas in chicken erythrocyte chromatin there are 0.4 moles of H1 and 0.9 moles of H5 per nucleosome (18). H1/H1 crosslinking in rat liver nucleosomes is also consistent with 1 mole of H1/nucleosome (19). A single copy of H1 per nucleosome can readily be accommodated in a number of models of chromatin structure (see below) but non-integral numbers present problems for which there is no obvious solution at present, particularly since H1 and H5 have been shown to be interspersed in chicken chromatin (20).

H1 Location

A characteristic feature of native chromatin is that when micrococcal nuclease digests down from monosomes to core particles, there is a pause at 168 bp (21). The digested chromatin is never fully in this form since there is always a considerable amount of 146 bp core-particle-length DNA also present. The difference between 168 bp and 146 bp is an extension of ~10 bp at each end of the core particle (22). However, prior removal of H1 totally removes this pause at 168 bp, which is thus a characteristic of the presence of H1. If H1-depleted chromatin is reconstituted with just the globular domain of H1 that lacks the two terminal domains, the 168 bp pause is still observed (14). It was, therefore, proposed that the globular domain is located between the two extra 10 bp DNA extensions beyond the core particle that, together with H1, represent the 168 bp "chromatosome" (see Figure 3). Such a symmetrical location should seal the core particle and help define the exit angles of the DNA. The binding pocket provided by the exiting DNA strands has a width close to that of the DNA superhelix pitch in the core particle, i.e., 28 Å, and so could accommodate the H1 globular domain very well. In this model of H1 binding to chromatin it is the globular domain which is responsible for defining the exact location of H1 in the nucleosome. Other possible locations of H1 can be envisaged that preserve the stoichiometry of one H1/nucleosome. Two such are shown in Figure 4. In both of these, each globular domain is in contact with 2 stretches of 10 bp core particle extensions, but these are on two different nucleosomes and not on the same nucleosome as in Figure 3.

The globular domain alone is not able to condense chromatin to the extent found for intact H1 (14), so the flanking domains must play this role. Histone/DNA crosslinking indicates that while there is a strong contact between H1 and the ends of chromatosome-length DNA, there is also weaker contact over the whole length of this 170 bp DNA (23). This could mean that the H1 flanking domains are wrapped around their "own" core particle.

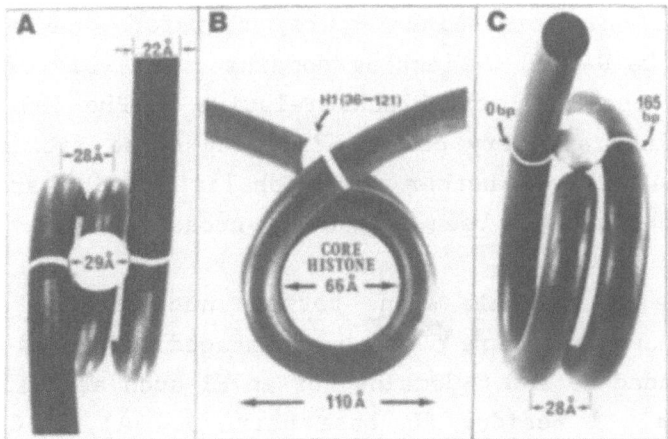

Figure 3. Proposed model for the location of the globular domain of H1-family molecules in a cage formed by two 10 bp DNA extensions to the core particle (-7 to -8 and +7 to +8). A third contact is with site 0.

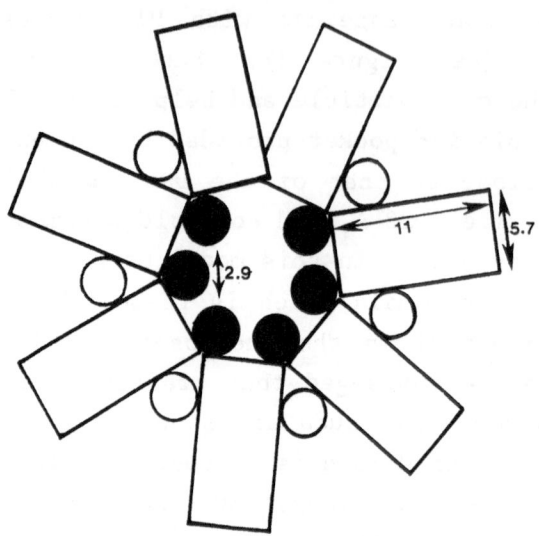

Figure 4. Two possible locations of the globular domain of H1 family molecules that maintain a stoichiometry of 1 H1/nucleosome.

H1/H1 protein crosslinking shows that H1s are readily polymerized and that this crosslinking does not disappear at low ionic strength when the chromatin is presumably extended to a beads-on-a-string structure (19). This suggests that H1 molecules are able to bridge between nucleosomes, i.e., stretch across the linker DNA. The crosslinks would, therefore, be expected to reside in the N- and C-flanking domains. Analysis of the positions of these H1-H1 crosslinks relative to Phe 106 shows that C-C crosslinks are very common, implying close proximity of C-domains (24, 25). Whether the globular domains are in close proximity has not yet been defined by crosslinking.

If the H1 molecule spans between nucleosomes, a model of the type shown in Figure 5 can be envisaged. This is drawn with fully extended N- and C-domains for an H1 such as that from calf thymus, i.e., 1 residue/DNA base pair (3.4 Å), a C-domain of ∿100 residues, and an N-domain of 35 residues. The function of the two "arms" of H1 would therefore be to neutralize the negative charge over a large region of the DNA and so act as a "cationic glue" to allow close approach of nucleosomes. This might be achieved by the C-domain acting to compact the linker DNA between adjacent chromatosomes.

The Supercoil

Chromatin higher order structure appears to be a cylinder of diameter ∿35 nm (26). The most attractive hypothesis is that this is a superhelix of ∿7 nucleosomes/turn with a pitch of ∿11 nm (3). Several lines of evidence indicate that the nucleosomes are fairly upright and have their 2-fold axes arranged radially with respect to the superhelix and not tangentially (26, 27). In such a model important questions are: 1) whether the nucleosomes are all identically arranged or are alternately arranged, and 2) whether the H1 molecules are located externally or internally. Ease of H1 removal and susceptibility to proteolysis might be taken to imply external location, but since the first stages of either of these processes would result in complete opening up of the superhelix, location cannot be defined from such considerations. Recent observations on essentially linkerless chromatin from neuronal nuclei (overall repeat ∿165 bp) containing about 0.4 H1s per nucleosome, suggest that a superhelix of fairly normal dimensions can be formed by such chromatin (28, 29). If such a structure were uniform and universal, it excludes a 180° alternation of adjacent nucleosomes since this requires a considerable length of linker DNA. It

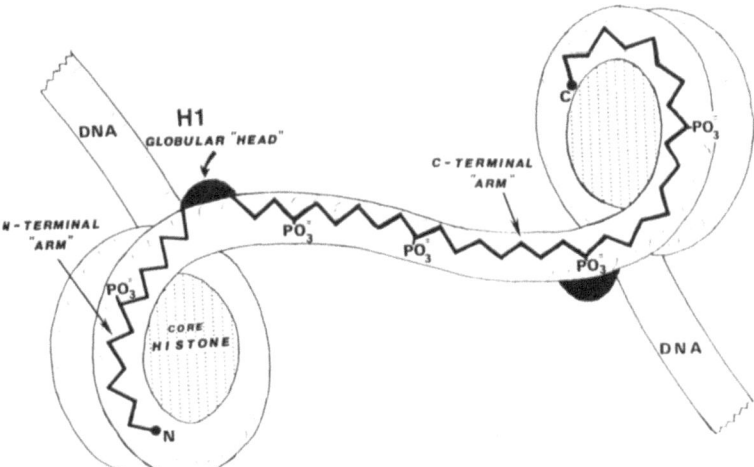

Figure 5. A possible mode of binding of histone H1 spanning 2 nucleosomes. Approximate sites of phosphorylation are indicated by $PO_3^=$.

also excludes a model in which all the H1 molecules (i.e., the site 0 end of the core particle) are uniformly on the periphery of the superhelix, since again linker DNA is needed to bridge between adjacent nucleosomes. The only model tenable on these results is one in which all nucleosomes have their "site 0 ends" and H1s to the center of the superhelix. Only then is it possible to envisage a linkerless chromatin with a "normal" 35 nm superhelix. In such a structure where significant lengths of linker DNA are present, these would have to be packed by some mechanism up the center of the superhelix. The density of negative charge so created would need neutralization and this could be the function of the flanking domains of H1. As seen from Figure 4, there seems no reason why the negative charge on the outside of the supercoil needs to be neutralized since the nucleosomes are well separated at this radius. H1 could there-fore be centrally located in the supercoil, albeit not folded as a single globule located at a single site, but broadly located so as to neutralize linker DNA charge.

REFERENCES

1. LITTAU, V.C., BURDICK, C.J., ALLFREY, V.G., and MIRSKY, A.E. (1965). The role of histones in maintenance of chromatin structure. Proc. Natl. Acad. Sci. USA 54, 1204-1212.

2. BRADBURY, E.M., CARPENTER, B.G., and RATTLE, H.W.E. (1973). Magnetic resonance studies of deoxyribonucleoprotein. Nature 241, 123-126.

3. THOMA, F., KOLLER, TH., and KLUG, A. (1979). Involvement of histone H1 in the organization of the nucleosome and of the salt-dependent superstructures of chromatin. J. Cell Biol. 83, 403-427.

4. KIMURA, T., MILLS, F.C., ALLAN, J., and GOULD, H. (1983). Selective unfolding of erythroid chromatin in the region of the active β-globin gene. Nature 306, 709-712.

5. KARPOV, V.L., PREOBRAZHENSKAYA, O.V., and MIRZABEKOV, A.D. (1984). Chromatin structure of hsp 70 genes, activated by heat shock: selective removal of histones from the coding region and their absence from the 5' region. Cell 36, 423-431.

6. SCHLISSEL, M.S. and BROWN, D.D. (1984). The transcriptional regulation of Xenopus 5S RNA genes in chromatin: the roles of active stable transcription complexes and histone H1. Cell 37, 903-913.

7. SOMMER, A. (1978). Yeast chromatin: search for histone H1. Molec. Gen. Genet. 161, 323-331.

8. PANYIM, S. and CHALKLEY, R. (1969). A new histone found only in mammalian tissues with little cell division. Biochem. Res. Commun. 37, 1042-1049.

9. STRICKLAND, W.N., STRICKLAND, M., BRANDT, W.F., VON HOLT, C., LEHMANN, A., and WITTMAN-LIEBOLD, B. (1980). The primary structure of histone H1 from sperm of the sea urchin Parechinus angulosus. 2. Sequence of the C-terminal CNBr peptide and the entire primary structure. Eur. J. Biochem. 104, 567-578.

10. KMIECIK, D. and SAUTIERE, P. (1985). Eur. J. Biochem. (in press).

11. GIANCOTTI, V., RUSSO, E., GASPARINI, M., SERRANO, D., DEL PIERO, D., THORNE, A.W., CARY, P.D., and CRANE-ROBINSON, C. (1983). Proteins from the sperm of the bivalve mollusc Ensis minor. Co-existence of histones and a protamine-like protein. Eur. J. Biochem. 136, 509-516.

12. HARTMAN, P.G., CHAPMAN, G.E., MOSS, T., and BRADBURY, E.M. (1977). Studies on the role and mode of operation of the very-lysine-rich histone H1 in eukaryote chromatin. The three structural regions of the histone H1 molecule. Eur. J. Biochem. 77, 45-51.

13. PUIGDOMENECH, P., PALAU, J., and CRANE-ROBINSON, C. (1980). The structure of sea-urchin-sperm histone 1 (H1) in chromatin and in free solution. Trypsin digestion and spectroscopic studies. Eur. J. Biochem. 104, 263-270.

14. ALLAN, J., HARTMAN, P.G., CRANE-ROBINSON, C., and AVILES, F.X. (1980). The structure of histone H1 and its location in chromatin. Nature 288, 675-679.

15. AVILES, F.J., CHAPMAN, G.E., KNEALE, G.G., CRANE-ROBINSON, C., and BRADBURY, E.M. (1978). The conformation of histone H5. Isolation and characterisation of the globular segment. Eur. J. Biochem. 88, 363-371.

16. BARBERO, J.L., FRANCO, L., MONTERO, F., and MORAN, T. (1980). Structural studies on histones H1. Circular dichroism and difference spectroscopy of histones H1 and their trypsin-resistant cores from calf thymus and from the fruit fly Ceratitis capitata. Biochemistry 19, 4080-4087.

17. TIKTOPULO, E.I., PRIVALOV, P., ODINTSOVA, T.I., ERMOKHINA, T.M., KRASHENINNIKOV, I.A., AVILES, F.X., CARY, P.D., and CRANE-ROBINSON, C. (1982). The central tryptic fragment of histones H1 and H5 is a fully compacted domain and is the only folded region in the polypeptide chain. A thermodynamic study. Eur. J. Biochem. 122, 327-331.

18. BATES, D.L. and THOMAS, J.O. (1981). Histones H1 and H5: one or two molecules per nucleosome? Nucleic Acids Res. 9, 5883-5894.

19. THOMAS, J.O. and KHABAZA, A.J.A. (1980). Cross-linking of histone H1 in chromatin. Eur. J. Biochem. 112, 501-511.

20. TORRES-MARTINEZ, S. and RUIZ-CARRILLO, A. (1982). Nucleosomes containing histones H1 and H5 are closely interspersed in chromatin. Nucleic Acids Res. 10, 2323-2335.

21. NOLL, M. and KORNBERG, R.D. (1977). Action of micrococcal nuclease on chromatin and the location of histone H1. J. Mol. Biol. 109, 393-404.

22. SIMPSON, R.T. (1978). Structure of the chromatosome, a chromatin particle containing 160 base pairs of DNA and all the histones. Biochemistry 17, 5524-5531.

23. BELYAVSKY, A.V., BAVYKIN, S.G., GOGUADZE, E.G., and MIRZABEKOV, A.D. (1980). Primary organization of nucleosomes containing all five histones and DNA 175 and 165 base-pairs long. J. Mol. Biol. 139, 519-536.

24. NIKOLAEV, L.G., GLOTOV, B.D., ITKES, A.V., and SEVERIN, E.S. (1981). Mutual arrangement of histone H1 molecules in chromatin of intact nuclei. FEBS Lett. 125, 20-24.

25. RING, D. and COLE, R.D. (1983). Close contacts between H1 histone molecules in nuclei. J. Biol. Chem. 258, 15361-15364.

26. SUAU, P., BALDWIN, J.P., and BRADBURY, E.M. (1979). Higher-order structures of chromatin in solution. Eur. J. Biochem. 97, 593-602.

27. MITRA, S., SEN, D., and CROTHERS, D.M. (1984). Orientation of nucleosomes and linker DNA in calf thymus chromatin determined by photochemical dichroism. Nature 308, 247-250.

28. PEARSON, F.C., BUTLER, P.J.G., and THOMAS, J.O. (1983). Higher-order structure of nucleosome oligomers from short-repeat chromatin. EMBO J. 2, 1367-1372.

29. ALLAN, J., RAU, D.C., HARBOURNE, N., and GOULD, H. (1984). Higher order structure in a short repeat length chromatin. J. Cell Biol. 98, 1320-1327.

STRUCTURAL STUDIES ON THE SECOND ORDER OF CHROMATIN ORGANIZATION

L. Wyns[1], S. Muyldermans[1], I. Lasters[1], and
J. Baldwin[2]

[1]Labo Algemene Biologie
Vrije Universiteit Brussel
Paardenstraat 65
1640 Sint Genesius Rode
Brussels, Belgium

[2]Department of Physics
Liverpool Polytechnic
Byrom Street
Liverpool L3 3AF   UK

INTRODUCTION

In the presence of histones of the H1 family, an increase in salt concentration to the region of 60-140 mM NaCl leads to a folding of 10 nm filaments of nucleosomes (which exist at 1-10 mM NaCl) into "thick" fibers of ∿30 nm diameter (1). It is this folding -- the next order above that of the nucleosome -- that we refer to as "second order chromatin structure." The second order structure is not uniform. At the level of the core particle (146 bp of DNA and the core histones), there is clearly a unique structure, although even at this level modifications of histones (and possibly DNA) may influence the higher order structure. Variability is found above the level of the core particles among chromatins from different sources. For example, species specific and cell-type specific differences in linker DNA length are observed: the nucleosomal repeat lengths reported vary from 154 base pairs in Aspergillus (2) up to 248 base pairs for sea urchin sperm (3). Most chromatins have repeat lengths between 180 and 210 base pairs. Also, differences are observed in H1 stoichiometry and the occurrence of different H1 subtypes and post-synthetical modifications (4).

These sorts of variability can be expected to introduce variation into the second order organization of chromatin.

ELECTRON MICROSCOPY STUDIES

There seems to be widespread agreement to consider the
nominal 30 nm fiber as the folded second order structure of
chromatin. However, the internal structure and overall organization of this fiber have not been determined with any cer-
tainty. Electron micrographs were the first and major source of
inspiration for the proposal of "thick-fiber" models. They have
led to two conflicting schools of thought: one that proposed
continuously coiled (solenoidal) structures, and the other that
favored a discontinuous structure (in the form of superbeads)
along the fiber's length.

Early studies on filaments spread from lysed nuclei established the existence of "thin" 10 nm and "thick" 25 nm fibers
(5). The thin fiber was observed in the presence of chelating
agents. Systematic analysis of thin sections of nuclei led
Davies et al. (6) to the concept of a "superunit thread" of
25-30 nm. This was manifested as a "dots and dashes" pattern
(transverse and longitudinal regions) in thin sections stained
with uranyl-lead. Most interestingly, one should not forget the
systematic presence in these studies of a 17 nm region that
stained densely with DNA-specific stains. This 17 nm region may
be related to the details of DNA organization within the 30 nm
fiber. More recently (i.e., since the recognition of nucleosomes), Davies et al. (7) have extended their findings, originally made on erythroblasts, to a variety of different tissues
and interpreted their pictures as sections through a helical
array of nucleosomes. Finch and Klug (8), on the basis of
freeze-etched and negatively stained images of polynucleosome
fragments (in 0.1 mM Mg^{++}), proposed a similar, solenoidal
arrangement of nucleosomes and specified ∿6 nucleosomes/turn and
a pitch of ∿11 nm. The latter figure would explain the systematic presence of a 11 nm reflection in fiber diffraction studies
discussed below.

Thoma et al. (1) performed a systematic electron microscopic analysis of the condensation of polynucleosomes obtained
by DNA cleavage with micrococcal nuclease. They continuously
increased the salt concentrations from ∿1 mM up to ∿100 mM NaCl.
The fibers folded with what seemed to be an increasing number of
nucleosomes, up to 6-8 nucleosomes per turn, and a nearly
constant pitch. Their homogeneously coiled model, based on
electron microscopic observations, allows for solenoidal structures which are both regular and not perfectly regular. Thus,
it can accommodate the linker length variation and the flexibility expected of a chromatin fiber. Handling of samples for
electron microscopic visualization certainly can increase any
pre-existing irregularity, or a fluctuating structure might be
trapped during fixation (see below).

Several electron microscopy studies have led to the concept
of the organization of nucleosomes into discrete-sized clusters
termed "superbeads" (9, 10) or "nucleomers" (11). In.their
studies, Davies et al. (6) had observed large chromatin granules
in thin sections of embedded intact nuclei, and Franke's group
has described the particularly striking arrangement of 18-26 nm
granules in cordlike threads in the peripheral chromatin adjacent to nuclear membranes (10). These fibers have a knobby
appearance and can also be seen upon lysis of nuclei from different sources. These authors associate their fibers with
globular arrangements of nucleosomes along the thick fiber.
More recently, they described clear differences of higher order
chromatin organization among different cell types (12).

There is a variety of possible artifacts in the preparation
of the especially fragile chromatin fiber for electron microscopy, and there is a tendency to rely too much on individual
images. It is necessary, therefore, to collect complementary
biochemical and physico-chemical information on chromatin in
solution.

NUCLEASE DIGESTION STUDIES

Discrete structures of higher order chromatin can be produced by partial nuclease digestion. However, the particles

produced in this way and the procedures used to generate them differ widely among laboratories and there is considerable confusion about the significance of particular particles. For instance, one should strictly distinguish reports (13, 14) of a _series_ of discrete nucleosome multimers from reports (12, 15, 16, 17, 18, 19) of a _single distribution_ of polynucleosomes clustered around one "stable peak" (Figure 1). By "stable peak" we mean that the peak position of the polynucleosomes' distribution persistently migrates to the same position in sucrose gradients containing sufficient NaCl (∿60-140 mM) to conserve the higher order structure.

A particle series from monomer multinucleosomes up to trimer, produced by nuclease digestion of nuclei (13, 14), was taken as strong evidence in favor of the existence of "super-beads." We still keep our earlier position (20), as confirmed by the work of Walker et al. (21), that the multiple maxima observed in these sucrose gradient separations result from an

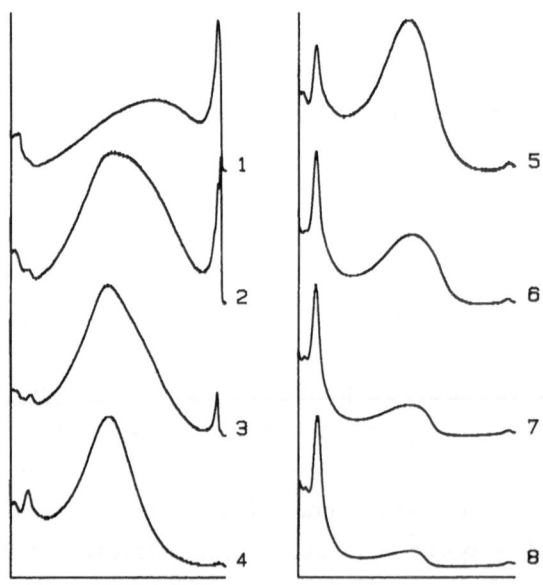

Figure 1. Sedimentation profiles of chromatin solubilized from chicken erythrocytes nuclei (2×10^9/ml) by digestion with micrococcal nuclease (150 u/ml) for 15 sec; 30 sec; 1; 2.5; 5; 7.5; 10; and 12.5 min (profiles 1 to 8, respectively). All sucrose gradients contained 80 mM NaCl. Mononucleosomes sediment at the position of the peak at the left end of the profiles. The peak in the middle of the sedimentation profile is referred to as the "stable peak."

isolation of RNP particles comigrating with a continuous DNP profile. Indeed, in one of the superbead papers it was stated that observations were "obfuscated by contaminating RNP's" (15).

In more recent work, Strätling and Klingholz described a quite different multinucleosomal entity in digests of rat liver chromatin. This took the form of a distribution of polynucleosome particles around a single stable peak (15) at the position of migration of intact 12-mers. Ruiz-Carrillo et al. (16) described a similar observation for chicken erythrocyte chromatin digests. Here the maximum occurred around the 25-mer position. Such distributions of particles, observed as we have said in sucrose gradients containing appropriate salt concentrations (60-140 mM), are produced early in digestion when little or no mononucleosomes have yet formed (Figure 1, profiles 2 and 3). Later in digestion, the bulk of the total chromatin can be solubilized into this sort of distribution (Figure 1, profile 4). Such a distribution about a stable peak does not even require intact DNA: the DNA may be internally nicked and cut, and protein-protein and/or protein-DNA contacts must hold the structure together (15, 18, 25). This indicates that the structures are not actually resistant to the action of the nuclease, but are resistant against unfolding at near physiological salt conditions (15). Moreover, the whole phenomenon is not nuclease specific: micrococcal nuclease, Ca-Mg dependent rat liver endonuclease, and DNase I digestions of chicken erythrocyte nuclei produce the same result.

In sucrose gradients at low ionic strength, the well-defined distribution falls apart and a continuous distribution of slower sedimenting fragments is observed with further digestion. The formation of particles in the well-defined distribution is clearly reversible: dialysis into buffers containing 5 mM NaCl and back into 80 mM NaCl quantitatively reforms the same distribution in sucrose gradient sedimentation. To our surprise, dialysis into 600 mM NaCl and back into 80 mM NaCl also reformed the distribution on sedimentation (25). (This, however, was not quantitative for long digestion times.)

Recently, Zentgraf and Franke (12) further extended the results to other cell types and compared differences in knobby fiber diameters and large granules, as observed by electron microscopy, with differences in position of migration of the "stable peak" in gradients: sea urchin spermatozoa gave rise to 260 S particles containing ∿48 nucleosomes and arise from 48 nm knobby fibers and chicken erythrocyte nuclei gave 105 S for ∿20 nucleosomes from 36 nm threads. This variability in sizes for these structures has been criticized (22), but we feel one should not be surprised by it. Differences in linker length and H1 stoichiometry clearly exist (23), and these could account for the variability.

Some authors have claimed that a massive rearrangement of H1/H5 histones is required to produce the fast sedimenting chromatin peak. Indeed, in an earlier report, Jorcano et al. (24) showed that upon prolonged digestion, when large quantities of mononucleosomes had been formed, a peak of up to 120 S developed in gradients containing 30 mM NaCl. These 120 S particles indeed required a massive accumulation of H1/H5 in order to be observed at this relatively low salt concentration. However, there is no observable accumulation of H1/H5 in similar "stable peak" structures obtained from early nuclease digestions (25). Independent of this, the data of Jorcano et al. (24) remain quite remarkable and seem to indicate that, in different ionic conditions, oligonucleosomes in the presence of H1/H5 are able to reassemble into similar higher order structures.

More recent experiments in our laboratory have strengthened this idea. Oligonucleosome fractions of defined length isolated from sucrose gradients in low salt (5 mM NaCl, in which the particles do not lose any H1 or H5 upon centrifugation (26)) reassociate into multimeric aggregates when redialyzed into 80 mM NaCl and re-analysed in 80 mM NaCl containing sucrose gradients (Figure 2). This figure clearly shows that several oligonucleosomal fractions differing in length from about 6 (profile 1) to 12 nucleosomes (profile 3) reassemble into a limit series of multimeric higher order chromatin structures. A fraction containing an average nucleosomal length of 25-mers did not aggregate into larger structures under similar conditions (profile 4). The existence of an assembly barrier around 25 nucleosomes could be inferred from these experiments, since we failed under any condition to observe a soluble, larger <u>assembly</u>

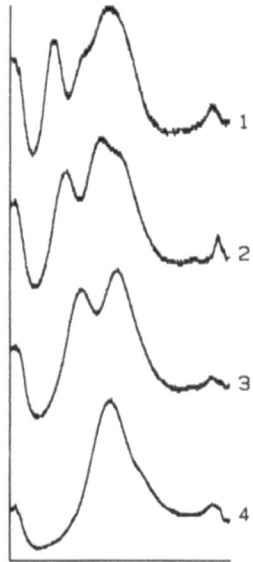

Figure 2. Sedimentation profiles of oligonucleosomal fractions, obtained from sucrose gradient sedimentation at 5 mM NaCl. The fractions have mean lengths of 6, 8, 12, and 20 nucleosomes (profiles 1 to 4, respectively). The slower migrating peak (at the left end of the profiles) is the monomer peak, while the other peaks arise from the assembly of this monomer. Note that no assembly products with a nucleosomal content of more than ∿24 nucleosomes are formed.

product. Quantitative gel electrophoresis of the multimer limit-assembly products revealed a similar H1/H5 stoichiometry to that obtained from a total nuclear chromatin extract (25). This indicates that the formation of multimeric assembly structures at near physiological ionic strengths is not dependent on a massive accumulation of lysine-rich histones.

In summary, our assembly experiments weaken the statement that the "stable peak" particles reflect a discontinuous stabilization of the native 30 nm fiber (15, 12). However, our results point toward the existence of an intrinsic property of oligonucleosomes to aggregate into more or less discrete multimeric structures at approximately physiological ionic strengths.

For 30 years it has been known that chromatin and even nuclei produce a "classic" series of reflections (27, 28). Since then, broad rings at ∿11, 5.5, 3.7, 2.7, and 2.1 nm (equivalent d-spacing) have been consistently observed from chromatin of many different sources (29, 30, 31). It became clear (32,33) that the 11 nm ring observed by neutron diffraction moved progressively to a spacing of ∿8 nm as chromatin was dried in D_2O and that the ring was present all the time during the drying procedure. The 11 nm ring behaves quite differently from the other rings in this respect.[*]

In pulled fibers, the 11 nm ring is highly meridionally oriented. It was shown that this diffraction-peak orientation was not truly meridional but in the form of a cross giving slightly off-meridional orientation (34). This could be interpreted in terms of a higher order arrangement of nucleosomes in the form of a flat helix which would give such a cross pattern. Tightening of the helix (to a smaller pitch) would explain the movement to 8 nm. This is entirely consistent with electron microscopy data of Finch and Klug (8); these data were taken from samples prepared only in 0.2 mM Mg^{++} conditions, which do not normally give a fully folded higher order chromatin structure (37). No doubt the higher order folding occurred during the extreme conditions used in drying fibers for electron microscopy.

For all chromatins and at all concentrations, the 3.5 nm ring remains fixed for neutron scattering in 65% D_2O, when the DNA is contrast-matched. This is observed for dilute solutions of chromatin right up to dry chromatin (in a vacuum). Studies

---

[*] When the drying procedure is in H_2O, there is a contrast variation phenomenon observed for neutron scattering (and for all drying experiments with X-ray diffraction). This results in a disappearance of the diffraction ring for chromatin at intermediate hydrations, e.g., equilibrated in ∿70% relative humidity. Many people have interpreted and, for some reason, still interpret this disappearance as a disappearance of structure (35) but we stress that the ring does not disappear for neutron scattering in D_2O. Moreover, the contrast variation effect has been fully explained and is entirely consistent with many data using deuterated chromatin (32, 34, 36).

of nucleosome core particles by scattering methods (38, 39) show that this maximum arises due to the Fourier transform of the protein core of the core particle. Thus there is no effect of hydration on the hydrophobic regions of the octamer histone core.

Scattering studies on core particles also give a clear insight into the origins of the 5.5 nm and 2.7 nm diffraction rings in polynucleosomal chromatin. These rings come from the the scattering contributions of the octamer core protein and the DNA of the core particle. For X-rays and neutrons, with samples in H_2O, a structure with lower scattering regions in the center (the protein) and higher scattering at the outside (the DNA) gives scattering patterns with just such diffraction rings (at 5.5 and 2.7 nm). If we look at the effect at 48% D_2O (the contrast match position for whole chromatin), we see the 5.5 nm ring move to the 7.0 nm ring observed in the internal structure of chromatin (38, 39, 40). Thus, we can explain the series of diffraction rings quite unambiguously. Changes in the 5.5 nm and 2.7 nm rings observed by X-rays or neutrons upon drying chromatin in H_2O are probably due to changes in DNA structure. The higher order structure simply tightens on drying chromatin and the octamer histone core does not change much. There is a change in DNA hydration and possibly in DNA conformation as the structure tightens. There is some enhancement of the ring at 5.5 nm, probably due to a second order reflection of the 11 nm higher order diffraction peak. Enhancement of the 2.7 nm peak in chromatin is sometimes observed in over-stretched or sheared material due to portions of DNA which are pulled out and pack together as DNA fibers.

Langmore and Paulson (35) have recently made a systematic analysis of the low-angle X-ray diffraction patterns of a wide range of materials: nuclei from many different cell types, metaphase chromosomes, and living chicken erythrocytes and lymphocytes. (Suitable background correction could be made for the whole cells.) A third type of reflection was systematically observed in erythrocytes centers around 40 nm. These authors interpret this as being due to periodical packing of thick fibers. This 40 nm ring moves to larger spacings upon removal of divalent ions, concomitant with disaggregation of fibers in

EM pictures. The other rings are completely unaltered upon this treatment. The 40 nm ring is the only one which differs among different sources: it is 31 nm in HeLa metaphase chromosomes, but 32-34 nm in lymphocytes. (Rat liver nuclei, containing nearly no chromocenters, do not show any such reflection at all.)

This study proved the existence in vivo of 30-40 nm fibers (35), and thus validates dilute solution studies in vitro. This is very reassuring. Diameters in the range 30-35 nm were obtained from solution studies, and both peak positions and relative magnitudes resembled closely those observed from nuclei and cells. Another very important finding, which is discomforting for electron microscopy work, was that low angle patterns from samples taken at different stages of the preparation procedure used for electron microscopy of erythrocytes (as described by Davies et al. (6, 7)) show that all reflections, except that at 40 nm, are abolished during the dehydration step (after fixation). From Langmore and Paulson's work (35) it is not quite clear, however, if this is due to the contrast effect described earlier. The 40 nm reflection shrinks to 31 nm, corresponding to the center to center spacing between fibers measured by these authors (35) and earlier by Davies et al. (6, 7) from electron micrographs.

One can therefore interpret diffraction from intact chromatin in fibers and gels, and -- now with Langmore and Paulson's recent work (35) -- in living cells. To proceed further it is necessary to cleave chromatin into fragments of 200 nucleosomes or less. This allows us to gain more information from solution scattering data and later from fibers. Parameters of the higher order coil can be obtained by cross-section analysis of solution scattering data from such polynucleosome chains. Such work has been reviewed recently (41).

SCATTERING STUDIES ON NUCLEASE-RELEASED MULTINUCLEOSOMAL PARTICLES

Recently, we compared the physico-chemical solution properties of two types of supranucleosomal particles containing about

23 nucleosomes. The first type moves as a single distribution around a "stable peak" position and contained cut DNA (as described above). The second type was polynucleosomes with intact DNA of the same size. The second type of particle came from very light micrococcal nuclease digestions followed by sucrose gradient centrifugation (with zonal rotors) in 5 mM NaCl. The low salt prevents H1/H5 redistribution (26, 42, 43). A sharp fraction corresponding to chains with 23 ± 2 nucleosomes (the length was taken from DNA gels) was used for further study, either as prepared or after overnight fixing in 0.1% glutaraldehyde (1). The first preparation, containing cut DNA, was obtained from similar runs in sucrose gradients containing 80 mM NaCl, but this material had been fixed in 0.1% glutaraldehyde after release from the nuclei and before further fractionation. Fixation was carried out again to prevent H1/H5 rearrangements or losses that occur upon centrifugation. The peak fraction from the resulting gradient profile was taken for further work. Hydrodynamic study of this fixed material led to a $s^0_{20,w}$ value of 101 ± 2 S (44). (Compare the value of 105 S of Zentgraf and Franke (12) with similar material.) This, together with a $D^0_{20,w}$ of (0.99 ± 0.02) x 10^{-7} cm^2/sec from photon correlation spectroscopy measurements, led to a molecular weight equivalent to 25 nucleosomes. Independent determinations by neutron scattering produced a figure of 23 nucleosomes (45).

A first and most important result from our scattering studies was that nearly identical scatter curves were obtained from all samples (i.e., from supranucleosomal particles -- either fixed or not fixed -- with intact DNA, and from particles aggregated from smaller oligonucleosomes). A comparison of theoretically predicted distance distribution functions with the experimentally derived ones allowed us to discard any model with spherical organization. Best fits were obtained with models having an overall cylindrical symmetry (Figure 3). (Both cylinders and solenoids of nucleosomes gave best fits.) An exercise in fitting models showed there was a rather sharp restriction in allowed diameter values. We also consistently observed that models lacking a central hole produced a better fit than did models with a hole. From a radius of gyration of 15.8 nm (obtained from the Guinier plots) and the assumption of a cylinder of about 30 nm diameter (which came from cross-section anal-

Our experiments show that intact 23-nucleosome chains fold in a way entirely consistent with the regular solenoidal model proposed from electron microscopy, fiber diffraction and cross-section analyses of solution scatter data. So far as we can tell at resolution of our data, the 23-nucleosome particles that have cleaved DNA (i.e., the multimeric aggregates) form an identical structure.

CONCLUSIONS

Size-fractionation of multinucleosomes produced by micro-coccal nuclease digestion of chicken erythrocyte chromatin has revealed a stable peak around 100 S (corresponding to 25 nucleo-somes) irrespective of the extent of digestion. Although one can obtain particles of this size (100 S) that contain intact DNA, particles of the same size can also be produced by sponta-neous reassociation of smaller oligonucleosomes. The inter-pretation of this reassembly must be that in avian erythrocytes there is a preference to form stable higher order chromatin particles containing 25 nucleosomes. Identification of these stable higher order chromatin particles as a structural repeat unit of the 300 Å fiber is, however, premature. Indeed, peaks at multiples of ∿25 nucleosomes are never observed on a sucrose gradient and there is no evidence for a preferential DNA cleav-age by the nuclease at multiples of ∿25 nucleosomes. We would rather favor the explanation that the 30 nm chromatin fiber is rather homogeneous but also has a fair amount of flexibility and that the superbeads observed on many electron micrographs find their origin in transient interruptions along the fiber axis. Fixation would generate multinucleosomal structures corre-sponding to cross-linked neighboring turns of the higher order coil. The behavior of chromatin in solution, as studied by sedimentation, is that of a fluctuating coil, with transient gaps between neighboring turns, rather than that of a stiff, non-bending solenoid (48).

Results of fiber diffraction and solution scattering from large multinucleosome fragments (reviewed in 41), of electron microscopy (1, 8), and of scattering from nuclei indicate the existence of a coiled higher order chromatin structure with a

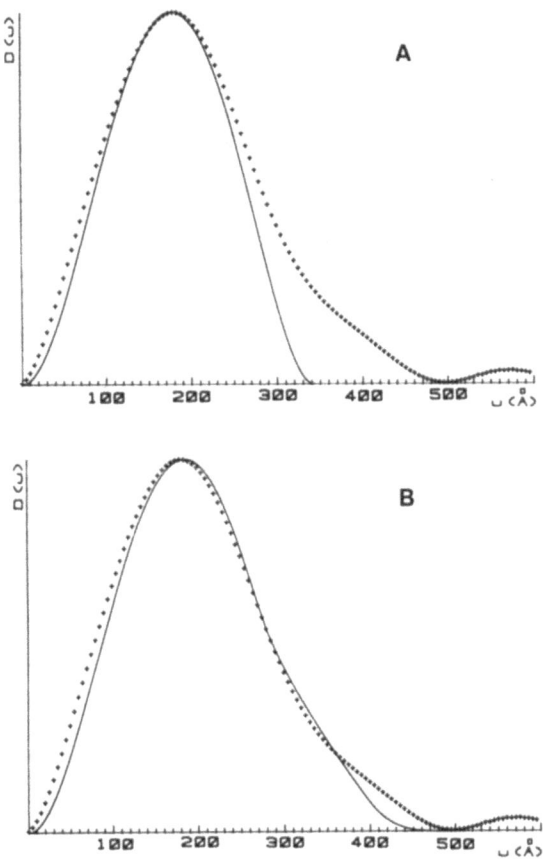

Figure 3. Comparison of the distance distribution function (D(u)) for the intact 23-mers (+ curve) with the D(u)'s for a 171 Å-radius sphere (solid curve in Å) and a solid cylinder of diameter 280 Å and height 400 Å (solid curve in B). The dimensions of these models were chosen to match the most probable vector length for the 23-mers. The D(u)'s for these models were analytically calculated whereas the D(u) for the 23-mers was derived by sine-Fourier transforming the X-ray scattered intensities. The data were extrapolated to zero-scattering angle using the Guinier approximation method (radius of gyration = 156 Å).

yses), a height of 40 nm was proposed (37,46,47). This agreed with the largest vector observed in the distance distribution function (48 nm). These proposed parameters would fit closely with the "classical" packing of nucleosomes in chicken erythrocyte 30 nm fibers: 6 nucleosomes/turn, a packing ratio of 0.55 nucleosomes/nm and an 11 nm pitch.

mass per unit length of 6-8 nucleosomes per nm and a diameter of ∿30-34 nm (the 30 nm fiber). Our solution scattering studies of nucleosomal (23 ± 2)-mers in a folded conformation show well-defined particles with radius of gyration 15.8 nm and maximum cord length 48 nm. These cannot have nearly spherical shape. The scattering data are consistent with models in which the 23-mers are considered as sections of the overall 30 nm fiber.

We must remember that chromatin is by its very nature heterogeneous, due to variation in linker length, histone H1 and non-histone protein contents, and histone modifications. Chromatin is also a very delicate structure, easily subject to mechanical and biochemical distortions. Therefore one should not be surprized that different sources of chromatin have different structures, which may or may not conform to the general scheme. This may perhaps account for the absence of the 40 nm packing reflection observed from rat liver nuclei (35), in which little or no heterochromatin is observed in the electron microscope, as opposed to the chicken erythrocyte material, where we are mainly studying heterochromatin.

ACKNOWLEDGEMENTS

The authors acknowledge the F.G.W.O.-I.K.W. (contract no. 3.0074.84) for grant support.

REFERENCES

1. THOMA, F., KOLLER, T., and KLUG, A. (1979). Involvement of histone H1 in the organization of the nucleosome and of the salt dependent superstructures of chromatin. J. Cell Biol. 83, 403-427.

2. HORGEN,P. and SILVER, J. (1978). Chromatin in eukaryotic microbes. Ann. Rev. Microbiol. 32, 249-284.

3. SAVIC, A., RICHMAN, P., WILLIAMSON, P., and POCCIA, D. (1981). Alteration in chromatin structure during early sea urchin embryogenesis. Proc. Natl. Acad. Sci. USA 78, 3706-3710.

4. HOHMAN, P. (1978). The H1 class of histone and diversity in chromosomal structure. Subcell. Biochem. 5, 87-127.

5. RIS, H. (1975). In: "Structure and Function of Chromatin," Ciba Found. Symp. Vol. 28 (D.W. Fitzsimmans and G.E. Wolstenholme, eds.) pp. 7-28. Elsevier, New York.

6. DAVIES, H.G., MURRAY, A.B., and WALMSLEY, M.E. (1974). Electron-microscope observations on the organization of the nucleus in chicken erythrocytes and a superunit thread hypothesis for chromosome structure. J. Cell Sci. 16, 261-299.

7. DAVIES, H.G. and HAYNES, M.E. (1976). Electron-microscope observations on cell nuclei in various tissues of a teleost fish: the nucleolus-associated monolayer of chromatin structural units. J. Cell Sci. 21, 315-327.

8. FINCH, J.T. and KLUG, A. (1976). Solenoidal model for superstructure in chromatin. Proc. Natl. Acad. Sci. USA 73, 1897-1901.

9. HOZIER, J., RENZ, M., and NEHLS, P. (1977). The chromosome fiber: evidence for an ordered superstructure of nucleosomes. Chromosoma 62, 301-317.

10. FRANKE, W.W., SCHEER, U., TRENDELENBURG, M., ZENTGRAF, H.W., and SPRING, H. (1978). Morphology of transcripitionally active chromatin. Cold Spring Harbor Symp. Quant. Biol. 42, 755-772.

11. KIRYANOV, G.J., SMIRNOVA, T.A., and POLYAKOV, V.Y. (1982). Nucleomeric organization of chromatin. Eur. J. Biochem. 124, 331-338.

12. ZENTGRAF, H.W. and FRANKE, W.W. (1984). Differences of supranucleosomal organization in different kinds of chromatin: cell type-specific globular subunits containing different numbers of nucleosomes. J. Cell Biol. 99, 272-286.

13. STRATLING, W.H., MULLER, U., and ZENTGRAF, H.W. (1978). The higher order repeat structure of chromatin is built up of globular particles containing eight nucleosomes. Exp. Cell Res. 117, 303-311.

14. BUTT, T.R., JUMP, D., and SMULSON, M. (1979). Nucleosome periodicity in HeLa cell chromatin as probed by micrococcal nuclease. Proc. Natl. Acad. Sci. USA 76, 1628-1632.

15. STRATLING, W. and KLINGHOLZ, R. (1981). Supranucleosomal structure of chromatin: digestion by calcium/magnesium dependent endonuclease proceeds via a discrete size class of particles with elevated stability. Biochemistry 20, 1386-1392.

16. RUIZ-CARRILLO, A., PUIGDOMEMECH, P., EDER, G., and LURZ, R. (1980). Stability and reversibility of higher ordered structure of interphase chromatin: continuity of deoxyribonucleic acid is not required for maintenance of folded structure. Biochemistry 19, 2544-2554.

17. RUIZ-CARRILLO, A. and PUIGDOMENECH, P. (1982). Effect of histone composition on the stability of chromatin structure. Biochem. Biochys. Acta 696, 267-274.

18. MUYLDERMANS, S., LASTERS, I., and WYNS, L. (1982). Observation and characterization of discrete supranucleosomal structures or minisolenoids. Arch. Int. Physiol. Biochim. 90, (4), B228.

19. RENZ, M. (1979). Heterogeneity of the chromosome fiber. Nucleic Acids Res. 6, 2761-2767.

20. MUYLDERMANS, S., LASTERS, I., WYNS, L., and HAMERS, R. (1980). Upon the observation of superbeads in chromatin. Nucleic Acids Res. 8, 2165-2172.

21. WALKER, B., LOTHSTEIN, L., BAKER, C., LE STOURGEON, W. (1980). The release of 40S hnRNP particles by brief digestion of HeLa nuclei with micrococcal nuclease. Nucleic Acids Res. 8, 3639-3657.

22. BUTLER, P. (1983). The folding of chromatin. CRC Crit. Rev. Biochem. 15, 57-91.

23. BATES, D. and THOMAS, J. (1981). Histones H1 and H5: one or two molecules per nuclesomes? Nucleic Acids Res. 9, 5883-5894.

24. JORCANO, J., MEYER, G., DAY, L., and RENZ, M. (1980). Aggregation of small oligonucleosomal chains into 300 A globular particles. Proc. Natl. Acad. Sci. USA 77, 6443-6447.

25. MUYLDERMANS, S., LASTERS, I., HAMERS, R., and WYNS, L. Assembly of oligonucleosomes into a limit series of multimeric higher order chromatin structures. Eur. J. Biochem. (in press).

26. CARON, F. and THOMAS, J. (1981). Exchange of histone H1 between segments of chromatin. J. Mol. Biol. 146, 518-537.

27. WILKINS, M. (1956). Cold Spring Harbor Symp. Quant. Biol. 21, 75-90.

28. LUZZATI, V. and NICOLAEV, A. (1959). Etude par diffusion des rayons X aux petits angles des gels d'acide desoxyribonucleique et de nucleoproteines. J. Mol. Biol. 1, 127-133.

29. LUZZATI, V. and NICOLAEV, A. (1963). The structure of nucleohistones and nucleoprotamines. J. Mol. Biol. 7, 142-163.

30. ZUBAY, G. and WILKINS, M. (1962). An X-ray diffraction study of histones and protamine in isolation and in combination with DNA. J. Mol. Biol. 4, 444-450.

31. PARDON, J., WILKINS, M., and RICHARDS, B. (1967). Superhelical model for nucleohistone. Nature 215, 508-509.

32. BALDWIN, J., BOSELEY, P., BRADBURY, E., and IBEL, K. (1975). The subunit structure of the eukaryotic chromosome. Nature 253, 245-249.

33. CARPENTER, B., BALDWIN, J., BRADBURY, E., and IBEL, K. (1976). Organization of subunits in chromatin. Nucleic Acids Res. 3, 89-100.

34. BALDWIN, J., CARPENTER, B., CREPI, H., HANCOCK, R., STEPHENS, R., SIMPSON, J., BRADBURY, E., and IBEL, K. (1978). Neutron scatter from chromatin in relation to higher order structure. J. App. Cryst. 11, 484-486.

35. LANGMORE, J. and PAULSON, J. (1983). Low angle X-ray diffraction studies of chromatin structure in vivo and in isolated nuclei and metaphase chromosomes. J. Cell Biol. 96, 1120-1131.

36. BRADBURY, E., BALDWIN, J., CARPENTER, B., HJELM, R., HANCOCK, R., and IBEL, K. (1975). Neutron scatter studies of chromatin. Brookhaven Symp. Biol. 27, IV, 97-117.

37. SUAU, P., BRADBURY, E., and BALDWIN, J. (1979). Higher order structures of chromatin in solution. Eur. J. Biochem. 97, 539-602.

38. SUAU, P., KNEALE, G., BRADDOCK, G., BALDWIN, J., and BRADBURY, E. (1977). A low resolution model for the chromatin core particle by neutron scattering. Nucleic Acids Res. 4, 3769-3786.

39. BRADDOCK, G., BALDWIN, J., and BRADBURY, E. (1981). Neutron-scattering studies of the structure of chromatin core particles in solution. Biopolymers 20, 327-343.

40. HJELM, R., KNEALE, G., SUAU, P., BALDWIN, J., BRADBURY, E., and IBEL, K. (1977). Small angle neutron scattering studies of chromatin subunits in solution. Cell 10, 139-151.

41. CRANE-ROBINSON, C., STAYNOV, D., and BALDWIN, J. (1984). Chromatin higher order structure and histone H1. Comments on Molecular and Cellular Biophysics 2, 219-265.

42. LASTERS, I., MUYLDERMANS, S., WYNS, L., and HAMERS, R. (1980). The role of H1 and H5 in the structure of chicken erythrocyte chromatin. Arch. Int. Physiol. Biochim. 88, (3), B140.

43. LASTERS, J., MUYLDERMANS, S., WYNS, L., and HAMERS, R. (1981). Differences in rearrangements of H1 and H5 in chicken erythrocyte chromatin. Biochemistry 20, 1104-1110.

44. WYNS, L., NIEUWENHUYSEN, P., MUYLDERMANS, S., and CLAUWAERT, J. (1982). Chromatin polynucleosomal fragments of well-defined size: their homogeneity, number of nucleosomes and solution structure. Arch. Int. Physiol. Biochim. 90, (2), BP10.

45. WYNS, L., MUYLDERMANS, S., POLAND, T., and BALDWIN, J. (1982). Neutron-scattering studies of chromatin minisolenoids. Arch. Int. Physiol. Biochim. 90, B226-227.

46. BRUST, R. and HARBERS, E. (1981). Structural investigation on isolated chromatin of higher order organization. Eur. J. Biochem. 117, 609-615.

47. LEE, K., MANDELKERN, M., and CROTHERS, D. (1981). Solution structural studies of chromatin fibers. Biochemistry 20, 1438-1445.

48. THOMAS, J. and BUTLER, P. (1980). Size dependence of a stable higher order structure in chromatin. J. Mol. Biol. 144, 89-93.

MITOTIC CHROMOSOME STRUCTURE: AN UPDATE, DECEMBER 1984

William C. Earnshaw

Department of Cell Biology and Anatomy
Johns Hopkins School of Medicine
725 North Wolfe Street
Baltimore, Maryland 21205

INTRODUCTION

Even though the study of mitotic chromosomes dates back to the 19th century, little is known about the details of chromosome architecture and the mechanism of chromosome condensation at mitosis (for reviews see references 1 and 2). In this review, I will briefly describe the current models of chromosome architecture and subsequently examine the current status of the chromosome scaffold model in greater depth.

Chromosome models are of two basic types: those which postulate the need of a non-histone scaffold for establishment and/or maintenance of the characteristic mitotic morphology; and those which hold that mitotic chromosome morphology is solely determined by properties of the nucleohistone fiber.

NON-SCAFFOLD MODELS

The Folded Fiber Model

This model was proposed by DuPraw (3) based on observation of the chromatin fiber at the periphery of chromosomes prepared for electron microscopy by spreading on an aqueous surface and critical point drying. The model was extended to suggest that

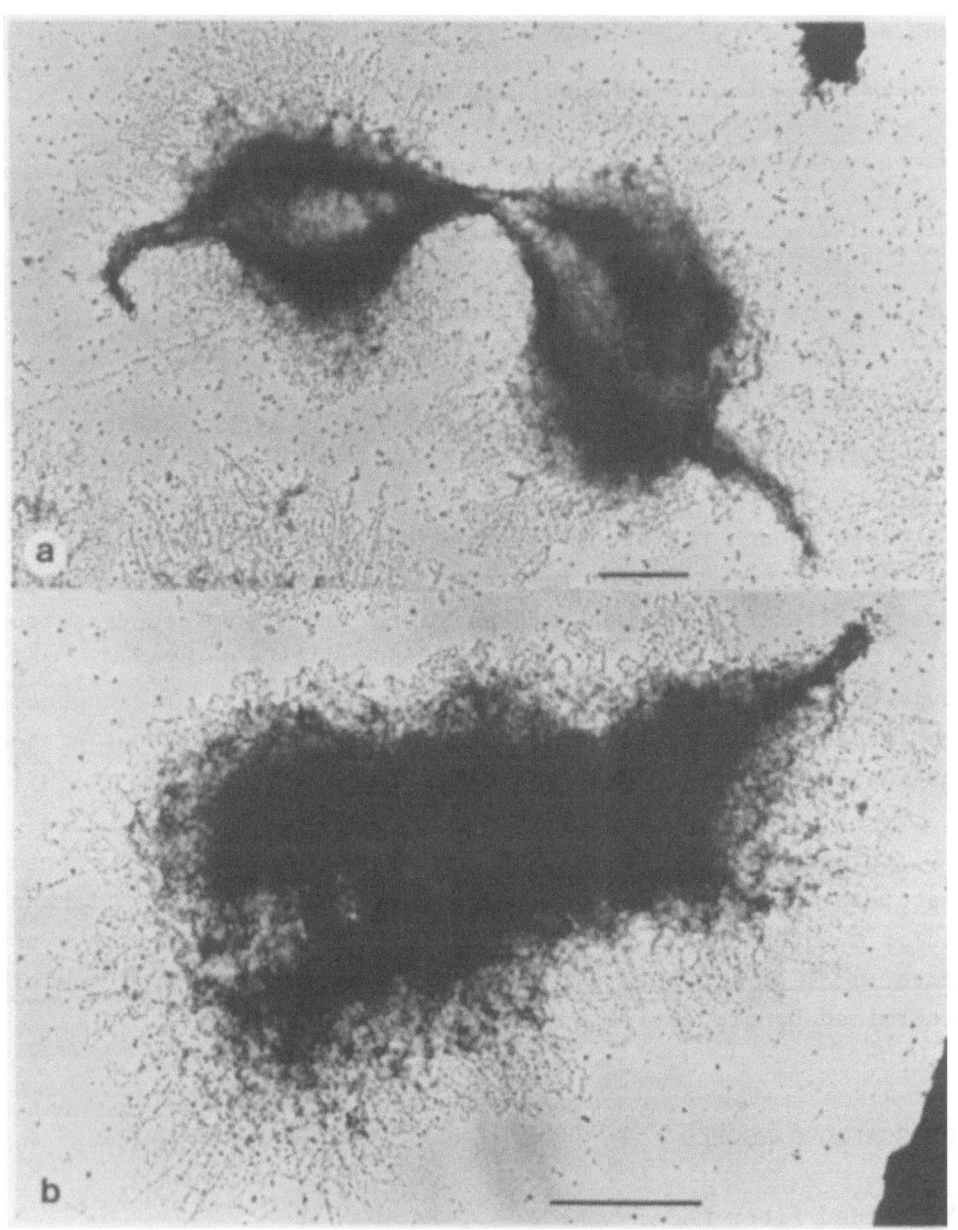

the randomly coiled chromatin fiber could be supertwisted to give the whole chromatid a helical appearance (4). These packing patterns were assumed to arise from intrinsic properties of the chromatin fiber itself.

This model is difficult to evaluate since even though the chromatin fiber does appear to be randomly folded in chromosomes prepared for electron microscopy in this way, the contribution of the specimen preparation procedure to this apparent disorder is unknown. In addition, specific banding patterns which are observed following a variety of treatments of mitotic chromosomes (e.g., G-banding, Q-banding, R-banding; reviewed in reference 5) are difficult to explain if the folding of the chromatin fibers is truly random.

Helical Models

Knowledge that the DNA follows a helical path about the nucleosome (6), coupled with the observation that isolated chromosomes may readily be induced to adopt a macroscopic helical coiling (7-9) have led to the proposal of helical chromosome models (10-12). The first of these, the "unit fiber" model (10, 11, 13-15), has been brought into question by recent results. This model proposed that the penultimate level of coiling of the

---

Figure 1.   Early stages in "unit fiber" formation. Isolated HeLa chromosomes were subjected to shear by centrifugation, aspiration of the supernatant, and vigorous vortexing of the pellet. The sample was then resuspended in RSB buffer (10 mM Tris-HCl pH 7.4, 10 mM NaCl, 5 mM $MgCl_2$) and processed for electron microscopy as described in detail for Figures 2 and 3 of reference 16. This procedure involved swelling the chromosomes in solution in the presence of dilute EDTA, centrifugation onto an electron microscope grid, fixation with glutaraldehyde, and positive staining with phosphotungstic acid while wet. After drying, the particles were then shadowed with Pt:Pd. This sample contained many "unit fibers" where no chromatin loops were seen (not shown; see reference 16). These images suggest strongly that the "unit fiber" is substantially more dense than the chromatin in the undistorted chromatid arms, and give the impression that the fiber is formed by twisting the chromatin. The magnification bar indicates 1 micron.

chromatin fiber of mitotic chromosomes is into a "unit fiber" of diameter 4000 Å. Thread-like structures (often seen in preparations of mitotic chromosomes which have been exposed to shear) were proposed to be formed by unwinding of the chromatids to generate "unit fibers" (10, 11, 13, 14).

Two observations suggest that these fibers are formed not by unwinding the chromatids, but as a result of a complex alteration of the chromatin packing. 1) When "unit fibers" are observed under conditions where intact chromosomes are expanded (so that radial chromatin loops are readily visible), the chromatin of the fiber remains highly compacted (16). Therefore, the chromatin in the "unit fibers" is packed differently from the chromatin in intact chromosomes. 2) One mitotic chromosome (two sister chromatids) gives rise to a single fiber (Figure 1). Two views of early stages in the formation of a "unit fiber" are shown. These images suggest that "unit fibers" are formed by a spinning out process, similar to twirling spaghetti around a fork. The explanation for why this transformation occurs is totally unknown.

In addition to the above experimental results, certain theoretical calculations performed for the number 4 human chromosome have suggested that helical coiling cannot account for the packing of the chromatin fiber into structures of the observed size (17). Although data in support of strictly helical models are limited, it must be remembered that certain gentle treatments result in pronounced helical macrocoiling of the chromatids (8, 9). One possible explanation for this phenomenon is suggested below.

SCAFFOLD MODELS

The Radial Loop Model

It has been known for many years that lampbrush chromosomes (transcriptionally active chromosomes found during the diplotene stage of meiotic prophase in oocytes of a number of species) are organized into an axial beaded (chromomeric) fiber with radially projecting chromatin loops (18, 19). In recent years a wide variety of studies has suggested that the DNA of interphase, meiotic, and mitotic chromosomes is also constrained, into topologically closed loop domains, each of which contains 50-200 kb of DNA.

The most comprehensive loop model for mitotic chromosome architecture was proposed as a result of experiments from Laemmli's laboratory (20-25). This model proposed that the shape of the mitotic chromosome is determined by a chromosomal substructure which consisted of non-histone proteins (21-23, 25). This structure, which was termed the "chromosome scaffold," was thought to contain the components responsible for binding the chromatin fiber into radial loops. Laemmli was not the first worker to propose the existence of either loops (26-28) or a core structure within mitotic chromosomes (29-31). However, he was the first to obtain experimental evidence that such a scaffold structure might be responsible for maintenance of the loop architecture of mitotic chromosomes.

A rather similar loop model for mitotic chromosomes has also been proposed by Comings (32). This model is based both on observation of histone-depleted mitotic chromosomes (28, 33-35) and on analogy with the chromomere/loop structure of lampbrush chromosomes (18, 19, 36). The model claims that the loops are tethered by nuclear matrix components (32), although no role is suggested for these components in determining higher order chromosome structure. In fact, the author later concluded that the chromosome scaffold was an artifact (37).

A brief summary of these measured loop lengths is compiled in Table 1. The extent to which the numbers shown in the table represent accurate estimates for the size of chromatin domains in vivo is difficult to judge. It is striking that the values obtained depend significantly on the measurement method employed. Even though nuclease digestion and electron microscopy measurements show an excellent agreement, there is no assurance that these numbers are intrinsically more trustworthy than the sedimentation values. If all values are included, the average size of a loop domain is calculated to be 144 $\pm$ 75 kb.

TABLE 1. Domain sizes in mitotic chromosomes and interphase nuclei.

MITOTIC/ INTERPHASE	AVERAGE DOMAIN SIZE (kb)		METHOD	SPECIES	REF
M	70		electron microscopy	human	20
M	83		"	"	16
M	64		"	"	24
I	90		"	mouse	38
I	53		"	"	39
	72 ± 13				
I	230		sedimentation	yeast	40
I	220		"	human	41
I	138		"	"	41a
I	323		"	" -normal	42
I	154		"	" -trans.	42
	213 ± 66				
I	85		nuclease digestion	Drosophila	43
I	150		"	rat liver	44
I	34	(75 max.)	"	"	45
M/I	62		"	mouse	46
I	80		"	rat liver	47
I	36	(80 max.)	"	chick	48
	75 ± 39				
I	100		DNase 1 sensitivity	chick	49
I	80		Topoisomerase II- binding periodicity	human	50

Recent experiments from Laemmli's laboratory suggest that the loop positions are precisely defined by specific sequences located 5' from the promoter elements (51). These sequences presumably bind to one of the nuclear matrix proteins. Data were presented for the _Drosophila_ histone and hsp70 (heat shock) loci. The histone genes were found in loops much smaller than those presented in Table 1 (4.8 - 5 kb), but the authors claim to have mapped much larger loops in _Drosophila_ Kc cells around the rosy locus (51).

It is important to note that the scaffold:loop model is not incompatible with the chromosomal macrocoiling alluded to above (7-9). Scaffold components might self-assemble into a helical structure along the chromosome axis. Each subunit of the scaffold helix would have attached to it a radially projecting chromatin loop. Recent micrographs of swollen helical chromosomes support the idea that radial chromatin loops project outwards from a helical chromatid axis (Figure 2) (Rattner and Lin, _Cell_, in press).

ASSESSMENT OF THE SCAFFOLD MODEL

A Warning: Methodological Differences Are Significant

The nuclear matrix-chromosome scaffold field has been plagued by the conflicting results obtained by different laboratories. This variation is likely to be influenced by three factors -- cell type under study, chromosome isolation procedure used, and the method used to prepare scaffolds.

Chromosomes used for studies of scaffolds have generally been isolated from Chinese hamster or HeLa cells, often with different results. To my knowledge, the methods developed by

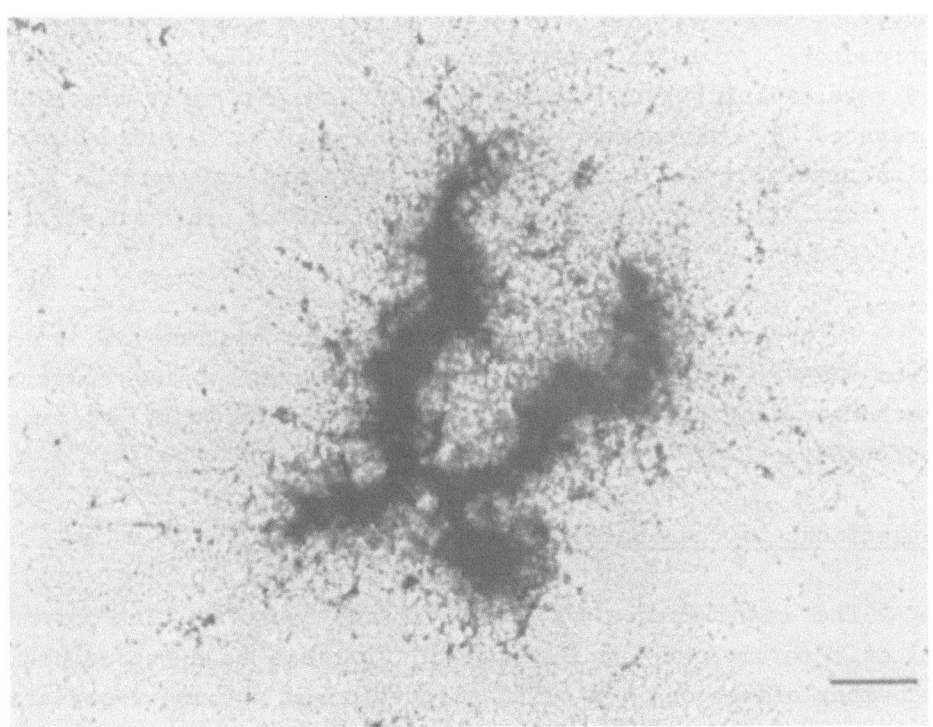

Figure 2. Loops projecting from a "spiralized" HeLa chromosome. This chromosome was found by chance in a sample prepared as in Figure 1 (except that the shear step was avoided). The magnification bar indicates 1 micron.

Laemmli for the isolation of chromosomes and chromosome scaffolds from HeLa cells (25) have not yet been applied to Chinese hamster material in a systematic way.

Chromosomes must be kept condensed during isolation in order to prevent damage caused by shear during cell lysis and centrifugation steps. The experimental conditions generally employed include: exposure to Ca^{++} and hexylene glycol (25, 52, 53); exposure to Mg^{++} or NaCl concentrations sufficient to condense isolated interphase chromatin (25, 54-57); and exposure to a combination of chelating agents and polyamines (25, 53, 55, 58). The latter method has the advantage that it minimizes endogenous nuclease digestion of the DNA, but the disadvantage that the chelating agents may disrupt certain aspects of chromosome structure (59). A useful study of the effect of various solution conditions on mitotic chromosome morphology is found in reference 60.

Finally, chromosome scaffolds may be isolated following nuclease digestion of isolated chromosomes by extraction of chromosomal proteins at either high (2 M NaCl) or low (dextran sulfate:heparin) ionic strength (21, 22, 25). For chromosomes prepared by a given method, the method used for scaffold production generally has no effect on the protein composition of the scaffold (21, 22, 25), though it does have a profound effect on scaffold morphology (16).

In practice, we have found that if chromosomes are isolated from such diverse species as man and chicken using comparable methods, scaffolds isolated from them have similar protein compositions (61).

Objections to the Scaffold Model

The significance of the scaffold proteins in governing mitotic chromosome architecture in vivo has been questioned for a number of reasons. A brief discussion of the major objections follows.

1) The scaffold has not generally been visualized directly in intact chromosomes (37, 62). The low mass of the scaffold

relative to that of the rest of the chromosome (less than 3%) (25) may partly explain this difficulty. However, under appropriate conditions putative scaffold components may be visualized in swollen chromosomes (16). In any case, recent studies using antibodies directed against a number of major scaffold proteins have permitted localization of these proteins along the axes of swollen chromosomes (79, 63).

2) Since the chromosome scaffold consists of the most insoluble protein components associated with mitotic chromosomes, it is possible that some of the scaffold proteins might derive from cytoskeletal contamination (25). In fact, vimentin has been found to be a major component of CHO metaphase scaffolds (57). However, one major chicken scaffold protein, characterized in detail, has been shown to be located solely in the nucleus of interphase cells (61).

3) Since scaffolds are often isolated following extraction of nuclease-digested chromosomes with 2 M NaCl, it has been suggested that the scaffold might be composed of a random set of proteins which precipitate when exposed to high ionic strength during the 2 M NaCl extraction step (36, 64). However, scaffolds may also be produced at an ionic strength equivalent to 20 mM NaCl (with the polyanion extraction method) (16). In fact, the polyanion extraction is inhibited by as little as 25-50 mM NaCl (Earnshaw and Laemmli, unpublished observation). Furthermore, it is now possible to produce scaffolds of a strictly defined and limited protein composition, suggesting that random precipitation models are incorrect (16, 25, 61). Finally, recent data indicate that at least one major scaffold protein is recovered in scaffolds with high efficiency (ca. 70%) (61). This would not be expected if the scaffold consists of a random subset of nuclear proteins. These points have previously been discussed at length (16, 25, 61, 65).

Scaffold Stabilization and the Question of Scaffold Integrity

It has been reported that the scaffold is stabilized by specific metalloprotein interactions involving Cu^{++} (25). An alternative explanation of these results is that the Cu^{++} effect is actually due to oxidation of scaffold components rather than specific metalloprotein interactions (J. Van Ness, personal communication; also see reference 66). This hypothesis has been the subject of a recent careful study (66a).

An alternative procedure for scaffold stabilization was found to be treatment of chromosomes with Ca^{++} at 37° for 10 min (25). If this treatment was done at 4°, no stabilization of the scaffold occurred. We have recently found, however, that the Ca^{++} can be omitted, provided that the 37° incubation is retained (Earnshaw, Heck, and Cooke, unpublished results). A similar finding was recently reported for the isolation of nuclear matrix (51). This again suggests some non-specific mechanism, such as oxidation, rather than a Ca^{++}-specific effect. Whatever the explanation, it appears that if chromosomes are isolated by the polyamine procedure (25, 58), the scaffolds must be stabilized or else they are destroyed during the extraction.

Two considerations arise:

1) When confronted with a given chromosome isolation procedure, one must ask if the procedure contains steps which either fortuitously or intentionally cause stabilization of the scaffold.

2) The scaffold may be a diffuse or disconnected protein network which does not possess structural integrity throughout the entire chromosome in the unstabilized (unoxidized?) state. However, the scaffold is composed of specific proteins which are closely enough associated that they can be readily crosslinked into a structure which has the size and shape of an intact chromosome. Thus, the scaffold may be more analogous to the many discrete spot welds which hold the components of an electric circuit in place than to the solid chassis on which the whole is mounted.

Several recent experiments have used a variety of disruptive treatments to claim that non-histone proteins serve no role in stabilization of chromosome structure (67-69). The observation that the scaffold proteins may function at many discrete sites rather than as a solid backbone for the chromosome means that these experiments will need to be re-interpreted.

RECENT ADVANCES IN THE STUDY OF CHROMOSOME SCAFFOLDS

During the past several years, research in several laboratories has focused on attempts to observe scaffold components in
intact mitotic chromosomes. The first apparent successes in
this area were achieved using cytological silver-staining procedures which specifically label the proteinaceous backbone of the
synaptonemal complex during meiotic prophase (70-72). When
applied to mitotic chromosomes, this technique revealed the
existence of a silver-binding "core" region along the chromatid
axis (73-76). It was proposed that this axial binding was an
artifactual consequence of elevated chromatin concentrations
along the chromatid axis (75-77), but it has recently been shown
that some as yet unidentified component(s) of the isolated
mitotic chromosome scaffold is intensely stained with silver
under these conditions (65).

Conditions used for silver-staining are harsh, involving
pretreatment of the chromosomes with 2.9 M $AgCl_2$ at elevated
temperatures (65). In addition, the silver binding components
have not as yet been identified. We have therefore adopted an
immunological approach to study of the chromosome scaffold.

Lewis and Laemmli (25) showed that isolated HeLa chromosome
scaffolds contained large amounts of two high molecular weight
proteins, Sc-1 (M_r = 170 kd) and Sc-2 (M_r = 135 kd). Antibodies
directed against Sc-1 from human (63) and chicken (61) chromosome scaffolds have recently been obtained, and it has been
shown that chicken Sc-1 is topoisomerase II (61), an abundant
nuclear enzyme which alters the linking number of covalently
closed DNA through the transient introduction of protein-linked
double-stranded DNA breaks. (For a review of topoisomerase II
see reference 78). The identification of chicken Sc-1 as topoisomerase II involved comparison of the binding of independently
prepared anti-chicken Sc-1 and anti-bovine topoisomerase II
antibodies to a series of peptide fragments produced by degradation of the scaffold protein by an endogenous protease (61).
This result was confirmed by showing that the anti-scaffold
antibody inhibited topoisomerase II activity in extracts
prepared from isolated mitotic chromosomes (61). In other
experiments, it has been shown that topoisomerase II of chicken

mitotic chromosomes is found along the chromatid axis and is not bound to the chromatin of the radial loops (79). This is in contrast with the distribution of DNA, HMG-17, and topoisomerase I, all of which were found in the loops (79). The implication of these results is that topoisomerase II is bound at the base of the radial chromatin loops in mitotic chromosomes.

THE FUNCTION OF SCAFFOLD COMPONENTS DURING INTERPHASE

While the loop model provides a convenient framework for understanding the structure and condensation of mitotic chromosomes, most interest on loop domains centers on interphase nuclei where the processes of replication, transcription, and recombination occur. A direct correlation between loop sizes and the length of replicons for several species has led to the proposal that each loop defines a single replicon (44, 47, 80, 81). Likewise, the role of closed domains in control of transcription has recently been discussed (49, 82-85). Clearly the presence of topoisomerase II as an integral part of the loop domain is of great significance for the above models.

Recent experiments have suggested that DNA supercoiling is required to activate certain promotors in microinjected frog oocytes (85). It has been suggested that the oocyte transcribes only DNA molecules which are kept under superhelical strain through the action of a novobiocin-sensitive DNA gyrase (85). These "dynamic" chromatin molecules appear to be in a different chromatin conformation from that of bulk chromatin. The obvious candidate for the gyrase is topoisomerase II, although as isolated the eukaryotic enzyme shows no gyrase function (78). It is possible that this function is lost during enzyme purification. Alternatively, Weintraub (84) has suggested that transient superhelical strain might be induced in circular DNA molecule through brief dissociation of nulceosomes.

What might superhelical strain accomplish? It is known that plasmids which are negatively supercoiled undergo the structural transition to left-handed Z-DNA if the appropriate alternating purine-pyrimidine sequences are present (86-88). This change might affect the activity of promotors by altering

DNA binding sites for either nucleosomes or regulatory proteins, or conversely, by altering the superhelicity of the loop (which would become more relaxed). There has been as yet no clear implication of a functional role of Z-DNA in vivo, however. It is also possible that changes in superhelicity might alter the DNA conformation at regulatory binding sites in some more subtle way.

A second possible role for topoisomerase II in metaphase chromosomes is resolution of knots which result from DNA replication. It has been shown that during SV40 replication, catenated dimers are generated, perhaps as a result of steric hindrance of topoisomerase I as the two converging replication complexes approach each other in formation of the so-called "last Cairns structure" (89, 90). These dimers are normally resolved quickly by a topoisomerase II activity in infected cells (90). Since eukaryotic replicons are also thought to be constrained into closed loop domains (see references in Table 1), the enzyme may be required to resolve catenated structures formed during cellular replication as well. This model is consistent with the observed phenotype of yeast topoisomerase II mutants -- i.e., catenated dimers of 2 micron plasmid are produced in S. cervisiae at restrictive temperature (91) and nuclear division is defective as an apparent result of failure of the chromosomes to separate (91). Similar results have also been obtained for S. pombe (92) and E. coli (93).

When chromosomes are examined by electron microscopy, the DNA of the two sister chromatids often appears intertwined. It has been proposed that unreplicated satellite DNA sequences hold the sister chromatids together until anaphase, with separation being triggered by replication of these regions (94, 95). It seems equally likely, however, that the onset of anaphase is triggered by activation of centromere-associated topoisomerases which would only then resolve replication-induced knots, permitting sister chromatid separation (90).

The enzyme might also function in pathways of recombination, if sister chromatid exchange can be taken as a valid model for this process. The epipodophyllotoxins VP-16 and VM-26, which do not bind to DNA, interact directly with topoisomerase

II (50). These drugs cause a drastic increase in sister chromatid exchange (96).

How would topoisomerase II function in sister chromatid exchange? The enzyme binds to DNA as a dimer (78). During the topoisomerization reaction the DNA is transiently cleaved by a mechanism which leaves it covalently bound to the enzyme. If, at this stage, subunit exchange occurred between topoisomerases on sister chromatids, then exchange would occur upon completion of the reaction. This mechanism requires that the sister chromatids be held in exact register during the event, a constraint which might be more efficiently accomplished by making the enzyme part of the chromosome scaffold.

The above model implies that sister chromatids must be tightly associated following replication. This is consistent with data which suggest that sister kinetochores become distinct from each other only in very late G2, immediately prior to prophase (97).

IMPLICATIONS OF THE NEW RESULTS

The results discussed above suggest that topoisomerase II defines a new class of nuclear proteins which are at the same time enzymatically active: structural proteins. Such proteins are not unknown, notable examples being myosin and dynein. The concept is a novel one when applied to chromatin structure, however.

It is hoped that as other components of the scaffold are identified, the role of this structure in the establishment and maintenance of the condensed mitotic chromosome morphology will be explained. It is also hoped that these studies of mitotic chromosomes will yield insight into the function of chromatin domains in the interphase nucleus, where visualization of the structures has so far proven impossible.

ACKNOWLEDGMENTS

In preparing Table 1, I made extensive use of references cited by R. Hancock (98). I would like to thank M. Heck for her comments on the manuscript. Experiments from my laboratory were supported by NIH grant GM 30985.

REFERENCES

1. WILSON, E.B. (1911). "The Cell in Development and Inheritance." Macmillan Co., London.

2. PAULSON, J.R. (1982). 3. Chromatin and chromosomal proteins. In: "Electron Microscopy of Proteins," Vol. 3 (R. Harris, ed.) pp. 77-134. Academic Press, New York.

3. DuPRAW, E.J. (1965). Macromolecular organization of nuclei and chromosomes: a folded fibre model based on whole-mount electron microscopy. Nature 206, 338-343.

4. DuPRAW, E.J. (1966). Evidence for a "folded-fibre" organization in human chromosomes. Nature 209, 577-581.

5. COMINGS, D.E. (1978). Mechanisms of chromosome banding and implications for chromosome structure. Ann. Rev. Genet. 12, 25- 46.

6. McGHEE, J.D. and FELSENFELD, G. (1980). Nucleosome structure. Ann. Rev. Biochem. 49, 1115-1156.

7. MANTON, I. (1950). The spiral structure of chromosomes. Biol. Rev. Cambridge Phil. Soc. 25, 486-508.

8. OHNUKI, Y. (1968). Structure of chromosomes I. Morphological studies of the spiral structure of human somatic chromosomes. Chromosoma (Berl.) 25, 402-428.

9. UTSUMI, K.R. and TANAKA, T. (1975). Studies on the structure of chromosomes I. The uncoiling of chromosomes revealed by treatment with hypotonic solution. Cell Struct. Funct. 1, 93-99.

10. BAK, A.L., ZEUTHEN, J., and CRICK, F.H.C. (1977). Higher-order structure of human mitotic chromosomes. Proc. Natl. Acad. Sci. USA 74, 1595-1599.

11. BAK, A.L. and ZEUTHEN, J. (1978). Higher-order structure of mitotic chromosomes. Cold Spring Harbor Symp. Quant. Biol. 42, 367-377.

12. SEDAT, J. and MANUELIDIS, L. (1978). A direct approach to the structure of eukaryotic chromosomes. Cold Spring Harbor Symp. Quant. Biol. 42, 331-350.

13. BAK, P., BAK, A.L., and ZEUTHEN, J. (1979). Characterization of human chromosomal unit fibers. Chromosoma (Berl.) 73, 301-315.

14. ZEUTHEN, J., BAK, P., and BAK, A.L. (1979). Chromosomal unit fibers in Drosophila. Chromosoma (Berl.) 73, 317-326.

15. JORGENSEN, A.L. and BAK, A.L. (1982). The last order of coiling in human chromosomes. Exp. Cell Res. 139, 447-451.

16. EARNSHAW, W.C. and LAEMMLI, U.K. (1983). Architecture of metaphase chromosomes and chromosome scaffolds. J. Cell Biol. 96, 84-93.

17. PIENTA, K.J. and COFFEY, D.S. (1985). The nuclear matrix: an organizing structure for the interphase nucleus and chromosome. J. Cell Sci. (in press).

18. GALL, J.G. (1955). Problems of structure and function in the amphibian oocyte nucleus. Symp. Soc. Exptl. Biol. 9, 358-370.

19. CALLAN, H.B. (1982). The Croonian lecture, 1981. Lampbrush chromosomes. Proc. Roy. Soc. Lond. B 214, 417-448.

20. PAULSON, J.R. and LAEMMLI, U.K. (1977). The structure of histone-depleted metaphase chromosomes. Cell 12, 817-828.

21. ADOLPH, K.W., CHENG, S.M., and LAEMMLI, U.K. (1977). Role of nonhistone proteins in metaphase chromosome structure. Cell 12, 805-816.

22. ADOLPH, K.W., CHENG, S.M., PAULSON, J.R., and LAEMMLI, U.K. (1977). Isolation of a protein scaffold from mitotic HeLa cell chromosomes. Proc. Natl. Acad. Sci. USA 74, 4937-4941.

23. LAEMMLI, U.K., CHENG, S.M., ADOLPH, K.W., PAULSON, J.R., BROWN, J.A., and BAUMBACH, W.R. (1978). Metaphase chromosome structure: the role of nonhistone proteins. Cold Spring Harbor Symp. Quant. Biol. 42, 351-360.

24. MARSDEN, M.P.F. and LAEMMLI, U.K. (1979). Metaphase chromosome structure: evidence for a radial loop model. Cell 17, 849-858.

25. LEWIS, C.D. and LAEMMLI, U.K. (1982). Higher order metaphase chromosome structure: evidence for metalloprotein interactions. Cell 29, 171-181.

26. DuPRAW, E.J. and RAE, P.M.M. (1966). Polytene chromosome structure in relation to the "folded fibre" concept. Nature 212, 598-600.

27. BAHR, G.F. (1970). Human chromosome fibers. Considerations of DNA-protein packing and of looping patterns. Exp. Cell Res. 62, 39-49.

28. COMINGS, D.E. and OKADA, T.A. (1973). Some aspects of chromosome structure in eukaryotes. Cold Spring Harbor Symp. Quant. Biol. 38, 145-153.

29. TAYLOR, J.H. (1957). The time and mode of duplication of chromosomes. Amer. Naturalist 91, 209-221.

30. MAIO, J.J. and SCHILDKRAUT, C.L. (1966). Isolated mammalian metaphase chromosomes. I. General characteristics of nucleic acids and proteins. J. Mol. Biol. 24, 29-39.

31. STUBBLEFIELD, E. and WRAY, W. (1971). Architecture of the Chinese hamster metaphase chromosome. Chromosome (Berl.) 32, 262-294.

32. COMINGS, D.E. (1977). Mammalian chromosome structure. In: "Chromosomes Today," Vol. 6 (A. de la Chapelle and M. Sorsa, eds.) pp. 19-26. Elsevier/North-Holland Biomedical Press, Amsterdam.

33. OKADA, T.A. and COMINGS, D.E. (1979). Higher order structure of chromosomes. Chromosoma (Berl.) 72, 1-14.

34. MULLINGER, A.M. and JOHNSON, R.T. (1979). The organization of supercoiled DNA from human chromosomes. J. Cell Sci. 38, 369-389.

35. MULLINGER, A.M. and JOHNSON, R.T. (1980). Packing DNA into chromosomes. J. Cell Sci. 46, 61-86.

36. ANGELIER, N., PAINTRAND, M., LAVAUD, A., and LECHAIRE, J.P. (1984). Scanning electron microscopy of amphibian lampbrush chromosomes. Chromosoma (Berl.) 89, 243-253.

37. OKADA, T.A. and COMINGS, D.E. (1980). A search for protein cores in chromosomes: is the scaffold an artifact? Am. J. Hum. Genet. 32, 814-832.

38. VOGELSTEIN, B., PRADOLL, D.M., and COFFEY, D.S. (1980). Supercoiled loops and eucaryotic DNA replication. Cell 22, 79-85.

39. HANCOCK, R. and HUGHES, M.E. (1982). Organization of DNA in the eucaryotic nucleus. Biol. Cell. 44, 201-212.

40. PINON, R. and SALTS, Y. (1977). Isolation of folded chromosomes from the yeast Saccharomyces cerevisiae. Proc. Natl. Acad. Sci. USA 74, 2850-2854.

41. COOK, P.R. and BRAZELL, I.A. (1978). Spectrofluorometric measurement of the binding of ethidium to superhelical DNA from cell nuclei. Eur. J. Biochem. 84, 465-477.

41a. HARTWIG, M. (1978). Organization of mammalian chromosomal DNA: supercoiled and folded circular DNA subunits from interphase cell nuclei. Acta Biol. Med. Germ. 37, 421-432.

42. HARTWIG, M. (1982). The size of independently supercoiled domains in normal human lymphocytes and leukemic lymphoblasts. Biochim. Biophys. Acta 698, 214-217.

43. BENYAJATI, C. and WORCEL, A. (1976). Isolation, characterization, and structure of the folded interphase genome of Drosophila melanogaster. Cell 9, 393-407.

44. WANKA, F., MULLENDERS, L.H.F., BEKERS, A.G.M., PENNINGS, L.J., AELEN, and EYGENSTEYN, J. (1977). Association of nuclear DNA with a rapidly sedimenting structure. Biochem. Biophys. Res. Comm. 74, 739-747.

45. IGO-KEMENES, T. and ZACHAU, H.G. (1978). Domains in chromatin structure. Cold Spring Harbor Symp. Quant. Biol. 42, 109-118.

46. RAZIN, S.V., MANTIEVA, V.L., and GEORGIEV, G.P. (1979). The similarity of DNA sequences remaining bound to scaffold upon nuclease treatment of interphase nuclei and metaphase chromosomes. Nucleic Acids Res. 7, 1713-1735.

47. BEREZNEY, R. and BUCHHOLTZ, L.A. (1981). Dynamic association of replicating fragments with the nuclear matrix of regenerating liver. Exp. Cell Res. 132, 1-13.

48. HYDE, J.E. (1982). Expansion of chicken erythrocyte nuclei upon limited micrococcal nuclease digestion. Exp. Cell Res. 140, 63-70.

49. LAWSON, G.M., KNOLL, B.J., MARCH, C.J., WOO, S.L., TSAI, M.-J., and O'MALLEY, B.W. (1982). Definition of 5' and 3' structural boundaries of the chromatin domain containing the ovalbumin multigene family. J. Biol. Chem. 257, 1501-1507.

50. CHEN, G.L., ROWE, T.C., HALLIGAN, B.D., TEWEY, K.M., and LIU, L.F. (1984). Non-intercalcative antitumor drugs interfere with the breakage-reunion reaction of mammalian topoisomerase II. J. Biol. Chem. 259, 13560-13566.

51. MIRKOVITCH, J., MIRAULT, M.-E., and LAEMMLI, U.K. (1984). Organization of the higher-order chromatin loop: specific DNA attachment sites on nuclear scaffold. Cell 39, 223-232.

52. WRAY, W. and STUBBLEFIELD, E. (1970). A new method for the rapid isolation of chromosomes, mitotic apparatus, or nuclei from mammalian fibroblasts at near neutral pH. Exp. Cell Res. 59, 469-478.

53. PAULSON, J.R. (1982). Isolation of chromosome clusters from metaphase-arrested HeLa cells. Chromosoma (Berl.) 85, 571-581.

54. THOMA, F., KOLLER, T.H., and KLUG, A. (1979). Involvement of histone H1 in the organization of the nucleosome and of the salt-dependent superstructures of chromatin. J. Cell Biol. 83, 403-427.

55. ADOLPH, K.W. (1980). Isolation and structural organization of human mitotic chromosomes. Chromosoma (Berl.) 76, 23-33.

56. DETKE, S. and KELLER, J.M. (1982). Comparison of the proteins present in HeLa cell interphase nucleoskeletons and metaphase chromosome scaffolds. J. Biol. Chem. 257, 3905-3911.

57. GOODERHAM, K. and JEPPESEN, P. (1983). Chinese hamster metaphase chromosomes isolated under physiological conditions. Exp. Cell Res. 144, 1-14.

58. BLUMENTHAL, A.B., DIEDEN, J.D., KAPP, L.N., and SEDAT, J.W. (1979). Rapid isolation of metaphase chromosomes containing high molecular weight DNA. J. Cell Biol. 81, 255-259.

59. RIS, H. and WITT, P.L. (1981). Structure of the mammalian kinetochore. Chromosoma (Berl.) 82, 153-170.

60. COLE, A. (1967). Chromosome structure. Theoret. Biophys. 1, 305-375.

61. EARNSHAW, W.C., HALLIGAN, B.H., COOKE, C.A., HECK, M.M., and LIU, L.F. (1985). Topoisomerase II is a major component of mitotic chromosome scaffolds. J. Cell Biol. 100, 1706-1715.

62. RATTNER, J.B., BRANCH, A., and HAMKALO, B.A. (1975). Electron microscopy of whole mount metaphase chromosomes. Chromosoma (Berl.) 52, 329-338.

63. VAN NESS, J. and LAEMMLI, U.K. Submitted.

64. HADLACZKY, G., SUMNER, A.T., and ROSS, A. (1981). Protein-depleted chromosomes. II. Experiments concerning the reality of chromosome scaffolds. Chromosoma (Berl.) 81, 557-567.

65. EARNSHAW, W.C. and LAEMMLI, U.K. (1984). Silver staining the chromosome scaffold. Chromosoma (Berl.) 89, 186-192.

66. KAUFMAN, S.H., COFFEY, D.S. and SHAPER, J.H. (1981). Considerations in the isolation of rat liver nuclear matrix, nuclear envelope, and pore complex lamina. Exp. Cell Res. 132, 105-123.

66a. JEPPESEN, P. and MORTEN, H. (1985). Effects of sulphydryl reagents on the structure of dehistonezed metaphase chromosomes. J. Cell Sci. 73, 245-260.

67. GOYANES, V.J., MATSUI, S.-I., and SNADBERG, A.A. (1980). The basis of chromatin fiber assembly within chromosomes studied by histone-DNA crosslinking followed by trypsin digestion. Chromosoma (Berl.) 78, 123-135.

68. LABHART, P. and KOLLER, T. (1982). Involvement of higher order chromatin structures in metaphase chromosome organization. Cell 30, 115-121.

69. WUNDERLI, H., WESTPHAL, M., ARMBRUSTER, B., and LABHART, P. (1983). Comparative studies on the structural organization of membrane-depleted nuclei and metaphase chromosomes. Chromosoma (Berl.) 88, 241-248.

70. PATHAK, S. and HSU, T.C. (1979). Silver-stained structures in mammalian meiotic prophase. Chromosoma (Berl.) 70, 195-203.

71. DRESSER, M.E. and MOSES, M.J. (1979). Silver staining of synaptonemal complexes in surface spreads for light and electron microscopy. Exp. Cell Res. 121, 416-419.

72. FLETCHER, J.M. (1979). Light microscope analysis of meiotic prophase chromosomes by silver staining. Chromosoma (Berl.) 72, 241-248.

73. HOWELL, W.M. and HSU, T.C. (1979). Chromosome core structure revealed by silver staining. Chromosoma (Berl.) 73, 61-66.

74. SATYA-PRAKASH, K.L., HSU, T.C., and PATHAK, S. (1980). Behavior of the chromosome core in mitosis and meiosis. Chromosoma (Berl.) 81, 1-8.

75. BURKHOLDER, G.D. and KAISERMAN, M.Z. (1982). Electron microscopy of silver-stained core-like structures in metaphase chromosomes. Can. J. Genet. Cytol. 24, 193-199.

76. ZHENG, H.-Z. and BURKHOLDER, G.D. (1982). Differential silver staining of chromatin in metaphase chromosomes. Exp. Cell Res. 141, 117-125.

77. BURKHOLDER, G.D. (1982). Dansyl chloride-stained nucleolar organizers and core-like structures in chinese hamster metaphase chromosomes. Exp. Cell Res. 142, 485-488.

78. LIU, L.F. (1983). DNA topoisomerases-enzymes that catalyse the breaking and rejoining of DNA. CRC Crit. Rev. Biochem. 15, 1-24.

79. EARNSHAW, W.C. and HECK, M.M.S. (1985). Localization of topoisomerase II in mitotic chromosomes. J. Cell Biol. 100 1716-1725.

80. BUONGIORNO-NARDELLI, M., MICHELI, G., CARRI, M.T., and MARILLEY, M. (1982). A relationship between replicon size and supercoiled loop domains in the eukaryotic genome. Nature 298, 100-102.

81. PARDOLL, D.M., VOGELSTEIN, B., and COFFEY, D.S. (1980). A fixed site of DNA replication in eukaryotic cells. Cell 19, 527-536.

82. LAWSON, G.M., TSAI, M-J., O'MALLEY, B.W. (1980). Deoxyribonuclease I sensitivity of the nontranscribed sequences flanking the 5' and 3' ends of the ovomucoid gene and the ovalbumin gene and its related X and Y genes in hen oviduct nuclei. Biochemistry 19, 4403-4422.

83. HARLAND, R.M., WEINTRAUB, H., and McKNIGHT, S.L. (1983). Transcription of DNA injected into Xenopus oocytes in influenced by template topology. Nature 302, 38-43.

84. WEINTRAUB, H. (1983). A dominant role for DNA secondary structure in forming hypersensitive structures in chromatin. Cell 32, 1191-1203.

85. RYOJI, M. and WORCEL, A. (1984). Chromatin assembly in Xenopus oocytes: in vivo studies. Cell 37, 21-32.

86. POHL, F.M. and JOVIN, T.M. (1972). Salt-induced co-operative conformational change of synthetic DNA: equilibrium and kinetic studies with poly(dG-dC). J. Mol. Biol. 67, 375-396.

87. NORDHEIM, A., PARDUE, M.L., LAFER, E.M., MOLLER, A., STOLLAR, B.D., and RICH, A. (1981). Antibodies to left-handed Z-DNA bind to interband regions of Drosophila polytene chromosomes. Nature 294, 417-422.

88. NORDHEIM, A., LAFER, E.M., PECK, L.J., WANG, J.C., STOLLAR, B.D., and RICH, A. (1982). Negatively supercoiled plasmids contain left-handed Z-DNA segments as detected by specific antibody binding. Cell 31, 309-318.

89. SUNDIN, O. and VARSHAVSKY, A. (1980). Terminal stages of SV40 DNA replication proceed via multiply intertwined catenated dimers. Cell 21, 103-114.

90. SUNDIN, O. and VARSHAVSKY, A. (1981). Arrest of segregation leads to accumulation of highly intertwined catenated dimers: dissection of the final stages of SV40 DNA replication. Cell 25, 659-669.

91. DiNARDO, S., VOELKEL, K., and STERNGLANZ, R. (1984). DNA topoisomerase II mutant of Saccharomyces cervisiae: topoisomerase II is required for segregation of daughter molecules at the termination of DNA replication. Proc. Natl. Acad. Sci. USA 81, 2616-2620.

92. UEMURA, T. and YANAGIDA, M. (1984). Isolation of type I and II DNA topoisomerase mutants from fission yeast: single and double mutants show current phenotypes in cell growth and chromatin organization. EMBO J. 3, 1737-1744.

93. STECK, T. and DRLICA, K. (1984). Bacterial chromosome segregation: evidence for DNA gyrase involvement in decatenation. Cell 36, 1081-1088.

94. PRYOR, A., FAULKNER, K., RHOADES, M.M., and PEACOCK, W.J. (1980). Asynchronous replication of heterochromatin in maize. Proc. Natl. Acad. Sci. USA 77, 6705-6709.

95. ALBERTS, B., BRAY, D., LEWIS, J., RAFF, M., ROBERTS, K., and WATSON, J.D. (1983). "Molecular Biology of the Cell," p. 6555. Garland, New York.

96. SINGH, B. and GUPTA, R.S. (1983). Mutagenic responses of thirteen anticancer drugs on mutation induction at multiple genetic loci and on sister chromatid exchanges in chinese hamster ovary cells. Cancer Res. 43, 577-584.

97. BRENNER, S., PEPPER, D., BERNS, M.W., TAN, E., and BRINKLEY, B.R. (1981). Kinetochore structure, duplication, and distribution in mammalian cells: analysis by human autoantibodies from scleroderma patients. J. Cell Biol. 91, 95-102.

98. HANCOCK, R. (1982). Topological organization of interphase DNA: the nuclear matrix and other skeletal structures. Biol. Cell. 46, 105-122.

DNase I HYPERSENSITIVE SITES: A STRUCTURAL FEATURE OF CHROMATIN
ASSOCIATED WITH GENE EXPRESSION

Graham H. Thomas, Esther Siegfried, and
Sarah C.R. Elgin

Department of Biology
Washington University
St. Louis, Missouri 63130

INTRODUCTION

Investigations in the field of gene regulation in eukaryotes have identified several conserved sequences in and around
genes. Such sequences are presumed, or have been demonstrated,
to be important in promoting, terminating, or otherwise regulating gene transcription. (For examples see references 1-3 and
the chapter by Wasylyk, this volume). Similarly, there has been
a growing realization of the importance of the role of chromatin
structure in gene expression (see reference 4 for an extensive
review). Specifically, potentially transcribed genes have been
shown to be preferentially sensitive to digestion by the enzyme
deoxyribonuclease I (DNase I) (5), and to lie within large
domains of chromatin containing 10-100 kilobase pairs of DNA
that are more sensitive to this enzyme than are the surrounding
regions (6). Within such regions exist more localized chromatin
sites which are hypersensitive to DNase I. These "DH sites" are
the subject of this review.

WHERE ARE DH SITES?

There is a strong correlation between the presence of a DH
site and the potential for transcriptional activity. Frequently, the DH sites associated with gene expression are located

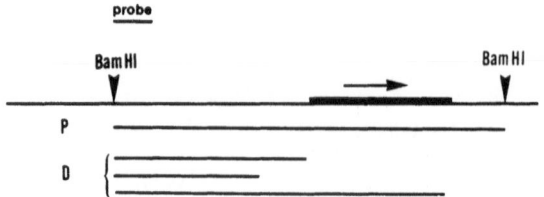

Figure 1. Diagram of the indirect end-labeling experiment. A stretch of DNA encompassing a region of interest, in this case a transcription unit (heavy line), is encompassed by two restriction enzyme sites (e.g., Bam H1). Nuclei are lightly digested with DNase I and the purified DNA is digested to completion by the restriction enzyme and the resultant fragments are separated by agarose gel electrophoresis. Fragments with an end at one of the restriction sites are visualized after Southern transfer by probing with a short cloned fragment as indicated ("probe" in figure). Examples of the fragments which might be seen are shown below the gene. P is the parental restriction fragment not cut by DNase I, D are fragments generated by DNase I which would be visualized with this probe. Frequent cutting by DNase I at one location would produce a discrete band smaller than P.

at or near the 5' ends of genes transcribed by RNA polymerases I and II but, in addition, DH sites may also be found 3' to the transcription unit. There are some cases where DH sites have been detected within introns (7-9). Table 1 lists a number of genes for which DH sites have been detected and mapped relative to the gene. To date no DH sites have been reported in plants, although it is premature to generalize from this due to the limited number of studies which have emerged.

A common method for mapping DH sites is the indirect end-labeling technique first described by Nedospasov and Georgiev (10) and Wu (11). This technique is described in Figure 1.

The _Drosophila_ heat shock genes hsp70, at locus 87 A and C, and hsp83, at locus 63 BC, were the first genes to be examined for DH sites using the method described above (11). In both cases, DH sites were mapped to regions 5' and 3' to the genes. The DH sites are present in the chromatin of both heat shocked and non-heat shocked cells. In the heat shocked nuclei, the DH sites are harder to detect, presumably because of the overall increased sensitivity of the gene when it is active. The DH sites 5' to hsp70 have been mapped more precisely (13, 14). Wu (13) has detected two sites 5' to hsp70 at +100 bp[*] to -8 bp and -38 bp to -215 bp, the sequence around -93 bp being the most sensitive and a region of relative resistance occurring at -8 bp to -38 bp. Costlow and Lis (14) define the DH sites over a range from +42 bp to -128 bp with peak intensities at +42 bp, -2 bp, and -128 bp. In addition to hsp70 and hsp83 in Drosophila, the cluster of small heat shock genes (hsp22, hsp23, hsp26, and hsp28) and one developmentally regulated gene (R) at locus 67B1 have been investigated. Each of these genes has a doublet of DH sites 5' to the gene (15), a major site at or near the 5' end and a weaker site ∿200 bp upstream. Costlow and Lis have mapped these sites for three of the four small heat shock genes to an average location of -17 bp and -115 bp from the start of transcription (14).

Many other genes from a variety of organisms have been examined for DNase I hypersensitivity, frequently revealing a correlation between DH sites and transcription. There are a number of genes for which DH sites are mapped to a region that contains sequences necessary for transcription. The herpes simplex virus thymidine kinase (tk) gene is constitutively expressed in transformed mouse tk⁻ fibroblasts. In these cells a DH site was mapped to a region from -4 bp to -182 bp by restriction endonucleases (16). The hypersensitivity entirely encompasses the three control signals (proximal, first and second distal) within the promoter region, as defined by mutational analysis (17). The rat preproinsulin gene has been investigated and, in an insuloma (a tumor derived from pancreatic B cells), a tissue-specific DH site was detected between -250 bp and -300 bp from the 5' end of the gene (18). This is approximately the region defined by deletion analysis and transformation experiments to be necessary for transcription in pancreatic cells (19). The expression of this gene is tissue-specific, as are the DH sites. The mouse metallothionein gene, MT-1, in a variety of tissues, has a DH region from -250 bp to +1 bp with peaks at -225 bp and -30 bp and a relatively resistant region centered around -140 bp (20). After induction by

---

[*]Unless otherwise stated, distances are given relative to the start point of transcription. bp = basepairs, kb = kilo-basepairs.

TABLE 1.   DNase 1 hypersensitive sites in chromatin

GENE	TISSUE EXAMINED	LOCATION	REFERENCES
CONSTITUTIVE GENES			
alcohol dehydrogenase ADC1 (Saccharomyces)	not applicable	5' (doublet)	21
ribosomal protein rp49 (Drosophila)	embryos	5' (5 sites)	22
thymidine kinase, tk (herpes simplex virus)	mouse L cell	5' (-4 to -182)	16
thymidine kinase, tk (chicken)	erythrocytes, bursal lymphocytes, RP9 – B cell line, MSB1 – T cell line, embryonic fibroblasts, brain	5'	23
c-myc (chicken)	erythrocytes embryonic fibroblasts, bursa	5' (-1.2 to - 1.3 kb, -2.5 kb) plus minor site in intron	9,24
rDNA (Tetrahymena thermophila)	not applicable	5' (3 sites from 0 to -1200, one at replication origin)	30,31,44
rDNA (Xenopus)	embryo, erythrocyte	5'	29
CELL-CYCLE REGULATED GENES			
Histone (sea urchin)	blastula	5', 3'	25
Histone (Drosophila)	embryo, tissue culture	5'	26
INDUCIBLE GENES			
Drosophila			
heat shock, hsp83	embryo, tissue culture	5' (+20, -20, -120, -213, -451), 3'	11,13,14
heat shock, hsp70	embryo, tissue culture	5' (-8 to +100, -38 to -215), 3'	14,15

GENE	TISSUE EXAMINED	LOCATION	REFERENCES
Drosophila			
heat shock, hsp28	embryo, tissue culture	5' (-18, -82, -293)	14,15
heat shock, hsp26	embryo, tissue culture	5' (-15, -118, -296, -374)	14,15
heat shock, hsp23	embryo, tissue culture	5' (-28, -125)	14,15
heat shock, hsp22	embryo, tissue culture	5' (doublet)	14,15
alcohol dehydrogenase ADR2, (Saccharomyces)	not applicable	5'	21
metallothionein, mt-1	liver, kidney sarcoma cell line	5' (-225 to -30 plus -30 to +60 after Cd induction)	20
TISSUE SPECIFIC			
α-globin (chicken)	erythrocytes	5' α^D, α^A 14 day embryo, 5' u only in 5 day embryo, 3' of α^A site distal to U	35
	brain		
α-globin (mouse)	erythroleukemia cells	5' (-100 to -200), additional site on induction with hexamethylene bisacetamide	90,96
β-globin (chicken)	erythrocytes	5' & 3' ρ and ε in 5 day embryos, 5' (-160 to +65) & 3' β^H and β^A in 14 day embryos	34,36,71
	brain	sites not seen	
β-globin (mouse)	erythroleukemia cells	5'	90
β-globin (human)	foetal erythroid liver erythroid bone marrow leukemia cells	5' Gγ, Aγ, δ, and β (-75 to -100), 5' δ and β, 5' ε (six sites 0 to -6.0 kb)	97, 98
glue protein gene, sgs4 (Drosophila)	salivary glands embryo	5' (+30, -70, -330, -405, -480), 3'	33
preproinsulin (rat)	pancreatic cells liver	5' (-250 to -300) 3', 5' sites not seen	18

(continued)

TABLE 1. DNase 1 hypersensitive sites in chromatin (continued)

GENE	TISSUE EXAMINED	LOCATION	REFERENCES
prolactin (rat)	pituitary tumors liver	5' (-150) sites not seen	40
tryptophan oxidase (rat)	liver kidney	5' (-130 to -180, -210 to -250, -420 to -470) sites not seen	100
tryptophan amino-transferase (rat)	liver kidney	5' (-2kb - dexamethasone dependent, -750, 2 in +1 to -200 region), internal (+700) sites not seen	100
collagen (chicken)	fibroblasts sperm, brain	5' (-100 to -300) sites not seen	37
lysozyme (chicken)	oviduct erythrocytes	5' (5 sites from -0.1 to -7.8 kb), 3' (3 sites from 0.1 to 4.2 kb 3' of poly(A) site) 3' (0.1 kb 3' of poly(A) site)	38
vitellogenin (chicken)	embryonic liver livers of egg laying hens, induced rooster livers, induced embryo livers oviduct erythrocytes, fibroblasts, brain	one within gene, 3' (3 sites) one within gene, 3' (3 sites) plus 5' (0, -300, -700) one within gene, 3', 5' (-700) sites not seen	8
ovalbumin (chicken)	oviduct of laying hens erythrocytes	5' (4 sites from +0.001 to -6.2 kb) sites not seen	39
chorion genes (Drosophila)	ovaries	5' & 3'	95
immunoglobulin kappa light chain gene (mouse)	myeloma cell line brain, liver	5' (-300), two sites in Jk - Ck intron sites not seen	7,47
immunoglobulin mu heavy chain gene (human)	B cell line nonlymphoid cell line	within Jh - Cμ intron site not seen	48

GENE	TISSUE EXAMINED	LOCATION	REFERENCES
VIRUSES			
mammary tumor virus (mouse)	L cell line	5' flanking regions, 5' & 3' LTR, gluco-corticoid inducible sites in 5' & 3' LTR	27
moloney murine leukemia virus	mouse tumors	5' LTR / cellular DNA boundary	86
avian sarcoma virus	rat L cells	5' LTR, 5' flanking region	84
ev3 (chicken)	erythrocytes	5' and 3' LTR	99
SV40	isolated minichromosome, african green monkey kidney cells, CV-1 cells	origin of replication, 5' to late and early promoters and at the enhancer region	41,42, 64—66
polyoma	NIH 3T6 mouse cells	origin of replication, 5' of late promoter	43
MISCELLANEOUS			
mating type loci (Saccharomyces)	not applicable	MATα 5' to active and inactive, 3' to active genes MATa 5' and 3' in active genes, only within genes at the Y, Z border	46
centromeres (Saccharomyces)	not applicable	flanking the centromeric region	45

cadmium, a new DH site appears between -30 bp and +60 bp. These DH sites encompass the sequence required for cadmium regulation (-95 bp to +63 bp) as indicated by gene transfection and oocyte injection experiments.

DH sites are thus associated with several different classes of genes: constitutively expressed genes (tk gene), tissue-specific genes (rat insulin), and inducible genes (mouse MT-1). Others among the class of constitutively expressed genes that have DH sites are: yeast alcohol dehydrogenase gene, ADC 1 (21); Drosophila ribosomal protein gene, rp49 (22); the endogenous chicken tk gene (23); and c-myc (24). In addition, DH sites have been detected for cell-cycle regulated genes, specifically the sea urchin and Drosophila histone genes (25, 26).

Genes that are inducible frequently have both a constitutive and an inducible DH site. As mentioned above, the mouse metallothionein gene has an additional DH site upon induction by cadmium (20). In mouse L cells harboring an integrated copy of the mouse mammary tumor virus promoter linked to the tk gene, both constitutive and dexamethasone inducible DH sites have been detected (27). One of the constitutive sites, C1, is within the upstream long terminal repeat (LTR), ~850 bp upstream from the start of transcription (this site appears to be duplicated in the downstream LTR). There is another constitutive site, C2, that is upstream of C1 and maps within the flanking rat DNA. Two inducible DH sites (I1 and I2) are detected within the upstream and downstream LTRs, respectively, mapping within the glucocorticoid receptor binding regions. Site I1 is induced within seven minutes of hormone addition, whereas, steady state levels of transcription are achieved fifteen to twenty minutes after addition of dexamethasone. A similar situation is seen for the chicken lysozyme gene (102). Here two constitutive DH sites are seen 5' to the gene in oviduct (at -100 bp and -2.4 kb) along with a third one at -1.9 kb which is dependent on induction with diethylstilboestrol. Hormone withdrawal is accompanied by loss of the third site. When this gene is expressed constitutively in macrophages, a different set of DH sites is seen, at -100 bp (in common with the other expressing tissues), -700 bp and -2.7 kb (102). Thus, depending on the mode of regulation, a specific set of sites is seen. It appears that the inducible DNase I hypersensitivity is established prior to, or coincident with, hormonal stimulation of transcription.

The link between the presence of a DH site and transcription has been further investigated using red cell precursors transformed by a temperature sensitive avian erythroblastosis virus (101). At the permissive temperature, when the transforming gene of the virus, erb, is active, no hemoglobin is synthesized; at the nonpermissive temperature hemoglobin is expressed. One line was established that, at the permissive temperature, lacked globin gene expression and had no DH sites associated with the globin genes. At the nonpermissive temperature DH sites appeared and the genes were expressed. A second line was established that, at the permissive temperature, did not express globin genes, but did possess DH sites. At nonpermissive temperatures this line expressed hemoglobin and the DH sites were maintained. It was suggested that this second line was "frozen" at a developmental time after the establishment of a DH site but before transcription is possible. This supports the idea that the formation of a DH site may occur prior to transcription initiation.

The above examples are in contrast to other inducible genes for which the pattern of DH sites does not change markedly upon induction. The inducible yeast alcohol dehydrogenase gene, ADR 2, has a 5' DH site that is present whether the gene is repressed or expressed (21). This is reminiscent of the heat shock genes of Drosophila for which DH sites are present prior to induction (11, 15, 28).

DH sites have been detected in the oncogene c-myc. In a Burkitt lymphoma cell line, BL31, one allele of the c-myc gene is juxtaposed to the immunoglobulin mu chain constant region; the other allele is in the germline location. In this cell line only the translocated copy is expressed. When the DH sites are mapped for the translocated allele a cluster of sites are detected 5' to the c-myc gene (9). There are three major sites, at -2 kb, at -900 bp, and at -70 bp from the start of transcription (I, II, and III, respectively). In the germline allele only DH site I is detected. This is in contrast to an Epstein-Barr virus-transformed B cell line, in which c-myc is expressed, where both alleles of the gene are located at their germline position and DH sites I, II, and III are detected.

All the genes described so far are transcribed by RNA polymerase II. Genes transcribed by RNA polymerase I also have 5' DH sites. In particular, such chromatin structures have been detected for the ribosomal RNA genes of Xenopus (29) and Tetrahymena (30, 31). Several genes transcribed by RNA polymerase III have been examined for DH sites. The Drosophila 5S gene repeat and two tRNA genes flanking this region have no distinguishable DH sites in chromatin, although sites sensitive to a chemical cleavage reagent, methidiumpropyl-EDTA.iron(II), are seen in this region (32).

The evidence obtained so far reveals a strong correlation between the presence of certain DH sites and the potential for transcription of a gene. The establishment of DH sites may also be a necessary step in the differentiation of a cell type and its subsequent commitment to the expression of a given gene. Expression of the sgs4 gene in Drosophila is limited to the salivary glands of third instar larvae and the DH sites 5' to this gene are specific for this tissue (33). The globin genes are also a clear example of the association of DH sites with tissue-specific expression. The α- and β-globin genes of chicken have a number of DH sites that appear in erythrocytes or embryonic red blood cells but not in brain cells (34-36). The one exception is a DH site that is present in brain nuclei 5' to the U-globin gene, an embryonic α-globin gene. However, this site is some 4 kb 5' and so its relevance to the U-globin gene must be questioned. In addition to being tissue-specific, the DH sites reflect the developmental expression of the globin gene family. In the case of the α- and β-globin genes, the DH sites 5' to the embryonic genes are present in five day old, but not in fourteen day old erythrocytes. At fourteen days DH sites appear 5' to the adult genes and disappear from the embryonic ones.

Among the other genes known to be tissue-specific for gene expression, the rat preproinsulin gene has a 5' DH site in a tumor line derived from pancreatic B cells. This site is absent from liver, spleen, kidney, and brain cells (18). However, a DH site 3' to the gene is observed only in liver nuclei. Tissue-specific DH sites are seen for the collagen (37), lysozyme (38), vitellogenin (8), and ovalbumin (39) genes of chicken and the rat prolactin gene (40). It is interesting to note that the DH site for the α(I) collagen gene in chick embryo fibroblasts is maintained in fibroblasts transformed with Rous sarcoma virus, which causes transcription of this gene to decrease five- to ten-fold. However, no DH sites are detected for the α(I) collagen gene in brain nuclei where there is no transcription.

Most of the DH sites described above have been associated with 5' ends of genes and correlated with transcriptional potential. There are other DH sites that fall in different locations and may be associated with other functions. DH sites have been detected within, or near, the origin of replication for a number of viruses, including SV40 (41, 42) and polyoma (43). In addition, the origin of replication of the extrachromosomal rDNA of Tetrahymena (30, 31, 44) has a DH site. The centromere region of yeast chromosomes is flanked by DH sites (45). The MAT locus of the mating type genes of yeast is associated with a DH site (46). This site falls close to, and may coincide with, the site of an endogenous cut is the DNA which may be involved in mating type switching. Similarly one of the DH sites in a human μ immunoglobulin gene lies adjacent to the Sμ heavy chain switching region and may be involved in gene rearrangement (48).

WHAT ARE DH SITES?

DH sites, then, appear to be symptomatic of more than one, perhaps many, genetic processes. However, more information is available regarding their role in transcriptional regulation than in other processes. We shall, therefore, limit ourselves to transcription in considering the fine structures and roles of DH sites.

It is now generally assumed that a DH site represents a region without a normal nucleosomal structure, possibly devoid of core histones. This is strongly suggested by the large range of DNA cleavage reagents which will contrast these regions from the surrounding chromatin. These include DNase I (see Table 1 for references), DNase II (36), micrococcal nuclease (36, 39), restriction enzymes (36, 16), and chemical cleavage reagents such as methidiumpropyl-EDTA.iron(II) (52). Electron microscopic studies on SV40 minichromosomes have revealed that the origin/promoter region, which is DNase I hypersensitive, contains no discernible nucleosomes in the (presumed) transcriptionally active subpopulation of molecules (41, 42, 53). Further evidence comes from protein crosslinking experiments on isolated tissue culture cell nuclei, which suggest that there is a histone-free region at the 5' end of the Drosophila hsp70 gene (54). This is not to say that these regions are necessarily entirely naked DNA. Evidence is accumulating that non-histone proteins are very specifically bound to the DNA in these sites and possibly at the 5' ends of active genes in general (13, 55-59). The cellular specificity of enhancer sequences (47-51, 60-62), often DH sites, could be explained through the interaction of host cell factors, presumably proteins, with these sites.

DH sites are almost certainly associated with specific DNA sequences. The clearest experimental evidence for this comes from studies on the SV40 origin/promoter region. Movement of this region to (63), or duplication of this region at (64, 65), other locations in the SV40 genome results in the production of DH sites over these sequences at their novel position. One example in which DNase I sensitivity was not detected also failed to have its normal enhancing effect (63), providing further evidence of functional association with these sensitivities. Both the 72 bp repeats (64) and the 21 bp repeats (64, 66) of the SV40 genome appear to have the ability to "create" DH sites independently. It is striking that the DH sites dependent upon the 21 bp repeats lie not over these sequences, but in an adjacent region whose sequence is also important for DH site formation (64). Similarly, the avian leukosis virus LTR can create a DH site on its own (24). However, this case is complex. When the proviral sequences are present as well, two sites are formed within the 5' end of the provirus (relative to c-myc) irrespective of the end of the virus at which the LTR resides (24). This case is also interesting since the virus-induced DH sites are the only ones seen. The c-myc sites which exist in non-transformed tissue are lost even though the sequences over which they form are still present.

In Drosophila, a series of five DH sites 5' of the sgs4 glue protein gene correlate with its activity and are also dependent on specific sequences (33, 67). A series of naturally occurring mutations 5' of the gene (Figure 2), which decrease its expression, also cause an alteration in the DH sites (33). Interestingly, these mutations indicate an interaction between the sites. Thus, one small deletion (52 bp) around -330 bp, which decreases expression 50 fold, abolishes a single DH site at -330 bp. However, a somewhat larger deletion (103 bp) which removes the DH sites at -405 bp and -480 bp and eliminates transcription, abolishes the other three DH sites in the chromatin, although the downstream sequences over which they form in the wild type are unaltered (33). Similarly, insertion of a 1.3 kb transposable element (hobo) in this 5' region moves three of the DH sites away form the gene, but they still form over the same sequences, and the developmental nature of the gene regulation is unaffected (67). However, the expression of this gene can also be reduced by small mutations in the DNase I sensitive regions without the loss of the DH sites themselves (68).

A specific association of DH sites with DNA sequences begs the question of whether it is the sequences per se which causes the sensitivity -- perhaps adopting a novel structure in response to an alteration in superhelical density of a domain --

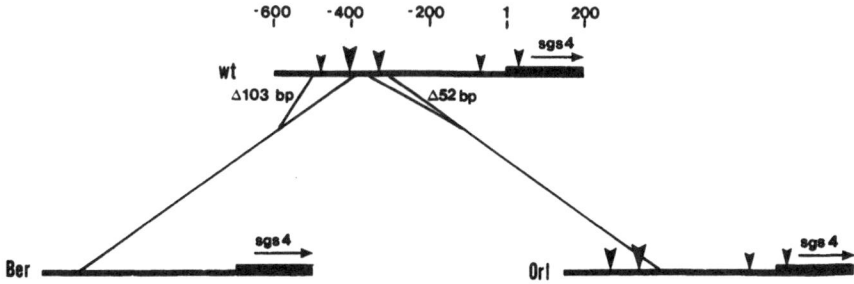

Figure 2.   Summary of the effects of the BER and ORL strain deletions on the DNase I hypersensitive sites associated with the Drosophila sgs4 gene. Vertical arrows indicate the locations and magnitudes of the DH sites located at (in wild type) +30 bp, -70 bp, -330 bp, -405 bp, and -480 bp respectively. The extent of the deletions are indicated by the converging lines below the wt gene (33).

or whether it is the interaction of proteins with those sequences which induces the transitions. It is well documented, for instance, that highly asymmetric sequences (i.e., polypyrimidine tracts) do not assemble well into nucleosomes (69, 70). These sequences are often associated with sensitivity to the single-strand-specific nuclease, S1, in a supercoil-dependent manner. Such sequences are found in the region of some DH sites (71-78). Regions of altered DNA structure may have some stability once established. An S1-sensitive site established to the 5' of the chicken β^A-globin gene in a supercoiled plasmid is maintained on reconstitution with histones despite substantial relaxation of the DNA by the reconstitution process. However, when the DNA is linearized with a restriction enzyme, the S1 sensitivity is lost, so some degree of local supercoiling is likely to be important (76). Z-DNA can also be formed in a supercoil-dependent manner (79) and is probably refractory to nucleosome formation (80). The presence of Z-DNA near the DH sites in the SV40 DNase I hypersensitive region has been suggested (81).

Recently it has become apparent that specific proteins are associated with DH sites and may well be the causative agents in maintaining them in an open conformation. Two factors (presumably proteins) have been described by their footprint at the 5' end of the Drosophila hsp70 genes in intact nuclei (13). The localization of this footprint in nuclei is possible through a novel use of exonuclease III. Nuclei are first digested with DNase I, as for the mapping of a DH site, and the free 3' ends that this produces are digested progressively by exonuclease III. Digestion with this enzyme halts when it encounters a "block" on the DNA molecule. After trimming the overhanging 5' end with S1 nuclease, the site of the blockage is mapped using the standard, indirect end-labeling technique. Using probes on either side of the DH site in question, the 5' and 3' limits of the footprint are mapped. In nuclei from non-heat shocked cells this footprint extends from -12 bp to -40 bp relative to the transcription start site, and in nuclei from heat shocked cells the protected region expands to cover sequences from -12 bp to -108 bp (13). The implication of this is that some macromolecule is occupying a site within the DH site prior to heat shock and that it is joined by a second molecule after heat shock. The second factor is obviously not involved in the formation of the DH site but the first factor might be. Two factors footprinting in a very similar region have also proved necessary for efficient in vitro transcription of these same genes (58). Whether these are the same proteins as those detected in nuclei remains to be seen. Both factors are present in both heat shocked and non-heat shocked cell extracts, but one becomes more stimulatory on heat shock.

More directly associated with DH sites is a factor which binds to the 5' region of the chicken adult β-globin gene (82). This factor (or factors), which behaves like a 60 kDa protein, will "create" a DNase I hypersensitive region on a plasmid carrying this gene when the plasmid is reconstituted with histones in vitro. However, the factor will create the DH site only if added before or during the reconstitution process, suggesting that some other condition(s) may be needed for this site's establishment in vivo. A similar pre- or co-assembly requirement is seen for the RNA polymerase III transcription factor TFIIIA (83).

In actuality it may well turn out that a DH site results from a combination of protein binding and DNA conformational change. One could provoke the other. There might be a conformational change on binding, or the protein(s) might specifically recognize an altered DNA conformation. However, the situation is more complicated than this. For example, there are several cases where the terminal repeats of viruses and transposable elements show differing chromatin structure despite an identity of DNA sequence (84-86). The DH sites 5' of Xenopus rDNA repeats are associated with only the most distal "Bam island," one of a series of repeated promoter elements (29). Perhaps the act of transcription precludes the establishment of DH sites at the other repeats.

DH sites tend to be reasonably broad structures, on the order of 50-200 or even 400 base pairs across (16, 18, 36, 87), and in the case of hsp70 (13) a reasonably large region surrounding the "protein" footprint is DNase I sensitive. A DNA structure induced by the DNA/protein combination may be important in excluding nucleosomes, since it seems unlikely that one protein could cover the entire region of the DH site. However, more than one protein may be involved.

An intriguing observation which associates DH sites with changes in DNA conformation comes from the rDNA minichromosomes in the macronucleus of _Tetrahymena_ _thermophila_. These linear chromosomes consist of a palindromic rDNA repeat with a replication origin in between. The whole region has a complex chromatin structure (44) including DH sites around the presumed origin sequences and at the start and finish of transcription (88). An endogenous nuclease is associated with all these sites (88) which has properties which are very reminiscent of a type I topoisomerase (89). SDS treatment of the nuclei leaves a series of single-stranded DNA nicks, specifically in the non-coding strand, with a protein covalently bound to the 3' ends of the DNA. These proteins, if they are topoisomerase molecules, could be involved in regulating the degree of supercoiling before and/or during transcription.

If a protein must associate with the DNA in order to establish a DH site but cannot "invade" the histone-associated DNA, as previously suggested (82), then some ancillary process is necessary to allow access to the DNA. One possibility is an alteration in DNA methylation, but this has not been consistently associated with the presence or absence of DH sites. An increase in DNA methylation of the embryonic chicken U-globin gene and the loss of its 5' DH site is seen on switching from U- to α-globin transcription (35). Similarly, the cell lines which fail to express the mouse metallothionein-1 gene (20) and the HSV thymidine kinase gene (16) are hypermethylated, and do not possess DH sites in the 5' regions, whereas their active counterparts do. In the latter case, activation of the gene by 5-azacytidine saw the re-establishment of a 5' DH site. However, no evidence of a methylation change in the globin genes during hexamethylene bisacetamide-induced differentiation of a murine erythroleukemia cell line was seen despite the appearance of DH sites 5' to these genes (90). Similarly, in _Xenopus_ _borealis_/_laevis_ hybrids only the _X_. _laevis_ genes possess DH sites and produce rRNA, although both genes are hypomethylated (91).

DNA replication represents another possibility whereby access to the DNA for DH site establishment might be achieved. Once established, DH sites appear to be stable through many rounds of replication (92, 28), but as with methylation, there is no strong correlation of replication with their establishment. A different pattern of DH sites is seen when the SV40 origin-promoter region (in a chimeric plasmid) is used to transfect CV1 cells, where they cannot replicate, compared to COS cells, where they can (93). This is correlated with a different pattern of transcription. After replication, transcription shifts to a predominantly late pattern. However, the α- and β-globin gene in a Rous sarcoma virus-transformed chicken embryo fibroblast cell line can form a detectable DH site 5' of the gene within thirty minutes of a shift to the permissive temperature (92). An even more rapid response is seen on administering dexamethasone to mouse cells transformed with mouse mammary tumor virus LTR-tk fusions. In this case, a DH site is established over the glucocorticoid receptor binding regions within seven minutes and can sometimes be detected as early as one minute after hormone addition (27). This site is lost on hormone withdrawal. There are also three DH sites established to the 5' of the major chicken vitellogenin gene (VTGII) in liver chromatin in response to estrogen, probably without replication. Two of these remain on hormone withdrawal, but one is completely dependent on the presence of estrogen (i.e., its establishment is reversible (94, 8)).

Thus, there is no real indication at the present time of what conditions must prevail in order to establish a DH site at a given sequence. The general "function" of DH sites may be to allow access to the DNA, but this covers a multitude of sins. At centromeres (45), it might allow access for microtubule attachment. At replication origins (44), it might facilitate the entry of the replication apparatus, and those at topoisomerase I sites (88, 89) may be involved in winding-state regulation.

Assuming that they are not merely a symptom of the binding of regulatory or other proteins, the role of DH sites in transcriptional regulation may be twofold. First, they may provide access for RNA polymerase initiation, and second, they may allow

the future access of other regulatory molecules. Examples of DH
site association with "housekeeping" genes which exhibit a constant transcription rate are given in Table 1, and may be
examples of the first possibility above. The footprint 5' of
the Drosophila hsp70 gene (in the DH site) enlarges upon heat
shock (13), presumably due to the binding of a second factor.
The mouse metallothionein 1 gene also shows an increase in its
5' DNase I hypersensitivity on induction (20, and see above).
These may be instances of the second possibility. Whatever
their precise role, DH sites provide a very useful assay for
identifying regions of interest and such analyses of chromatin
structure should be most fruitful in dissecting associated
regulatory processes.

REFERENCES

1. GROSSCHEDL, R., WASYLYK, B., CHAMBON, P., and BIRNSTEIL,
    M.L. (1981). Point mutations in the TATA box curtails the
    expression of sea urchin H2A histone gene in vivo. Nature
    294, 178.

2. STRUHL, K. (1983). Promoter elements, regulatory elements,
    and chromatin structure of the yeast HIS 3 gene. Cold
    Spring Harbor Symp. Quant. Biol. 47, 901.

3. McKNIGHT, S.L. (1983). Constitutive transcriptional con-
    trol signals of the herpes simplex virus tk gene. Cold
    Spring Harbor Symp. Quant. Biol. 47, 945.

4. CARTWRIGHT, I.L., KEENE, M.A., HOWARD, G.C., ABMAYR, S.M.,
    FLEISCHMANN, G., LOWENHAUPT, K., and ELGIN, S.C.R. (1982).
    Chromatin structure and gene activity: the role of nonhis-
    tone chromosomal proteins. CRC Crit. Rev. Biochem. 13, 1.

5. WEINTRAUB, H. and GROUDINE, M. (1976). Chromosome subunits
    in active genes have an altered conformation. Science 193,
    848.

6. LAWSON, G.M,. KNOLL, B.J., MARCH, C.M., WOO, S.L.C., TSAI,
    M.J., and O'MALLEY, B.J. (1982). Definition of 5' and 3'
    structural boundaries of the chromatin domain containing
    the ovalbumin multigene family. J. Biol. Chem. 257, 1501.

7. PARSLOW, T.G. and GRANNER, D.K. (1983). Structure of a
    nuclease sensitive region inside the immunoglobulin kappa
    gene: evidence for a role in gene regulation. Nucleic
    Acids Res. 11, 4775.

8. BURCH, J.B.E. and WEINTRAUB, H. (1983). Temporal order of
    chromatin structural changes associated with activation of
    the major chicken vitellogenin gene. Cell 33, 65.

9. SEIBENLIST, U., HENNIGHAUSEN, L., SATTEY, J., and LEDER, P.
    (1984). Chromatin structure and protein binding in the
    putative regulatory region of the c-myc gene in Burkitt
    lymphoma. Cell 37, 381.

10. NEDOSPASOV, S.A. and GEORGIEV, G.P. (1980). Non-random cleavage of SV40 DNA in the compact minichromosome and free in solution by micrococcal nuclease. Biochem. Biophys. Res. Comm. 92, 532.

11. WU, C. (1980). The 5' ends of Drosophila heat shock genes in chromatin are hypersensitive to DNase I. Nature 286, 854.

12. SOUTHERN, E.M. (1975). Detection of specific sequences among DNA fragments separated by gel electrophoresis. J. Mol. Biol. 98, 503.

13. WU, C. (1984). Two protein-binding sites in chromatin implicated in the activation of heat shock genes. Nature 309, 229.

14. COSTLOW, N. and LIS, J.T. (1984). High resolution mapping of DNase I hypersensitive sites of Drosophila heat shock genes in Drosophila melanogaster and Saccharomyces cervisiae. Mol. Cell. Biol. 4, 1853.

15. KEENE, M.A., CORCES, V., LOWENHAUPT, K., and ELGIN, S.C.R. (1981). DNase I hypersensitive sites in Drosophila chromatin occur at the 5' ends of regions of transcription. Proc. Natl. Acad. Sci. USA 78, 143.

16. SWEET, R.W., CHAO, M.V., and AXEL, R. (1982). The structure of the thymidine kinase promoter: nuclease hypersensitivity correlates with expression. Cell 31, 347.

17. McKNIGHT, S.L. and KINGSBURY, R. (1982). Transcriptional control signals of a eukaryotic protein coding gene. Science 217, 316.

18. WU, C. and GILBERT, W. (1981). Tissue-specific exposure of chromatin structure at the 5' terminus of the rat pre-proinsulin II gene. Proc. Natl. Acad. Sci. USA 78, 1577.

19. WALKER, M.D., EDLUND, T., BOULET, A.M., and RUTTER, W.J. (1983). Cell-specific expression controlled by the 5' flanking region of insulin and chymotrypsin genes. Nature 306, 577.

20. SENEAR, A.W. and PALMITER, R.D. (1983). Expression of the mouse metallothionein-1 gene alters the nuclease hypersensitivity of its 5' regulatory region. Cold Spring Harbor Symp. Quant. Biol. 47, 539.

21. SLEDZIEWSKI, A. and YOUNG, E.T. (1982). Chromatin conformational changes accompany transcriptional activation of a glucose-repressed gene in Saccharomyces cervisiae. Proc. Natl. Acad. Sci. USA 79, 253.

22. WONG, Y-C., O'CONNELL, P., ROSBASH, M., and ELGIN, S.C.R. (1981). DNase I hypersensitive sites of the chromatin for Drosophila melanogaster ribosomal protein 49 gene. Nucleic Acids Res. 9, 6979.

23. GROUDINE, M. and CASIMIR, C. (1984). Post-transcriptional regulation of the chicken thymidine kinase gene. Nucleic Acids Res. 12, 1427.

24. SCHUBACH, W. and GROUDINE, M. (1984). Alteration of c-myc chromatin structure by avian leukosis virus integration. Nature 307, 702.

25. BRYAN, P.N., OLAH, J., and BIRNSTEIL, M.L. (1983). Major changes in the 5' and 3' chromatin structure of sea urchin histone genes accompany their activation and inactivation in development. Cell 33, 843.

26. SAMAL, B., WORCEL, A., LOUIS, C., and SCHEDL, P. (1981). Chromatin structure of the histone genes of D. melanogaster. Cell 23, 401.

27. ZARET, K.S. and YAMAMOTO, K.R. (1984). Reversible and persistent changes in chromatin structure accompanying activation of a glucocorticoid-dependent enhancer element. Cell 38, 29.

28. LOWENHAUPT, K., CARTWRIGHT, I.L., KEENE, M.A., ZIMMERMAN, L.M., and ELGIN, S.C.R. (1983). Chromatin structure in pre- and post-blastula embryos of Drosophila. Devel. Biol. 99, 194.

29. LaVOLPE, A., TAGGERT, M., McSTAY, B., and BIRD, A. (1983). DNase I hypersensitive sites at promoter-like sequences in the spacer of Xenopus laevis and Xenopus borealis ribosomal DNA. Nucleic Acids Res. 11, 5361.

30. BORCHSENIUS, S., BONVEN, B., LEER, J.C., and WESTERGAARD, O. (1981). Nuclease-sensitive regions on the extrachromosomal r-chromatin from Tetrahymena pyriformis. Eur. J. Biochem. 117, 245.

31. PALEN, T., GOTTSCHLING, D.S., and CECH, T. (1982). Transcribed and non-transcribed regions of the ribosomal RNA gene of Tetrahymena exhibit different chromatin structures. J. Cell Biochem. Suppl. 6, 336.

32. CARTWRIGHT, I.L. and ELGIN, S.C.R. (1984). Chemical footprinting of 5S RNA chromatin in embryos of Drosophila melanogaster. EMBO J. 3, 3101.

33. SHERMOEN, A.W. and BECKENDORF, S.K. (1982). A complex of interacting DNase I hypersensitive sites near the Drosophila glue protein gene, sgs 4. Cell 29, 601.

34. STALDER, J., LARSEN, A., ENGEL, J.D., DOLAN, M., GROUDINE, M., and WEINTRAUB, H. (1980). Tissue-specific DNA cleavages in the globin chromatin domain introductd by DNase I. Cell 20, 451.

35. WEINTRAUB, H., LARSEN, A., and GROUDINE, M. (1981). α-globin gene switching during the development of chicken embryos: expression and chromosome structure. Cell, 24, 333.

36. McGHEE, J.D., WOOD, W.I., DOLAN, M., ENGEL, J.D., and FELSENFELD, G. (1981). A 200 base pair region at the 5' end of the chicken adult β-globin gene is accessible to nuclease digestion. Cell 27, 45.

37. McKEON, C., PASTAN, I., and DE CROMBRUGGHE, B. (1984). DNase I sensitivity of the α2(I) collagen gene: correlation with its expression but not its methylation pattern. Nucleic Acids Res., 12, 3491.

38. FRITTON, H.P., SIPPEL, A.E., and IGO-KEMENES, I. (1983). Nuclease hypersensitive sites in the chromatin domain of the chicken lysozyme gene. Nucleic Acids Res. 11, 3467.

39. KAYE, J.S., BELLARD, M., DRETZEN, G., BELLARD, F., and CHAMBON, P. (1984). A close association between sites of DNase I hypersensitivity and sites of enhanced cleavage by micrococcal nuclease in the 5' flanking region of the actively transcribed ovalbumin gene. EMBO J. 3, 1137.

40. DURRIN, L.K., WEBER, J.L., and GORSKI, J. (1984). Chromatin structure, transcription, and methylation of the prolactin gene domain in pituitary tumors of Fischer 344 rats. J. Biol. Chem. 259, 7086.

41. VARSHAVSKY, A.J., SUNDIN, O., and BOHN, M. (1979). A stretch of "late" SV40 viral DNA about 400 bp long, which includes the origin of replication, is specifically exposed in SV40. Cell 16, 453.

42. SARAGOSTI, S., MOYNE, G., and YANIV, M. (1980). Absence of nucleosomes in a fraction of SV40 chromatin between the origin of replication and the region coding for the late leader RNA. Cell 20, 65.

43. HERBOMEL, P., SARAGOSTI, D., BLANGY, D., and YANIV, M., (1981). Fine structure of the origin-proximal DNase I hypersensitive region in wild-type and EC mutant polyoma. Cell 25, 651.

44. PALEN, T.E. and CECH, T.R. (1984). Chromatin structure at the replication origins and transcription-initiation region of the ribosomal RNA genes of Tetrahymena. Cell 36, 933.

45. BLOOM, K.S. and CARBON, J. (1982). Yeast centromere DNA is in a unique and highly ordered structure in chromosomes and small circular minichromosomes. Cell 29, 305.

46. NASMYTH, K.A. (1982). The regulation of yeast mating-type chromatin structure by SIR: an action at a distance affecting both transcription and transposition. Cell 30, 567.

47. CHUNG, S-Y., FOLSOM, V., and WOOLEY, J. (1983). DNase I hypersensitive sites in the chromatin of immunoglobulin k light chain genes. Proc. Natl. Acad. Sci USA 80, 2427.

48. MILLS, F.E., FISHER, L.M., KURODA, R., FORD, A.M., and GOULD, H.J. (1983). DNase I hypersensitive sites in the chromatin of human u immunoglobulin heavy-chain genes. Nature 306, 809.

49. QUEEN, C. and BALTIMORE, D. (1983). Immunoglobulin gene transcription is activated by downstream sequence elements. Cell 33, 741.

50. BANERJI, J., OLSEN, L., and SCHAFFNER, W. (1983). A lymphocyte specific cellular enhancer is located downstream of the joining region in immunuglobulin heavy chain genes. Cell 33, 729.

51. GILLES, S.D., MORRISON, S.L., OI, V.I., and TONEGAWA, S. (1983). A tissue-specific transcription enhancer element is located in the major intron of a rearranged immunoglobulin heavy chain gene. Cell 33, 717.

52. CARTWRIGHT, I.L., HERTZBERG, R.P., DERVAN, P.B., and ELGIN, S.C.R. (1983). Cleavage of chromatin with methidium propyl-EDTA·iron(II). Proc. Natl. Acad. Sci. USA 80, 3213.

53. JAKOBOVITIS, E.B., BRATOSIN, S., and ALONI, Y. (1980). A nucleosome-free region in SV40 minichromosomes. Nature 285, 263.

54. KARPOV, V.L., PREOBRAZHENSKAYA, O.V., and MIRZABECKOV, A.D. (1984). Chromatin structure of hsp 70 genes activated by heat shock: selective removal of histones from the coding region and their absence from the 5' region. Cell 36, 423.

55. JACK, R.S., GEHRING, W.J., and BRACK, C. (1981). Protein component from Drosophila larval nuclei showing sequence specificity for a short region near a major heat shock protein gene. Cell 24, 321.

56. DAVISON, B.L., EGLY, J.M., MULVIHILL, E.R., and CHAMBON, P. (1983). Formation of stable preinitiation complexes between eukaryotic class B transcription factors and promoter sequences. Nature 301, 680.

57. DAVISON, B.L., MULVIHILL, E.R., EGLY, J.M., and CHAMBON, P. (1983). Interaction of eukaryotic class B transcription factors and chick progesterone-receptor complex with conalbumin promoter sequences. Cold Spring Harbor Symp. Quant. Biol. 47, 9.

58. PARKER, C.S. and TOPOL, J. (1984). A Drosophila RNA polymerase II transcription factor binds to the regulatory site of an hsp 70 gene. Cell 37, 273.

59. DYNAN, W.S. and TJIAN, R. (1983). The promoter-specific transcription factor Sp1 binds to upstream sequences in the SV40 early promoter. Cell 35, 79.

60. SCHOLER, H.R. and GRUSS, P. (1984). Specific interaction between enhancer-containing molecules and cellular components. Cell 36, 403.

61. DE VILLERS, J., OLSEN, L., TYNDALL, C., and SCHAFFNER, W. (1982). Transcriptional "enhancers" from SV40 and polyomavirus show a cell type preference. Nucleic Acids Res. 10, 7965.

62. BYRNE, B.J., DAVIS, M.S., YAMAGUCHI, J., BERGSMA, D.J., and SUBRAMANIAN, K.N. (1983). Definition of the simian virus 40 early promoter region and demonstration of a host range bias in the enhancement effect of the simian virus 40 72 base pair repeat. Proc. Natl. Acad. Sci. USA 80, 721.

63. FROMM, M. and BERG, P. (1983). Transcription in vivo from SV40 early promoter deletion mutants without repression by large T antigen. Mol. Cell. Biol. 3, 991.

64. JONGSTRA, J., REUDELHUBER, T.L., OUDET, P., BENOIST, C., CHAE, C-B., JELTSCH, J-M., MATHIS, D.J., and CHAMBON, P. (1984). Induction of altered chromatin structures by simian virus 40 enhancer and promoter elements. Nature 307, 708.

65. JAKOBOVITIS, E.B., BRATOSIN, S., and ALONI, Y. (1982). Formation of a nucleosome free region in SV40 minichromosomes is dependent upon a restricted segment of DNA. Virology 120, 340.

66. GERARD, R.D., WOODWORTH-GUTAI, M., and SCOTT, W.A. (1982). Deletion mutants which affect the nuclease sensitive site in simian virus 40 chromatin. Mol. Cell. Biol. 2, 782.

67. McGINNIS, W., SHERMOEN, A.W., and BECKENDORF, S.K. (1983). A transposable element inserted just 5' to a Drosophila glue protein gene alters gene expression and chromatin structure. Cell 34, 75.

68. McGINNIS, W., SHERMOEN, A.W., HEEMOKERK, J., and BECKENDORF, S.K. (1983). DNA sequence changes in an upstream DNase I hypersensitive region are correlated with reduced gene expression. Proc. Natl. Acad. Sci. USA 80, 1063.

69. SIMPSON, R.T. and KUNZLER, P. (1979). Chromatin core particles formed from the inner histones and synthetic polydeoxyribonucleotides of defined sequence. Nucleic Acids Res. 6, 1387.

70. RHODES, D. (1979). Nucleosome cores reconstituted from poly(dA-dT) and the octamer of histones. Nucleic Acids Res. 6, 1805.

71. LARSEN, A. and WEINTRAUB, H. (1982). An altered DNA conformation detected by S1 nuclease occurs at specific regions in active chick globin chromatin. Cell 29, 609.

72. NICKOL, J.M. and FELSENFELD, G. (1983). DNA conformation at the 5' end of the chicken adult β-globin gene. Cell 35, 467.

73. SCHON, E., EVANS, T., WELSH, J., and EFSTRATIADIS, A. (1983). Conformation of promoter DNA: fine mapping of S1-hypersensitive sites. Cell 35, 837.

74. KOHWI-SHIGEMATSU, T., GELINAS, R., WEINTRAUB, H. (1983). Detection of an altered DNA conformation at specific sites in chromatin and supercoiled DNA. Proc. Natl. Acad. Sci. USA 80, 4389.

75. SHEN, C.K. (1983). Superhelicity induces hypersensitivity of a human polypyrimidine-polypurine DNA sequence in the human a2-a1 globin intergenic region to S1 nuclease digestion - high resolution mapping of the clustered cleavage sites. Nucleic Acids Res. 11, 7899.

76. WEINTRUAB, H. (1983). A dominant role for DNA secondary structure in forming hypersensitive structures in chromatin. Cell 32, 1191.

77. MACE, H.A.F., PELHAM, H.R.B., and TRAVERS, A.A. (1983). Association of an S1 nuclease sensitive structure with short direct repeats 5' of Drosophila heat shock genes. Nature 304, 555.

78. SELLECK, S.B., ELGIN, S.C.R., and CARTWRIGHT, I.L. (1984). Supercoil-dependent features of DNA structure at Drosophila locus 67B1. J. Mol. Biol. 178, 17.

79. NORDHEIM, A. and RICH, A. (1983). The sequence (dC-dA)$_n$-(dG-dT)$_n$ forms left handed Z-DNA in negatively supercoiled plasmids. Proc. Natl. Acad. Sci. USA 80, 1821.

80. NICKOL, J., BEHE, J., and FELSENFELD, G. (1982). Effect of the B-Z transition in poly(dG-m^{5}dC)-poly(dG-m^{5}dC) on nucleosomal formation. Proc. Natl. Acad. Sci. USA 79, 1771.

81. NORDHEIM, A. and RICH, A. (1983). Negatively supercoiled simian virus 40 DNA contains Z-DNA segments within transcriptional enhancer sequences. Nature 303, 674.

82. EMERSON, B.M. and FELSENSELD, G. (1984). Specific factor conferring nuclease hypersensitivity at the 5' end of the chicken adult β-globin gene. Proc. Natl. Acad. Sci. USA 81, 95.

83. WOODLAND, H.R. (1982). Stable gene expression in vitro. Nature 297, 457.

84. EISSENBERG, J.C., KIMBRELL, D.A., FRISTROM, J.W., and ELGIN, S.C.R. (1984). Chromatin structure at the 44D larval cuticle gene locus in Drosophila: the effect of a transposable element insertion. Nucleic Acids Res. 12, 9025.

85. CHISWELL, D.J., GILLESPIE, D.A., and WYKE, J.A. (1982). The changes in proviral chromatin that accompany morphological variation in avian sarcoma virus infected rat cells. Nucleic Acids Res. 10, 3967.

86. VAN DER PLUTTEN, H., QUINT, W., VERMA, I.M., and BERNS, A. (1982). Moloney murine leukemia virus-induced tumors: recombinant proviruses in active chromatin regions. Nucleic Acids Res. 10, 577.

87. SARAGOSTI, S., CEREGHINI, S., and YANIV, M. (1982). Fine structure of the regulatory region of simian virus 40 minichromosomes revealed by DNase I digestion. J. Mol. Biol. 160, 133.

88. BONVEN, B.J. and WESTERGAARD, O. (1982). DNase I hypersensitive regions correlate with site specific endogenous nuclease activity on the r-chromatin of Tetrahymena. Nucleic Acids Res. 10, 7593.

89. GOCKE, E., BONVEN, B.J., and WESTERGAARD, O. (1983). A site and strand specific nuclease activity with analogies to topoisomerase I frames the rRNA gene in Tetrahymena. Nucleic Acids Res. 11, 7661.

90. SHEFFERY, M., RIFKIND, R.A., and MARKS, P.A. (1982). Murine erythroleukemia cell differentiation: DNase I hypersensitivity and DNA methylation near the globin genes. Proc. Natl. Acad. Sci. USA 79, 1180.

91. MACLEOD, D. and BIRD, A. (1982). DNase I sensitivity and methylation of active versus inactive rRNA genes in Xenopus species hybrids. Cell 29, 211.

92. GROUDINE, M. and WEINTRAUB, H. (1982). Propagation of globin DNase I hypersensitive sites in absence of factors required for induction: a possible mechanism for determination. Cell 30, 131.

93. CEREGHINI, S. and YANIV, M. (1984). Assembly of transfected DNA into chromatin: structural changes in the origin-promoter-enhancer region upon replication. EMBO J. 3, 1243.

94. BURCH, J.B.E. (1984). Indentification and sequence analysis of the 5' end of the major chicken vitellogenin gene. Nucleic Acids Res. 12, 1117.

95. GRIFFIN-SHEA, R., THIREOS, G., and KAFATOS, F.C. (1982). Organization of a cluster of four chorion genes in Drosophila and its relationship to developmental expression and amplification. Developmental Biol. 91, 325.

96. SHEFFERY, M., MARKS, P.A., and RIFKIND, R.A. (1984). Gene expression in murine erythroleukemia cells, transcriptional control and chromatin structure of the al-globin gene. J. Mol. Biol. 172, 417.

97. GROUDINE, M., KOHWI-SHIGEMATSU, T., GELINAS, R., STAMATOYANNOPOULOS, G., and PAPAYANNOPOULOS, T. (1983). Human fetal to adult hemoglobin switching: changes in chromatin structure of the β-globin gene locus. Proc. Natl. Acad. Sci. USA 80, 7551.

98. TUAN, D. and LONDON, I.M. (1984). Mapping of DNase I hypersensitive sites in the upstream DNA of human embryonic e-globin gene in K562 leukemia cells. Proc. Natl. Acad. Sci. USA 81, 2718.

99. GROUDINE, M., EISENMAN, R., and WEINTRAUB, H. (1981). Chromatin structure of endogenous retroviral genes and activation by an inhibitor of DNA methylation. Nature 292, 311.

100. BECKER, P., RENKAWITZ, R., and SCHUTZ, G. (1984). Tissue specific DNase I hypersensitive sites in the 5'-flanking sequences of the tryptophan oxygenase and the tryptophan aminotransferase genes. EMBO J. 3, 2015.

101. WEINTRAUB, H., BEUG, G., GROUDINE, M., and GRAF, T. (1982). Temperature-sensitive changes in the structure of a globin chromatin in lines of red cell precursors transformed by a ts-AEV virus. Cell 28, 931.

102. FRITTON, H.P., IGO-KEMENES, T., NOWOCK,MJ., STRECH-JURK, U., THEISEN, M., and SIPPEL, A.E. (1984). Alternative sets of DNase I hypersensitive sites characterize the various functional states of the chicken lysozyme gene. Nature 311, 163.

PROMOTER ELEMENTS OF EUKARYOTIC PROTEIN-CODING GENES

B. Wasylyk

Laboratoire de Génétique Moléculaire
des Eucaryotes de CNRS
Unité 184 de Biologie Moléculaire
et de Génie Génétique de l'INSERM
Faculté de Médecine
11, rue Humann
67085 Strasbourg-Cedéx, France

INTRODUCTION

In prokaryotes, initiation of transcription is controlled by specific DNA regions called promoters. Promoters were first defined on a genetic basis (1) as cis-acting regions indispensable for the expression of bacterial genes. Biochemical studies have shown that prokaryotic promoters are located 5' to the transcribed genes and are composed of multiple elements (2-5) as shown in Figure 1. One of these elements is involved in RNA polymerase binding. It includes the RNA start-site (consensus sequence 5'-CAT-3'), the Pribnow-Schaller box located 10 bp upstream from the start-site (consensus sequence 5'-TATAAT-3'), and frequently a third region, located in the -35 region (consensus sequence 5'-TTGACA-3'). The spatial relationship between the Pribnow-Schaller box and the -35 region is important because the insertion or deletion of a single base-pair can lead to a dramatic alteration of transcription (6). Other promoter elements, located either further upstream or downstream from the RNA start-site, interact with positive and negative regulatory proteins, which control the efficiency of transcription initiation (7-12).

The promoter regions of eukaryotic genes are defined as any DNA sequence required for accurate and efficient initiation of

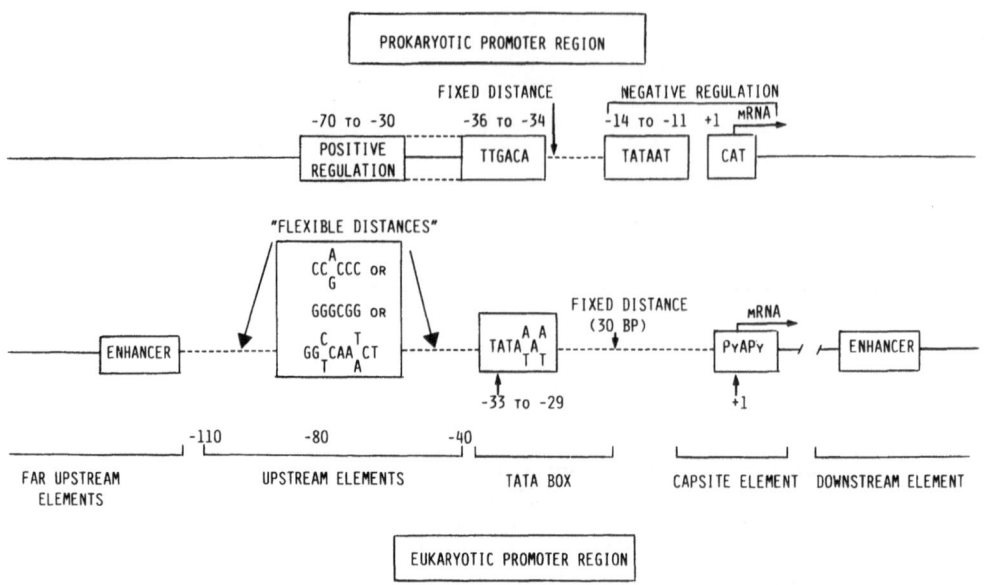

Figure 1. Comparison of prokaryotic and eukaryotic promoter regions. The boxes represent promoter elements. Some conserved sequence motifs are shown in the boxes.

transcription. For mRNA-coding genes, which are transcribed by RNA polymerase B (or II), the promoter region can be divided into several components: the cap-site element, the TATA box or proximal upstream elements, the distal upstream elements, and the enhancer (Figure 1). The characteristics of these sequences, and their interaction with the transcription machinery, have little similarity with their prokaryotic counterparts.

THE TATA BOX AND THE CAP-SITE

The TATA box is a highly conserved A-T rich region, with the consensus sequence $5'\text{-TATA}^{A}_{T}A^{A}_{T}\text{-}3'$, located approximately 28 bp upstream from the start-site of transcription of most eukaryotic protein-coding genes. In the few cases where it is absent, it may be replaced by a substitute TATA-like sequence, which plays the same role (13). The TATA box is a positioning element which directs the transcription machinery to initiate RNA synthesis 30 nucleotides downstream at the mRNA start-site (14, 15). Its integrity is also required to achieve maximum RNA synthesis from a given promoter region (16-18).

The cap-site element or the RNA start-site, which contains the bases coding for the 5' terminal nucleotides of the mRNA, is less conserved than the TATA box. Deletion of the cap-site element results in RNA chain initiation at a new site also located approximately 30 nucleotides downstream from the TATA box. Such a deletion is usually accompanied by a decrease in the amount of RNA chain initiation (reviewed in 14 and 15).

In vitro studies have shown that at least one factor interacts specifically with the TATA box and cap-site region of the DNA to form a stable preinitiation complex (19-25). This interaction occurs in the absence of RNA polymerase B and is apparently stabilized by a factor which has been extensively purified and which possesses actin-like properties (26). At this stage the promoter is committed, so that subsequently other factors and RNA polymerase can specifically initiate RNA synthesis (19, 21, 27). ATP hydrolysis, at the β-γ bond, is absolutely required for specific transcription initiation (28, 29).

THE UPSTREAM PROMOTER ELEMENTS

Several sequence elements, located in the -40 to -110 region upstream from the TATA box, are required for the efficiency but not the accuracy of initiation of transcription. These upstream promoter elements are polymorphic, and their number and position relative to the mRNA start-site is variable from one promoter to another. Although not as well conserved as the TATA box, some conservation of sequence is apparent. Several types of motifs have been identified. One of these, with the consensus sequence 5'-GGPyCAA^T_ACT-3', is located approximately 70-90 bp upstream from the mRNA start-site (30) and is found ubiquitously in both animal and plant promoters (31). Mutations in this sequence reduce specific transcription from the rabbit β-globin, adenovirus-2 EIIaE, and major-late promoters (17, 32-36). A second sequence, 5'-CCPuCCC-3' or the complementary sequence 5'-GGGCGG-3', is important for transcription from the SV40 early, Herpes thymidine kinase and glycoprotein-D, and rabbit β-globin genes (16, 17, 37-41).

Another class of distal upstream elements appears to be restricted to members of a gene family which are co-ordinately expressed either in terminally differentiated cells or in response to a defined signal such as heat shock. In the immuno-globulin gene family, the variable region promoters of the light chain genes have a conserved sequence 5'-ATTTGCAT-3', whilst the heavy chain promoters have the complementary sequence 5'-ATGCAAAT-3' (42, 43). Sequences containing this element are important for κ-light chain gene expression in lymphoid cells (43) and are probably non-functional in other cell lines (44, see also 45). The Drosophila heat-shock genes have a common distal upstream element 5'-CTNGAANTTCNAG-3' which is important for regulation of transcription in response to heat shock (46).

Upstream elements interact specifically with transcription factors. Several have been characterized and partially puri-fied, including those for the SV40 early 21 bp repeat (47), the Drosophila hsp70 heat-shock gene (24, 25, 48) and the adenovirus major-late promoter (35). Despite a strong sequence resemblance between the adenovirus major-late and EIIaE distal upstream ele-ments, they appear to interact with different factors (35). The upstream elements of the SV40 21 bp repeat, the herpes thymidine kinase gene and a monkey cellular gene have upstream GGGCGG motifs. The 21 bp repeat binding factor stimulates transcrip-tion from the SV40 and the cellular promoters (49) but not from the thymidine kinase promoter (50). These results suggest that the specificity of interaction between the upstream elements and transcription factors requires more than just a consensus sequence.

Upstream elements are interchangeable between different promoters. Cochran and Weissman (51) have shown that the two distal upstream elements of the thymidine kinase and the rabbit β-globin genes are interchangeable. They found that the natural β-globin and thymidine kinase promoters, as well as mosaic promoters containing all possible combinations of the upstream elements with the β-globin TATA box region, were about equally active. However, mosaic promoters with the thymidine kinase TATA box element were all about 10-fold less active. Miyamoto et al. (35) have shown that the SV40 21 bp repeat can function-ally replace the adenovirus major-late promoter upstream element to stimulate transcription through the major-late TATA box element. In these recombinants, it is the SV40 factor which stimulates transcription. The distance between the upstream element and the TATA box appears to be flexible within certain limits. Introducing up to about 20 bp of heterologous spacer DNA between upstream elements and the TATA box of the thymidine kinase or SV40 early promoters has little effect on transcription efficiency (40, 51, 52). In several cases, upstream elements have been shown to stimulate transcription in both their natural and inverse orientation (37, 50). These results suggest that efficient promoters can be assembled from a variety of heterologous elements and that considerable flexibility exists in the way that transcription factors binding to these regions may interact to stimulate transcription.

ENHANCER ELEMENTS

Discovery and Properties

Promoter sequences in far upstream sequences, extending beyond about -110 bp, were first identified in the SV40 early promoter (53, 54) and the sea-urchin H2A gene (55). The SV40 far upstream element, the 72 bp repeat, was later called an enhancer because of its ability to stimulate transcription from heterologous natural and substitute promoter elements, even when separated from them by several thousand base pairs of DNA (38, 52, 56, 57). Comparable promoter elements have been identified in several viral and cellular genes (for reviews see 58-65).

In vivo studies, using transfection of chimeric recombinants containing mainly the SV40 72 bp repeat enhancer, have shown that enhancers have the following properties. They appear to be quantitatively the most important promoter element. In cis location, they stimulate transcription from heterologous natural or substitute promoter elements, whether or not the promoters contain a TATA box or a functional upstream element. Enhancers stimulate transcription even when they are separated from the potentiated promoter element by several thousand base pairs of DNA. However, the efficiency of stimulation decreases as the length of the interposing sequence is increased. Stimulation is particularly efficient when the enhancer is located directly upstream from the stimulated promoter elements. Finally, enhancers are bidirectional, so that they stimulate transcription equally in either orientation relative to the potentiated promoter element (57, 66, 67). Enhancers appear to have no fixed position relative to other promoter elements. They are naturally located either upstream (the SV40 early promoter), within (in the intron separating the variable and constant regions of the immunoglobulin heavy and κ-light chain genes), or downstream (early promoter of bovine papilloma virus) from a transcription unit. Several enhancers (including those from SV40, the Moloney leukemia and sarcoma viruses, and the adenovirus EIa gene) which are naturally located upstream from the potentiated promoter elements, are less active when they are moved away from these promoter elements (66-69). This raises the question whether naturally downstream enhancers are in an unfavorable position for enhancement and, if not, what can account for this difference in properties.

Several sequence similarities have been identified among enhancers. Weiher et al. (70) showed that a critical sequence in the SV40 72 bp repeat, $5'-GTGG^{TTT}_{AAA}G-3'$, was also found in other enhancers. Similarly, there are important sequences in the adenovirus-5 EIa (5'-CAGGAGTGA-3'; reference 71) and the histone H2A enhancers (5'-GAGCCACCAACAGATGG-3'; reference 72) which are similar to sequences in other enhancers (for reviews see references 60, 65). In addition, several enhancers have short potential Z-forming stretches of alternating purines and pyrimidines, separated by 60-80 bp of DNA (73). These are not the only important sequences in enhancers since deletions outside the homologous sequences can reduce enhancer activity (69, 74, 75). Sequence repetition may affect the strength of an enhancer. Many enhancers are found naturally to have repeated sequences. Furthermore, duplication of non-enhancer sequences can generate an active enhancer (76, 77), suggesting that at least in certain cases repetition in itself can be critical.

The Entry-Site Model

The observations that proximal potential promoter elements are preferentially activated, and that activation of transcription decreases as the length of interposing sequences increases, led to the hypothesis that enhancers may act as an efficient entry site for the transcription machinery. A transcription factor (not necessarily RNA polymerase) might bind to the enhancer, and then scan the DNA until it finds a promoter element (57, 52). This model is supported by the experiments of de Villiers et al. (78) and Kadesh and Berg (79). For example, de Villiers et al. showed that a promoter element containing a TATA box, when placed between the SV40 enhancer and the β-globin gene decreases globin transcription. A point mutation in the TATA box of the inserted promoter decreases the inhibition.

There are several possible mechanisms by which an enhancer might act as an entry-site. 1) Enhancers may either adopt a particular DNA conformation (such as a change in superhelicity) or induce a conformational change at a distant location, which would facilitate an interaction with the transcription machinery. Changes in supercoiling are known to affect both gene expression in prokaryotes (for a review see reference 80) and the interaction of eukaryotic RNA polymerases with DNA (for a review see reference 81). 2) Enhancers may direct DNA to specialized regions of the nucleus, such as the matrix, which may be rich in transcription factors (see references 82-85). 3) Enhancers could organize chromatin structure to facilitate the access of transcription factors to this region of the DNA. 4) Finally, they may just interact directly and strongly with transcription factors. These mechanisms are not mutually exclusive and may all contribute to enhancer activity. Little evidence exists for mechanisms 1 and 2. Evidence for mechanisms 3 and 4 will now be discussed.

Enhancers Can Alter Chromatin Structure

Late in SV40 viral infection, the region between the origin of replication and the late start-sites of transcription is sensitive to digestion with a variety of nucleases (86-90) and in about 25% of extracted minichromosomes there is a nucleosome free gap (91, 92). Experiments using double origin SV40 viruses, in which a second normal or mutated origin region is inserted in a non-essential region of the virus, have shown that both the 72 bp repeat and the 21 bp repeat can induce DNase I

hypersensitivity. Insertion into the second origin of two repeats of the 72 bp repeats induces a sufficiently long stretch
of DNase I sensitive chromatin to generate a visible nucleosome-
free gap (93). In several cases increased transcription from a
promoter can be directly correlated with the appearance of a
DNase I hypersensitive site. In the 70Z/3 pre-B cell line,
treatment with mitogens leads to both induction of DNase I
hypersensitivity over the κ-light chain gene enhancer and
increased transcription of the gene (94). Induction of the
proviral MMTV promoter with glucocorticoids leads to both increased DNase I sensitivity over the enhancer and increased
transcription of the gene (95). These results show that enhancers can generate an "open window" in chromatin, and that
this is directly correlated with enhancement of transcription.
The significance of the observation that the 21 bp repeat, which
is quantitatively a less important element than the enhancer,
can also induce an altered chromatin structure (93, 96) is not
understood at present.

Enhancers Bind Transcription Factors

Some enhancers, especially those from the immunoglobulin
heavy and light chain genes, show marked tissue specificity,
suggesting that they could interact with specific regulatory
proteins (for a recent review see reference 65). More direct
evidence comes from both _in vivo_ and _in vitro_ experiments.
Scholer and Gruss (97), using an _in vivo_ assay, showed that
there is competition between enhancer-containing molecules for
cellular components. Specific competition involves the enhancers and not other promoter elements such as the SV40 21 bp
repeat and the TATA box. In addition, point mutations in the
SV40 enhancer, which diminish enhancer activity, also reduce the
ability to compete for cellular functions. _In vitro_ the SV40
72 bp repeat stimulates transcription from both the homologous
SV40 early promoter (98) as well as from heterologous promoter
elements (99). Competition assays have shown that, _in vitro_,
there is rapid and stable binding of the trans-acting factor(s)
with both the 5' and 3' domains of the enhancer sequence, and
that this factor(s) is different from those that interact with
the TATA box and the distal upstream elements of the SV40 early
and adenovirus major-late promoters. In addition, the same

factor(s) interacts with other enhancer elements (98, 100). The MMTV promoter region contains a glucocorticoid responsive element, which appears to be an enhancer because it stimulates transcription from heterologous promoters in an apparently orientation-independent manner (101). This element binds the purified glucocorticoid receptor in vitro (102).

CONCLUDING REMARKS

Despite considerable progress in the last few years, it appears that more questions are being raised than are being answered about how eukaryotic RNA polymerase B promoters function.

Several of these questions are:

1. How many transcription factors are we likely to find? We might expect there to be just a limited number of general transcription factors which appear to act at the level of the conserved TATA box and cap-site region of the DNA. Several have already been identified and there could be others for non-TATA box containing promoters. Distal upstream elements and enhancers are highly polymorphic although they contain some conserved sequence elements. Does each consensus sequence correspond to a binding site for a transcription factor which is common to many genes? The answer seems to be that the factors that bind to conserved sequences on different genes are not the same (see above), suggesting that there are whole families of factors even in the cases where the sequences are similar.

2. Upstream promoter elements and enhancers in eukaryotes are modular, there being apparently little constraint on the way they are mixed, the distance between them, and their orientation. How do the factors binding to these elements interact with each other and with the DNA to stimulate transcription?

3. Why do prokaryotes and eukaryotes use different mechanisms for promoter recognition by RNA polymerase? In prokaryotes RNA polymerase specifically recognizes promoter sequences before initiation of transcription. In eukaryotes RNA

polymerase B acts through stable binding factors, which them-selves are sequence specific. Is this difference due to the complexity of the eukaryotic genome? (For discussions see references 11, 103, and 104.)

4. How important are enhancer elements in controlling transcription of cellular genes? At the present time it is not known whether enhancers are a generalized feature of cellular promoters or whether they are restricted to actively transcribed viral and cellular genes. However, enhancers have many attrac-tive features for a general control element in cellular genes (see references 45 and 105). Answers to these and other ques-tions will be forthcoming in the next few years.

ACKNOWLEDGMENTS

I thank P. Chambon for many helpful discussions, G. Albrecht, J-M. Egly, N. Miyamoto, and M. Zenke for critically reading the manuscript, C. Aron for typing the manuscript, B. Boulay and C. Werlé for preparing the figures. This work was supported by grants from the CNRS, the INSERM, the MIR (82V1283), the Fondation pour la Recherche Médicale Française, the Association pour le Développement de la Recherche sur le Cancer, and the Fondation Simone et Cino del Duca.

REFERENCES

1. JACOB, F., ULLMAN, A., and MONOD, J. (1964). C.R. Acad. Sci. 258, 3125.

2. ROSENBERG, M. and COURT, D. (1979). Regulatory sequences involved in the promotion and termination of RNA tran-scription. Ann. Rev. Genet. 13, 319.

3. LOSICK, R. and CHAMBERLIN, M. (1976). In: "RNA Polym-erase", pp. 285-329. Cold Spring Harbor Laboratory, Cold Spring Harbor, New York.

4. RODRIGUEZ, R.L. and CHAMBERLIN, M.J., eds. (1982). "Promo-ters: Structure and Function." Praeger, New York.

5. SEIBENLIST, U., SIMPSON, R.B., and GILBERT, W. (1980). E. coli RNA polymerase interacts homologously with two dif-ferent promoters. Cell 20, 269-281.

6. STEFANO, J.E. and GRALLA, J.D. (1982). Spacer mutations in the lac ps promoter. Proc. Natl. Acad. Sci. USA 79, 1069-1072.

7. GUARENTE, L., NYE, J.S., HOCHSCHILD, A., and PTASHNE, M. (1982). Mutant lambda phage repressor with a specific defect in its positive control function. Proc. Natl. Acad. Sci. USA 79, 2236-2239.

8. HOCHSCHILD, A., IRWIN, N., and PTASHNE, M. (1983). Repressor structure and the mechanism of positive control. Cell 32, 319-325.

9. HAWLEY, D.K. and McCLURE, W.R. (1983). Compilation and analysis of Escherichia coli promoter DNA sequences. Cell 32, 327-333.

10. De CROMBRUGGHE, B., BUSBY, S., and BRUC, H. (1984). In: "Biological Regulation and Development," Vol. 3B (R.F. Goldberger and K.R. Yamamoto, eds.) pp. 129-167.

11. VON HIPPEL, P.H., BEAR, K.G., MORGAN, W.D., and McSWIGGEN, J.A. (1984). Protein-nucleic acid interactions in transcription: a molecular analysis. Ann. Rev. Biochem. 53, 389-446.

12. MAJUMDAR, A. and ADHYA, S. (1984). Demonstration of two operator elements in gal: in vitro repressor binding studies. Proc. Natl. Acad. Sci. USA 81, 6100-6104.

13. BRADY, J., RADONOVICH, M., VODKIN, M., NATARAJAN, V., THOREN, M., DAS, G., JANIK, J., and SALZMAN, N.P. (1982). Site-specific base substitution and deletion mutations that enhance or suppress transcription of the SV40 major late RNA. Cell 31, 625-633.

14. CORDEN, J., WASYLYK, B., BUCHWALDER, A., SASSONE-CORSI, P., KEDINGER, C., and CHAMBON, P. (1980). Expression of cloned genes in new environment. Science 209, 1406-1414.

15. BREATHNACH, R. and CHAMBON, P. (1981). Organization and expression of eucaryotic split genes coding for proteins. Ann. Rev. Biochem. 50, 349-383.

16. McKNIGHT, S.L. and KINGSBURY, R. (1982). Transcriptional control signals of a eukaryotic protein-coding gene. Science 217, 316-325.

17. DIERKS, P., VAN OOYEN, A., COCHRAN, M., DOBKIN, C., REISER, J., and WEISSMANN, C. (1983). Three regions upstream from the cap site are required for efficient and accurate transcription of the rabbit beta-globin gene in mouse 3T6 cells. Cell 32, 695-706.

18. WASYLYK, B., WASYLYK, C., MATTHES, H., WINTZERITH, M., and CHAMBON, P. (1983). Transcription from the SV40 early-early and late-early over-lapping promoters in the absence of DNA replication. EMBO J. 2, 1605-1611.

19. DAVISON, B.L., EGLY, J.M., MULVIHILL, E.R., and CHAMBON, P. (1983). Formation of stable preinitiation complexes between eukaryotic class B transcription factors and promoter sequences. Nature 301, 680-686.

20. CONCINO, M.F., LEE, R.F., MERRYWEATHER, J.P., and WEINMANN, R. (1984). The adenovirus major late promoter TATA box and initiation site are both necessary for transcription in vitro. Nucleic Acids Res. 12, 7423-7433.

21. FIRE, A., SAMUELS, M., and SHARP, P.A. (1984). Inter-actions between RNA polymerase II, factors, and template leading to accurate transcription. J. Biol. Chem. 259, 2509-2516.

22. HIROSE, S., TAKEUCHI, K., HORI, H., HIROSE, T., INAYAMA, A., and SUZUKI, Y. (1984). Contact points between tran-scription machinery and the fibroin gene promoter deduced by functional tests of single-base substitution mutants. Proc. Natl. Acad. Sci. USA 81, 1394-1397.

23. PARKER, C.S. and TOPOL, J. (1984). A Drosophila RNA polym-erase II transcription factor contains a promoter region specific DNA-binding activity. Cell 36, 357-369.

24. WU, C. (1984). Two protein-binding sites in chromatin implicated in the activation of heat-shock genes. Nature 309, 229-234.

25. WU, C. (1984). Activating protein factor binds in vitro to upstream control sequences in heat shock gene chromatin. Nature 311, 81-84.

26. EGLY, J.M., MIYAMOTO, N.G., MONCOLLIN, V., and CHAMBON, P. (1984). Is actin a transcription initiation factor for RNA polymerase B? EMBO J. 3, 2363-2371.

27. MATSUI, T., SEGALL, J., WEIL, P.A., and ROEDER, R.G. (1980). Multiple factors required for accurate initiation of transcription by purified RNA polymerase II. J. Biol. Chem. 255, 11992-11996.

28. BUNICK, D., ZANDOMENI, R., ACKERMAN, S., and WEINMANN, R. (1982). Mechanism of RNA polymerase II-specific initiation of transcription in vitro: ATP requirement and uncapped runoff transcripts. Cell 29, 877-886.

29. SAWADOGO, M. and ROEDER, R.G. (1984). Energy requirement for specific transcription initiation by the human RNA polymerase II system. J. Biol. Chem. 259, 5321-5326.

30. BENOIST, C., O'HARE, K., BREATHNACH, R., and CHAMBON, P. (1980). The ovalbumin gene -- sequence of putative control regions. Nucleic Acids Res. 8, 127-142.

31. SHAW, C.H., CARTER, G.H., WATSON, M.D., and SHAW, C.H. (1984). A functional map of the mopaline synthase pro-moter. Nucleic Acids Res. 12, 7831-7846.

32. GROSVELD, G.C., ROSENTHAL, A.I., and FLAVELL, R.A. (1982). Sequence requirements for the transcription of the rabbit beta globin gene in vivo the -80 region. Nucleic Acids Res. 10, 4951-4971.

33. ELKAIM, R., GODING, C., and KEDINGER, C. (1983). The adenovirus-2 EIIa early gene promoter: sequences required for efficient in vitro and in vivo transcription. Nucleic Acids Res. 11, 7105-7117.

34. ZAJCHOWSKI, D., BOEUF, H., and KEDINGER, C. (1984). The adenovirus-2 early EIIa transcription unit possesses two overlapping promoters with different sequence requirements for EIa-dependent stimulation. EMBO J., in press.

35. MIYAMOTO, N.G., MONCOLLIN, V., WINTZERITH, M., HEN, R., EGLY, J.M., and CHAMBON, P. (1984). Stimulation of in vitro transcription by the upstream element of the adeno-

virus-2 major late promoter involves a specific factor. Nucleic Acids Res. 12, 8779-8799.

36. HEN, R., WINTZERITH, M., MIYAMOTO, N., and CHAMBON, P. Manuscript in preparation.

37. EVERETT, R.D., BATY, D., and CHAMBON, P. (1983). The repeated GC-rich motifs upstream from the TATA box are important elements of the SV40 early promoter. Nucleic Acids Res. 11, 2447-2464.

38. FROMM, M. and BERG, P. (1983). Simian virus 40 early- and late-region promoter functions are enhanced by the 72-base pair repeat inserted at distant location and inverted orientations. J. Mol. Appl. Genet. 2, 127-135.

39. BATY, D., BARRERA-SALDANA, H.A., EVERETT, R.D., VIGNERON, M., and CHAMBON, P. (1984). Mutational dissection of the 21 bp repeat region of the SV40 early promoter reveals that it contains overlapping elements of the early-early and late-early promoters. Nucleic Acids Res. 12, 915-932.

40. McKNIGHT, S.L. (1982). Functional relationships between transcriptional control signals of the thymidine kinase gene of herpes simples virus. Cell 31, 355-365.

41. EVERETT, R. (1983). DNA sequence elements required for regulated expression of the HHSV-i glycoprotein D gene lie within 83 bp of the RNA capsites. Nucleic Acids Res. 11, 6647-6666.

42. PARSLOW, T.G., BLAIR, D.G., MURPHY, W., and GRANNER, D.K. (1984). Structure of the 5' ends of immunoglobulin genes a novel conserved sequence. Proc. Natl. Acad. Sci. USA 81, 2650-2654.

43. FALKNER, F.G. and ZACHAU, H.G. (1984). Correct transcription of an immunoglobulin κ gene requires an upstream fragment containing conserved sequence elements. Nature 310, 71-74.

44. FALKNER, F.G. and ZACHAU, H.G. (1982). Expression of mouse immunoglobulin genes in monkey cells. Nature 298, 286-288.

45. NORTH, G. (1984). Multiple levels of gene control in eukaryotic cells. Nature 312, 308-309.

46. PELHAM, H.R.B. and BIENZ, M. (1982). A synthetic heat-shock promoter element confers heat-inducibility on the herpes simplex virus thymidine kinase gene. EMBO J. 1, 1473-1477.

47. DYNAN, W.S. and TJIAN, R. (1983). The promoter-specific transcription factor Spl binds to upstream sequences in the SV40 early promoter. Cell 35, 79-87.

48. PARKER, C.S. and TOPOL, J. (1984). A Drosophila RNA polymerasee II factor binds to the regulatory site of an HSP70 gene. Cell 37, 273-283.

49. GIDONI, D., DYNAN, W.S., and TJIAN, R. (1984). Multiple specific contacts between a mammalian transcription factor and its cognate promoters. Nature 312, 409-413.

50. McKNIGHT, S.L., KINGSBURY, R.C., SPENCE, A., and SMITH, M. (1984). The distal transcription signals of the herpes virus TK gene share a common hexanucleotide control sequence. Cell 37, 253-262.

51. COCHRAN, M.D. and WEISSMANN, C. (1984). Modular structure of the β-globin and the TK promoters. EMBO J. 3, 2453-2459.

52. MOREAU, P., HEN, R., WASYLYK, B., EVERETT, R., GAUB, M.P., and CHAMBON, P. (1981). The SV40 72 base pair repeat has a striking effect of gene expression both in SV40 and other chimeric recombinants. Nucleic Acids Res. 9, 6047-6068.

53. BENOIST, C. and CHAMBON, P. (1981). In vivo sequence requirements of the SV40 early promoter region. Nature 290, 304-310.

54. GRUSS, P., DHAR, R., and KHOURY, G. (1981). Simian virus 40 tandem repeated sequences as an element of the early promoter. Proc. Natl. Acad. Sci. USA 78, 943-947.

55. GROSSCHEDL, R. and BIRNSTIEL, M.L. (1982). Delimitation of far upstream sequences required for maximal in vitro transcription of an H2A histone gene. Proc. Natl. Acad. Sci. USA 79, 297-301.

56. BANERJI, J., RUSCONI, S., and SCHAFFNER, W. (1981). Expression of a beta-globin gene is enhanced by remote SV40 DNA sequences. Cell 27, 299-308.

57. WASYLYK, B., WASYLYK, C., AUGEREAU, P., and CHAMBON, P. (1983b). The SV40 72 bp repeat preferentially potentiates transcription starting from proximal natural or substitute promoter elements. Cell 32, 503-514.

58. YANIV, M. (1982). Enhancing elements for activation of eukaryotic promoters. Nature 297, 17-18.

59. KHOURY, G. and GRUSS, P. (1983). Enhancer elements. Cell 33, 313-314.

60. GLUZMAN, Y. and SHENK, T., eds. (1983). "Enhancers and Eukaryotic Gene Expression." Cold Spring Harbor Laboratory, Cold Spring Harbor, New York.

61. GUARENTE, L.P. (1984). Yeast promoters: positive and negative elements. Cell 36, 799-800.

62. SHIMIZU, A. and HONJO, T. (1984). Immunoglobulin class switching. Cell 36, 801-803.

63. DUNN, A.R. and GOUGH, N.M. (1984). Tissue-specific enhancers. TIBS 11, 81-82.

64. KOLATA, G.B. (1984). New clues to gene regulation enhancer sequences seem to be involved in turning on genes and may themselves be regulated by a small group of proteins. Science 224, 588-589.

65. YANIV, M. (1984). Regulation of eukaryotic gene expression by trans-activating proteins and cis-acting DNA elements. Biol. Cell 50, 203-216.

66. WASYLYK, B., WASYLYK, C., and CHAMBON, P. (1984). Short and long range activation by the SV40 enhancer. Nucleic Acids Res. 12, 5589-5608.

67. AUGEREAU, P. and WASYLYK, B. (1984). The MLV and SV40 enhancers have a similar pattern of transcriptional activation. Nucleic Acids Res. 12, 8801-8818.

68. HEN, R., BORRELLI, E., SASSONE-CORSI, P., and CHAMBON, P. (1983). An enhancer element is located 340 base pairs up-

stream from the adenovirus-2 E1A capsite. Nucleic Acids Res. 11, 8747-8760.

69. LAIMINS, I.A., GRUSS, P., POZZATTI, R., and KHOURY, G. (1984). Characterization of enhancer elements in the long terminal repeat of moloney murine sarcoma virus. J. Virol. 49, 183-189.

70. WEIHER, H., KOENIG, M., and GRUSS, P. (1983). Multiple point mutations affecting the simian virus 40 enhancer. Science 219, 626-631.

71. HEARING, P. and SHENK, T. (1983). A duplicated enhancer element within the adenovirus type-5 E1A transcriptional control region. In: "Enhancers and Eukaryotic Gene Expression," (Y. Gluzman and T. Shenk, eds.) pp. 91-94. Cold Spring Harbor Laboratory, Cold Spring Harbor, New York.

72. GROSSCHEDL, R., MAECHLER, M., ROHRER, U., and BIRNSTIEL, M.I. (1983). A functional component of the sea urchin H2A gene modulator contains an extended sequence homology to a viral enhancer. Nucleic Acids Res. 11, 8123-8136.

73. NORDHEIM, A. and RICH, A. (1983). Negatively supercoiled simian virus 40 DNA contains Z-DNA seqments within transcriptional enhancer sequences. Nature 303, 674-679.

74. BANERJI, J., OLSON, L., and SCHAFFNER, W. (1983). A lymphocyte-specific cellular enhancer is located downstream of the joining region in immunoglobulin heavy chain genes. Cell 33, 729-740.

75. HERBOMEL, P., BIUCHAROT, B., and YANIV, M. (1984). Two distinct enhancers with different cell specificities coexist in the regulatory region of polyoma. Cell, in press.

76. WEBER, F., DeVILLIERS, J., and SCHAFFNER, W. (1984). An SV40 "enhancer trap" incorporates exogenous enhancers or generates enhancers from its own sequences. Cell 36, 983-992.

77. SWIMMER, C. and SHENK, T. (1984). A viable simian virus 40 variant that carries a newly generated sequence reiteration in place of the normal duplicated enhancer element. Proc. Natl. Acad. Sci. USA 81, 6652-6656.

78. DeVILLIERS, J., OLSON, L., BANERJI, J., and SCHAFFNER, W. (1983). Cold Spring Harbor Symp. on Quant. Biol., Vol. XLVII, pp. 911-919. Cold Spring Harbor Laboratory, Cold Spring Harbor, New York.

79. KADESCH, T.R. and BERG, P. (1983). Effects of the position of the 72-bp enhancer segment on transcription from the SV40 early region promoter. In: "Enhancers and Eukaryotic Gene Expression," (Y. Gluzman and T. Shenk, eds.) pp. 21-27. Cold Spring Harbor Laboratory, Cold Spring Harbor, New York.

80. WANG, J.C. (1983). In: "Genetics Rearrangement," (K.F. Chater, C.A. Cullis, D.A. Hopwood, A.W.B. Johnston, and H. Wolhouse, eds.) pp. 1-26. Sinauer, Amherst, Massachusetts.

81. CHAMBON, P. (1975). Eukaryotic RNA polymerases. Ann. Rev. Biochem. 44, 613-633.

82. MIRKOVITCH, J., MIRAULT, M.E., and LAEMMLI, U.K. (1984). Organization of the higher-order chromatin loop: specific

DNA attachment sites on nuclear scaffold. Cell 39, 223-232.

83. CIEJEK, E.M., TSAI, M., and O'MALLEY, B.W. (1983). Actively transcribed genes are associated with the nuclear matrix. Nature 306, 607-609.

84. ROBINSON, S.T., NELKIN, B.D., and VOGELSTEIN, B. (1982). The ovalbumin gene is associated with the nuclear matrix of chicken oviduct cells. Cell 28, 99-106.

85. COOK, P.R. and BRAZELL, I.A. (1976). Conformational constraints in nuclear DNA. J. Cell. Sci. 22, 287-302.

86. SCOTT, W.A. and WIGMORE, D.J. (1978). Sites in simian virus 40 chromatin which are preferentially cleaved by endonucleases. Cell 15, 1511-1518.

87. WALDECK, W., FOHRING, B., CHOWDHURY, K., GRUSS, P., and SAUER, G. (1978). Origin of DNA replication in papovavirus chromatin is recognized by endogenous endonuclease. Proc. Natl. Acad. Sci. USA 75, 5964-5968.

88. WIGMORE, D.J., EATON, R.W., and SCOTT, W.A. (1980). Endonuclease-sensitive regions in SV40 chromatin from cells infected with duplicated mutants. Virology 104, 462-473.

89. SUNDIN, O. and VARSCHAVSKY, A. (1979). Staphylococcal nuclease makes a single non-random cut in the simian virus 40 viral minichromosome. J. Mol. Biol. 132, 535-546.

90. VARSHAVSKY, A.J., SUDIN, O., and BOHN, M. (1979). A stretch of "late" SV40 viral DNA about 400 bp long which includes the origin of replication is specifically exposed in SV40 minichromosomes. Cell 16, 453-466.

91. SARAGOSTI, S., MOYNE, G., and YANIV, M. (1980). Absence of nucleosomes in a fraction of SV40 chromatin between the origin of replication and the region coding for the late leader RNA. Cell 20, 65-73.

92. JAKOBOVITS, E.B., BRATOSIN, S., and ALONI, Y. (1980). A nucleosome-free region in SV40 minichromosomes. Nature 285, 263-265.

93. JONGSTRA, J., REUDELHUBER, T.L., OUDET, P., BENOIST, C., CHAE, C.B., JELTSCH, J.M., MATHIS, D., and CHAMBON, P. (1984). Induction of altered chromatin structures by simian virus 40 enhancer and promoter elements. Nature 307, 708-714.

94. PARSLOW, T.G., and GRANNER, D.K. (1983). Structure of a nuclease-sensitive region inside the immunoglobin kappa gene: evidence for a role in gene regulation. Nucleic Acids Res. 11, 4775-4792.

95. ZARET, K.S. and YAMAMOTO, K.R. (1984). Reversible and persistent changes in chromatin structure accompany activation of a glucocorticoid-dependent enhancer element. Cell 38, 29-38.

96. INNIS, J. and SCOTT, W.A. (1984). DNA replication and chromatin structure of simian virus 40 insertion mutants. Mol. Cell. Biol. 4, 1499-1507.

97. SCHOLER, H.R. and GRUSS, P. (1984). Specific interaction between enhancer-containing molecules and cellular components. Cell 36, 403-411.

98. WILDEMAN, A.G., SASSONE-CORSI, P., GRUNDSTROM, T., ZENKE, T., and CHAMBON, P. (1984). Stimulation of in vitro transcription from the SV40 early promoter by the enhancer involves a specific trans-acting factor. EMBO J. 3, 3129-3133.

99. SASSONE-CORSI, P., DOUGHERTY, J., WASYLYK, B., and CHAMBON, P. (1984). Stimulation of in vitro transcription from heterologous promoters by the SV40 enhancer. Proc. Natl. Acad. Sci. USA 81, 308-312.

100. SASSONE-CORSI, P., WILDEMAN, A., and CHAMBON, P. (1984). A trans-acting factor is responsible for the SV40 enhancer activity. Nature 313, 458-463.

101. CHANDLER, V.L., MALER, B.A., and YAMAMOTO, K.R. (1983). DNA sequences bound specifically by glucocorticoid receptor in vitro render a heterologous promoter hormone responsive in vivo. Cell 33, 489-499.

102. SCHEIDEREIT, C. and BEATO, M. (1984). Contacts between hormone receptor and DNA double helix within a glucocorticoid regulatory element of mouse mammary tumor virus. Proc. Natl. Acad. Sci. USA 81, 3029-3033.

103. TRAVERS, A. (1983). Protein contacts for promoter location in eukaryotes. Nature 303, 755.

104. PTASHNE, M. (1984). DNA-binding proteins. Nature 308, 753-754.

105. CHAMBON, P., DIERICH, A., GOUB, M.P., JAKOWLEV, S.B., JONGSTRA, J., KRUST, A., LePENNEC, J.P., OUDET, P., and REUDELHUBER, T. (1984). Promoter elements of genes coding for proteins and modulation of transcription by oestrogens and progesterone. In: "Recent Progress in Hormone Research, The Proceedings of the Laurentian Hormone Conference," Vol. 40 (R.O. Greep, ed.) pp. 1-42. Academic Press, New York.

REGULATORY ELEMENTS IN STEROID HORMONE INDUCIBLE GENES: STRUCTURE AND EVOLUTION OF DNA SEQUENCES RECOGNIZED BY STEROID HORMONE RECEPTORS

M. Beato, C. Scheidereit, P. Krauter, D. von der Ahe, S. Janich, A.C.B Cato, G. Suske, and H.M. Westphal

Institut für Physiologische Chemie
der Phillipps-Universität
Deutschhausstrasse 1-2
D-3550 Marburg, FRG

INTRODUCTION

The mechanism by which steroid hormones control the synthesis of specific proteins in target cells has attracted the attention of molecular biologists because it represents a useful model system for studying regulation of gene expression in terminally differentiated cells of higher organisms. The cellular effects of steroid hormones are mediated by their corresponding receptors that, after binding the hormone, interact with specific sites in chromatin (1, 2). Considerable effort has been devoted to elucidate the mechanism of this interaction. Until recently a widespread hypothesis assumed the existence of specific "acceptor" sites in the chromatin of target cells, composed essentially of non-histone chromosomal proteins (3). This idea was based on the observation of a saturable preferential binding of hormone receptor to isolated nuclei or chromatin from target cells (4-8). However, some of these observations were later questioned, and they probably reflected in vitro artifacts (9-12). On the other side, a direct binding of hormone-receptor complexes to naked DNA has been observed in several systems (4, 12-14) and therefore the possibility was considered that the receptors could recognize specific DNA sequences (15, 16).

Initial experiments, using total cellular DNA coupled to cellulose and either crude cytosol or partially purified receptor preparations, showed similar binding to DNA of animal and bacterial origin (12, 17, 18). In fact, given the complexity of the animal genome, theoretical considerations show that even if a small number of high affinity sites do exist in DNA, they could not be detected using total cellular DNA and conventional techniques (15). This type of experiments showed, nevertheless, that the interaction of the hormone receptor with both total cellular DNA and chromatin exhibits similar kinetics and salt dependence, suggesting that DNA is indeed the acceptor for the hormone-receptor complex in the cell nucleus (12). Indeed, masking of "free" DNA phosphates by titration with poly-L-lysine leads to almost complete inhibition of receptor nuclear binding (12).

The availability of cloned inducible genes and progress in the purification of functional hormone receptors have recently allowed investigators to test directly the hypothesis of a specific interaction between receptors and DNA. In this chapter we will summarize evidence from our and other laboratories demonstrating that defined nucleotide sequences close to the promoters of hormonally regulated genes are involved both in binding of the hormone receptor and in hormonal regulation of transcription. Whether other chromatin components play a role in gene regulation by steroid hormones is a question not addressed in this paper.

THE BINDING ASSAYS

In principle, either crude extracts or more or less purified preparations of activated hormone-receptor complexes can be used for studying specific binding to DNA. Using crude extracts, two approaches are possible. In the first method, the steroid-receptor complex is bound to calf thymus or salmon sperm DNA immobilized on cellulose, and then released from the matrix by addition of individual DNA fragments. The ability of a given fragment to compete with calf thymus DNA is a measure of its affinity for the receptor (19, 20). In this assay one uses radioactive hormone to detect hormone-receptor complexes, and therefore a purification of the receptor is not necessary. The procedure, however, is indirect and many parameters can influence receptor release. In the second approach, a mixture of labeled DNA fragments is incubated with the crude extract and the receptor-bound fragments are isolated from the mixture by immunoprecipitation, using antibodies against the receptor that do not interfere with receptor binding to DNA (21). The radioactive DNA fragments are then separated from the protein and analyzed in agarose gels. This method is very specific and can be used with hormone-free receptors or with receptors that are occupied by unphysiological ligands such as antihormones.

Experiments with purified receptors are usually based on incubations with a mixture of radioactively labeled restriction fragments and on isolation of the protein-bound DNA fragments by

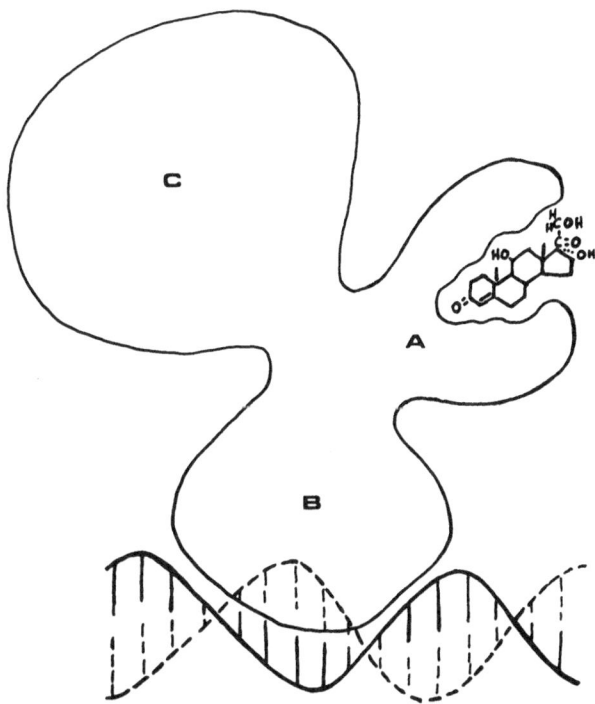

Figure 1.   Domain model of the activated glucocorticoid receptor. The model represents the 90,000 M_r form of the activated rat liver glucocorticoid receptor with domains for: (A) steroid binding; (B) DNA binding; and (C) antibody binding. The C domain is not functionally defined and could be heterogeneous. The 40,000 M_r form of the receptor, as obtained from frozen rat liver or by digestion with α-chymotrypsin (27), is composed of domains A and B.

a)

```
        - 220              - 200            - 180              - 160              - 140
CAACCTTGCGGTTCCCAGGGTTTAAATAAGTTTATGGTTACAAACTGTTCTTAAAACAAGGATGTGAGACAAGTGGTTTCCTGAGTTGGTTT
GTTGGAACGCCAAGGGTCCCAAATTTATTCAAATACCAATGTTTGACAAGAATTTTGTTCCTACACTCTGTTCACCAAAGGACTCAACCAAA

        - 120        Sau III A I      - 100                        - 60              - 40
GGTATCAAATGTTCTGATCTGAGCTCTTAGTGTTCTTATTTTCCTATGTTCTTTTGGAATCTATCCAAGTCTTATGTAAATGCTTATGTAAAC
CCATAGTTTACAAGACTAGACTCGAGAATCACAAGATAAAAGGATACAAGAAAACCTTAGATAGGTTCAGAATACATTTACGAATACATTTG

        - 20                              cap
CATAATATAAAAGAGTGCTGATTTTTTGAGTAAACTTGCAACAGTCCTAACATTCTTCTCTCGTGTGTTTGTGTCTGTTCGC-3'
GTATTATATTTTCTCACGACTAAAAAACTCATTTGAACGTTGTCAGGATTGTAAGAAGAGAGCACACAAACACAGACAAGCG-5'
```

b)

```
            - 320              - 300              - 280              - 260
TTCGAGTACAGGACAGGAGGGAGGGGAGCTGTGCACACGGCGGAGGCGCACGGCGTGGGCACCCAGCACCCGGTACACTGTGTCCTCCCGCT
AAGCTCATGTCCTGTCCTCCCTCCCCTCGACACGTGTGCCGCCTCCGCGTGCCGCACCCGTGGGTCGTGGGCCATGTGACACAGGAGGGCGA

    - 240              - 220              - 200              - 180              - 160
GCACCCAGCCCCTTCAGCGCCGAGGCGTCCCCGAGGCGCAAGTGGGCCGCCTTCAGGGAACTGACCGCCCGCGGCCCGTGTGCAGAGCCGGG
CGTGGGTCGGGGAAGTCGCGGCTCCGCAGGGGCTCCGCGTTCACCCGGCGGAAGTCCCTTGACTGGCGGGCGCCGGGCACACGTCTCGGCCC

    - 140              - 120              - 100              - 80              - 60
TGCGCCCGGCCCAGTGCGCGCGGCCGGGTGTTTCCCTTGGAGCCGCAAGTGACTTCTAGCGCGGGGCGTGTGCAGGCACGGCCGGGGCGGGG
ACGCGGGCCGGGTCACGCGCGCCGGCCCACAAAGGGAACCTCGGCGTTCACTGAAGATCGCGCCCCGCACACGTCCGTGCCGGCCCCGCCCC

    - 40              - 20              CAP
CTTTTGCACTCGTCCCGGCTCTTTCTAGCTATAAACACTGCTTCCCGCGCTGCACTCCACCACG-3'
GAAAACGTGAGCAGGGCCGAGAAAGATCGATATTTATGACGAAGGGCGCGACGTGAGGTGGTGC-5'
```

c)

```
        - 260              - 240              - 220              - 200
TTTTATGGCGGTTAGTAGTGGTACACTGATGATGAACAATGGCTATGCAGTAAAATCAAGACTGTAGATATTGCAACAGACTATAAAAATTCC
AAAATACCGACAATCATCACCATGTGACTACTACTTGTTACCGATACGTCATTTTAGTTCTGACATCTATAACGTTGTCTGATATTTTAAGG

      - 180              - 160              - 140              - 120              - 100
TCTGTGGCTTAGCCAATGTGGTACTTCCCACATTGTATAAGAAATTTGGCAAGTTTAGAGCAATGTTTGAAGTGTTGGGAAATTTCTGTATA
AGACACCGAATCGGTTACACCATGAAGGGTGTAACATATTCTTTAAACCGTTCAAATCTCGTTACAAACTTCACAACCCTTTAAAGACATAT

        - 80              - 60              - 40              - 20              cap
CTCAAGAGGGCGTTTTTGACAACTGTAGAACAGAGGAATCAAAAGGGGGTGGGAGGAAGTTAAAAGAAGAGGCAGGTGCAAGAGAGCTTGCA-3
GAGTTCTCCCGCAAAAACTGTTGACATCTTGTCTCCTTAGTTTTCCCCCACCGTCCTTCAATTTTCTTCTCCGTCCACGTTCTCTCGAACGT-5
```

filtration through nitrocellulose filters (22). Only the glucocorticoid receptor of rat liver and the progesterone receptors of avian oviduct and rabbit uterus have been used for this type of binding study. Different forms of the glucocorticoid receptor of rat liver have been partially purified (Figure 1). The 90,000 M_r form is obtained from fresh liver and appears to be composed of at least three domains: a steroid binding domain, a DNA binding domain, and a third domain of unknown function with strong antigenic properties (23-26). A 40,000 M_r form of the receptor can be purified from frozen liver (27), or obtained by limited chymotryptic digestion of the 90,000 M_r form (25). This smaller form still contains the steroid and DNA binding sites but has lost the antigenic domain (25). The progesterone receptor of chicken oviduct has been claimed to be composed of two different subunits, A and B, that have been purified to homogeneity (28). Only subunit A has been used for binding studies with cloned inducible genes. The progesterone receptor of rabbit endometrium has also been purified and used for binding studies with either avian or mammalian genes (29).

Since the receptor preparations are never completely homogeneous (generally around 50% pure), it is important to devise control experiments that eliminate artifacts due to contaminating proteins. In the case of the glucocorticoid receptor,

Figure 2. Nucleotide sequence around the regulated promoters: a) MMTV-LTR; b) human metallothionein II_A; and c) chicken lysozyme. The regions protected against DNase I in the presence of the receptor are indicated by thick lines. In b) only the lower strand has been tested with DNase I. Broken lines denote uncertain ties in establishing the limits of the protected regions. Sites of enhanced sensitivity to DNase I in the presence of the receptor are indicated by arrows. Purine residues undermethylated in the presence of the receptor are marked by an open triangle, whereas those in which methylation is enhanced by the receptor are indicated by dark triangles. Open arrows point to G-residues that, if methylated by dimethyl sulfate prevent receptor binding. Other relevant sequences are indicated by pointed lines. Numbers refer to positions upstream of the "cap" site (+1). For the chicken lysozyme gene neither footprint data nor methylation data are available.

monoclonal antibodies can be used to fulfill this requirement (see below).

MOUSE MAMMARY TUMOR VIRUS

Glucocorticoids enhance the transcription of mouse mammary tumor virus (MMTV) DNA in a variety of cell lines containing integrated proviral genomes (30). Gene transfer experiments suggest that the glucocorticoid regulatory element is located in the long terminal repeat (LTR) regions of the viral genome (32-35). Both with crude extracts of rat liver and with purified preparations of glucocorticoid receptor, a preferential binding of the receptor to restriction fragments containing LTR sequences has been detected (36-41). In addition, preferential binding of the receptor to other regions of the proviral genome and to the mouse flanking sequences was also observed (36, 41, 42). Interestingly, both forms of the partially purified glucocorticoid receptor (the 90,000 M_r form and the 40,000 M_r form) display similar patterns of sequence recognition, although the extent of specific binding is higher with the native 90,000 M_r form (36, 43). These results suggest that the antigenic domain of the receptor may modulate the affinity of the DNA binding domain for specific DNA sequences (44). Using a set of LTR deletion mutants and monoclonal antibodies to the receptor we have delimited a region of DNA between positions -50 and -202 upstream of the transcription initiation site that is responsible for glucocorticoid receptor binding (21). This region

```
                  1  2  3  4  5  6  7  8  9 10 11 12 13 14 15 16 17

MMTV I    -186    T  G  G  T  T  A  C  A  A  A  C  T  G  T  T  C  T

MMTV IIa  -129    T  G  G  T  A  T  C  A  A  A  .  T  G  T  T  C  T

hMTIIA    -263    C  G  G  T  .  A  C  A  C  T  G  T  G  T  C  C  T

LysI       -50    T  T  G  A  T  T  C  .  C  T  C  T  G  T  T  C  T

CONSENSUS:        T  G  G  T  T  C  A  C  T  C  T  G  T  T  C  T
                  c  t     a  A  A     .  A  A  G           c
```

Figure 3. Consensus sequence for the glucocorticoid regulatory element. The nucleotide sequence of the two main binding sites in the LTR region of MMTV, of the strong binding site of the human metallothionein II_A gene, and the equivalent site in the lysozyme gene are aligned to yield maximal similarity.

represents a glucocorticoid regulatory element since it has been shown to be required for glucocorticoid inducibility in gene transfer experiments (45). DNase I protection experiments (46) with the purified glucocorticoid receptor yield a complex pattern of footprints in this region of the LTR (21). Two sites, between positions -192 and -163 and between -124 and -105 are strongly protected against DNase I digestion, whereas two other smaller sites, between -101 and -85, and between -83 and -71 are only weakly protected by the receptor against nuclease digestion (Figure 2a). Although the two strong binding sites show considerable sequence similarity (Figure 3), the limits of the footprints are different for each site. This could be due to small differences in sequence or reflect an interaction of the bound receptor molecules with each other. The hexanucleotide 5'-TGTTCT-3' is common to all four subregions, suggesting that it plays an important role in binding. Similar results have been obtained by Yamamoto's group, who also detect additional footprints in the LTR region and within the viral genome (41).

Additional evidence for direct contacts between the receptor and specific base pairs in the binding sites comes from methylation interference and methylation protection experiments (47, 48). In the first type of experiment, a fragment containing the upstream binding site (-192 to -163) was methylated with dimethyl sulfate (under conditions leading to alkylation at the N-7 position of guanine) and incubated with the receptor. Protein- bound and free DNA fragments were separated by nitrocellulose filters, subjected to strand cleavage reaction at the modified bases, and analyzed in polyacrylamide sequencing gels. If methylation at a particular residue interferes with receptor binding, the corresponding band in the autoradiograms should be underrepresented in the population of receptor bound fragments and overrepresented in the population of free DNA fragments (49). We find that methylation of either or both G residues in the hexanucleotide $\begin{array}{l} 5'\text{-TGTTCT-}3' \\ 3'\text{-ACAAGA-}5' \end{array}$ (positions -174 in the sense strand and -171 in the antisense strand), as well as methylation at position -180 in the antisense strand, interferes with receptor binding (Figure 2a). This indicates that the receptor interacts with these G residues in such a way that each interaction in essential for binding and can be prevented by methylation at position N-7. If more than one binding site was present in the DNA fragment, no clear-cut results were obtained, suggesting that one intact site is sufficient for receptor binding in filter binding studies.

In methylation protection experiments, a labeled DNA fragment is treated with dimethyl sulfate in the absence or in the presence of the receptor. Purine residues in direct contact with the protein are protected against methylation, thus leading to a decrease in the intensity of the corresponding band in a subsequent chemical strand-scission reaction (50). The results of these experiments are summarized in Figure 2a. The G residues in each strand of the conserved hexanucleotide are protected by the receptor against methylation by dimethyl sulfate in all four binding sites (48). A quantitative analysis of these changes as a function of receptor concentration shows that the receptor binds to all four binding sites with affinities of the same order of magnitude. Within the footprint regions, additional G residues are protected by the receptor against methylation, whereas, at other residues, hypermethylation by dimethyl sulfate is observed in the presence of the receptor (Figure 2a).

Interestingly, under very stringent conditions we also detect receptor dependent alterations in the methylation pattern outside the footprint regions. We interpret this as a consequence of changes in the helix conformation due to receptor binding to adjacent sites. Considerable methylation changes are detected around position -150 in a region overlapping the octanucleotide 5'-GTGGTTTC-3' that exhibits extensive similarity to a consensus sequence detected in several viral and cellular enhancer cores: 5'-GTGG$^{TTT}_{AAA}$G-3' (51). Enhancers or activators are DNA elements that activate a promoter independently of their position and orientation (52-56). Since the glucocorticoid regulatory element is able to activate, in a hormone dependent manner, heterologous promoters located at different distances and even in opposite orientation (45,57), one can imagine that it acts as a receptor-activated enhancer. Possibly the changes in methylation pattern induced by the receptor around position -150 may represent the structural correlate of this functional activation. Alternatively, changes in methylation pattern outside of the footprint regions could reflect cryptic receptor binding sites or the binding of contaminating proteins present in the receptor preparation.

METALLOTHIONEIN

The human metallothionein II$_A$ gene (hMTII$_A$) has been shown to be induced by glucocorticoids in gene transfer experiments (58). We have used this gene to determine whether the glucocorticoid regulatory element found in the proviral genome is also responsible for hormonal regulation of a cellular structural gene. If this is the case, a comparison of the nucleotide sequence of both regulatory elements should give information both on the relevant aspects of the sequence and on the degree of evolutionary conservation of the elements between rodents and human.

In gene transfer experiments with a chimeric gene containing the promoter of the hMTII$_A$ gene and the tk gene of HSV, the region responsible for glucocorticoid induction was localized between -236 and -268 base pairs upstream of the transcription initiation site by comparison of different deletion mutants in the hMTII$_A$ promoter (59). In filter binding studies with the rat liver glucocorticoid receptor, the same DNA region was shown to be responsible for specific binding (59). DNase I protection experiments yielded a clear footprint between positions -242 and -267, and a weaker protection between -318 and -332 (Figure 2b). Common to both sites is the hexanucleotide 5'-TGTCCT-3' that is found in the sense strand at -253 and in the opposite strand at -319 (Figure 2b). This sequence is very similar to the hexanucleotide found in the binding sites of MMTV-LTR. In addition, further similarities are detected between the strong binding sites of hMTII$_A$ and MMTV-LTR that fit into a consensus sequence for the glucocorticoid regulatory element (Figure 3). A similar sequence is also found at position -234 in the mouse metallothionein gene I that is also regulated by glucocorticoids (60). The receptor binding site at -324 is not required for hormonal inducibility in gene transfer experiments and may represent a non-functional interaction (59).

Methylation protection experiments performed around the strong binding site show that a G at -252 in the sense strand and two residues at -249 and -258 in the opposite strand are protected by the receptor, whereas the G-residue at -250 in the antisense strand is hypermethylated in the presence of the receptor (Figure 2b). The data and the changes in methylation protection around the weak binding site (Figure 2b) suggest that the hexanucleotide 5'-TGTCCT-3' in the hMTII$_A$ promoter plays the same role as does the 5'-TGTTCT-3' sequence in the MMTV-LTR with respect to receptor binding. Thus, it appears that the glucocorticoid regulatory element has been evolutionarily conserved between rodents and human, and is common to both viral and cellular genes. Either a common ancestoral element or convergent evolution could account for our findings. The functional significance of the difference in the number of receptor binding sites in MMTV-LTR and hMTII$_A$ remains to be established.

EGG WHITE PROTEINS

The expression of the genes for the egg white proteins in chicken oviduct is under stringent control by steroid hormones, and this system has been extensively used for studying the molecular mechanisms of hormone action. In fact, one of the first reports on specific sequence recognition by steroid hormone receptors was based on a competition assay using crude progesterone receptor from chicken oviduct and cloned fragments of these genes (20). Selective binding of the purified A subunit of the receptor to the promoter region of the ovalbumin gene has been reported by other groups (61-63). The recognition sequences found by these two groups are multiple, however, and footprinting experiments are not yet available.

In an attempt to further analyze the evolutionary conservation of the glucocorticoid regulatory element we have used the chicken lysozyme gene and the purified rat liver glucocorticoid receptor for _in vitro_ binding studies. We have delimited a region between -39 and -74 upstream of the transcription initiation site of the lysozyme promoter that is mainly responsible for specific receptor binding (Figure 2c) (64, 64a). Within this region we find the hexanucleotide 5'-TGTTCT-3' at position

-60 in the antisense strand, confirming previous sequence data (65). In fact, there is further sequence similarity between the binding site in the lysozyme gene and those reported above for mouse and human regulated genes (Figure 3), suggesting that the glucocorticoid regulatory element has been conserved in avian genes. In addition, the regulatory element in the lysozyme promoter is in opposite orientation to those found in MMTV and hMTII$_A$. It seems that both orientations of the element are compatible with hormonal inducibility of natural promoters, thus confirming results with chimeric constructions (57). Using 3' deletions of the lysozyme promoter, we have detected a weak binding site for the glucocorticoid receptor between -161 and -208 that does not contain the hexanucleotide 5'-TCTTCT-3' (64, 64a).

Microinjection into oviduct cells of a chimeric gene containing lysozyme promoter sequences linked to the gene for the SV40 T antigen results in hormone dependent expression of the T antigen, but only if at least 208 base pairs upstream of the lysozyme initiation site are preserved in the constructions (64, 64a). Deletions containing only 164 base pairs of the lysozyme promoter are not expressed, although they still contain the strong binding site around -60. Unfortunately, data on the functional relevance of this binding site for glucocorticoid regulation are not available and will be difficult to obtain, as the binding site overlaps essential promoter elements. Possibly both binding sites, at around -180 and -60, have to be occupied for glucocorticoid inducibility.

Since the lysozyme gene is also induced by progesterone in vitro and in microinjection experiments, and inducibility is lost by deletion of sequences between -208 and -164 (64, 64a), we have analyzed the binding of a partially purified progesterone receptor from rabbit uterus to the promoter region of the lysozyme gene. Two progesterone receptor binding sites were detected, a weak one between -39 and -74 coinciding with the strong binding site for the glucocorticoid receptor and strong binding site between -164 and -200 overlapping the weak binding site for the glucocorticoid receptor (66). These results suggest that the binding sites for both receptors are similar but not identical. Interestingly, the decanucleotide

5'-ATTCCTCTGT-3' is present in both binding sites: between -53 and -62 in the sense strand, and between -192 and -181 in the antisense strand (Figure 2c). Whether this sequence is involved in recognition by the progesterone receptor remains to be established.

CONCLUSIONS

The results of binding data reported above allow certain conclusions about the structure, evolution, and function of the steroid hormone regulatory elements in hormone inducible genes.

1. Analysis of the binding sequence of four strong glucocorticoid binding sites in the promoter region of three different genes, defines a consensus sequence:

$$5'-\overset{1\ 3}{TGGT}\cdot\overset{T}{\underset{A}{}}CA\overset{CTC}{\underset{AAG}{}}\overset{13\ 17}{TGT}\overset{T}{\underset{C}{}}CT-3'$$

that has been preserved in evolution between rodents, avian, and human genes (Figure 3). There is a relatively high degree of variability, but the hexanucleotide $5'-TGT^{T}_{C}CT-3'$ appears in all strong binding sites, and in two weaker sites detected in MMTV-LTR as well as in the weak site of hMTII$_A$.

2. The receptor contacts the double helix through the major groove at least at four sites within the consensus sequence: the G-residues at positions 3 and 13 on the top strand, and at positions 7 and 16 on the bottom strand. These positions are well preserved in all strong binding sites. Interestingly, the contact sites are separated by either 9 or 10 base pairs, and therefore, are located on the same site of the helix on two consecutive turns (Figure 4). This structure of the binding sites is compatible with a model in which a dimer of the receptor molecule interacts with each of the strong binding sites. In the weaker binding site only two consecutive contacts are detected, which is compatible with binding of a receptor monomer (Figure 4).

3. The distance between the receptor binding site and the initiation of transcription varies between 60 and 250 bp in the four binding sites studied, and both orientations of the binding sequence are found. These findings are reminiscent of the properties of so-called enhancer or activator elements (51-56), and suggest that the steroid hormone regulatory elements could act as hormone dependent enhancers. The detection of receptor induced changes in the accessibility to dimethyl sulfate of purines in a sequence similar to the enhancer's "core" further supports this concept.

4. There is a striking similarity between the properties of the glucocorticoid regulatory element described here and the binding site for the cAMP receptor protein (CRP) of E. coli (67). Both proteins contain two interacting domains, one for ligand binding and another for DNA binding. Both appear to bind as a dimer to a sequence that is not strictly conserved among different genes and is located at variable distances from the

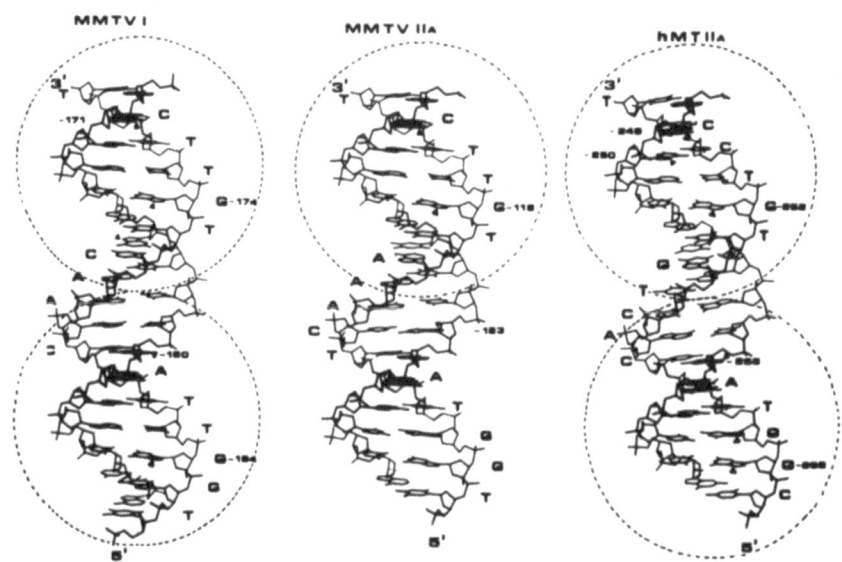

Figure 4.   Structure of the glucocorticoid binding sites in MMTV and hMTII$_A$. Computer graphic representation of the DNA double helix containing the sequence of MMTV I, MMTV IIa, and hMTII$_A$ shown in Figure 3. The sites of contact with the receptor are indicated by open triangles. Those positions hypermethylated in the presence of the receptor are marked by dark triangles. The receptor molecules are represented as contour circles. Numbers refer to the distance from the "cap" site.

transcription initiation site. Moreover, both regulatory proteins can act as positive as well as negative modulators of
transcription (68, 68a). These similarities suggest that the
mechanisms regulating gene expression in terminally differentiated eukaryotic cells are not basically different from those
found in bacteria and phages. Since, however, different cells

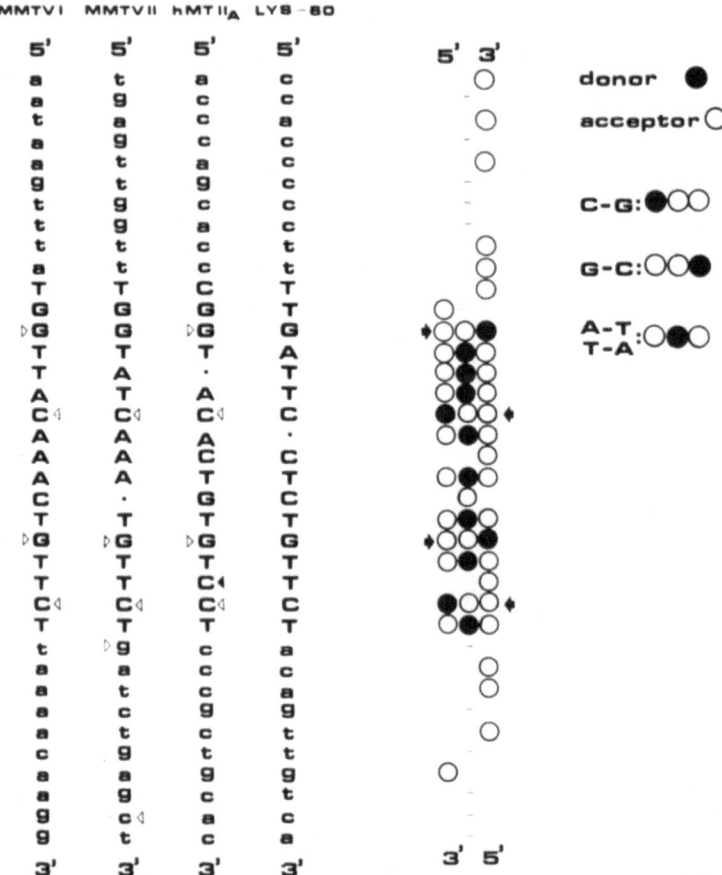

Figure 5. Pattern of hydrogen bond donor/acceptor sites in the
major groove of the DNA double helix in and around
the receptor binding sites. The sequence of 37 nucleotides comprising each of the four strong binding
sites for the glucocorticoid receptor described in
this paper is shown on the left. Hydrogen bond donor
and acceptor sites in the major groove are indicated
by closed and open circles respectively, and those
sites showing 100% conservation in the four sequences
are shown on the right of the figure. Those positions in the sequence where there is no conserved
hydrogen bond site are indicated by a hyphen. Arrows
point to the conserved N-7 positions of guanines that
have been found to be protected against methylation
by DMS in more than one of the binding sites. Other
symbols are as in Figure 2.

of an organism respond differently to the same hormone, and probably contain the same receptor and DNA, additional mechanisms must be responsible for differential gene regulation. These mechanisms should operate during cell differentiation and could involve chromatin organization changes leading to differential masking of regulatory elements.

5. The variability in the nucleotide sequence between different steroid receptors and CRP binding sites suggests that features of the double helix other than its linear sequence could be relevant for recognition by regulatory proteins. One possibility is that the pattern of hydrogen bond donor and acceptor sites characteristic of each base pair is an important component of the recognition mechanism (69-70). If we assume that the receptor contacts the double helix through the major groove, and that A-T and T-A base pairs are similar in terms of the sequence of hydrogen bond donor and acceptor sites (69), we obtain the results shown in Figure 5. Obviously, this pattern of donor-acceptor sites is better preserved within the binding sites than the direct nucleotide sequence. Of course, other parameters such as base inclination, propeller twist and base roll are influenced by the specific nucleotide sequence (71) and should be considered in this context. Possibly, this information could serve to define a regulatory code that is read by the amino acid side chains in the DNA binding sites of regulatory proteins and that regulates the expression of adjacent segments of the DNA containing genetic code information. The elucidation of the structural and functional features of the regulatory code is one of the main tasks to be solved in the near future.

ACKNOWLEDGMENTS

We thank B. Groner and N. Hynes, Karlsruhe, F.R.G., for the MMTV-LTR constructions. The hMTII$_A$ chimeric gene and deletions thereof were provided by M. Karin, Los Angeles, California, and the lysozyme gene constructions were from R. Renkawitz and G. Schutz, Heidelberg, F.R.G. The computer graphic pictures of Figure 4 were prepared by Heinz Bosshardt, EMBL, Heidelberg, F.R.G. The experimental work was supported by the Deutsche Forschungsgemeinschaft and Fonds der Chemischen Industrie.

REFERENCES

1. HIGGINS, S.J. and GEHRING, U. (1978). Molecular mechanisms of steroid hormone action. Adv. Cancer Res. 28, 313-397.

2. GRONEMEYER, H. and PONGS, O. (1980). Localization of ecdysterone on polytene chromosomes of D. melanogaster. Proc. Natl. Acad. Sci. USA 77, 2108-2112.

3. SPELSBERG, T.C., LITTLEFIELD, B.A., SEELKE, R., DANI, G.M., TOYODA, H., BOYD-LEINEN, P., THRALL, C., and KON, O.L. (1983). Role of specific chromosomal proteins and DNA sequences in the nuclear binding sites for steroid receptors. Recent Progr. Hormone Res. 39, 463-513.

4. BAXTER, J.D., ROUSSEAU, G.G., BENSON, M.C., GARCEA, R.L., ITO, J., and TOMKINS, G.M. (1972). Role of DNA and specific cytoplasmic receptors in glucocorticoid action. Proc. Natl. Acad. Sci. USA 69, 1892.

5. SCHRADER, U.T., TOFT, D.O., and O'MALLEY, B.W. (1972). Progesterone-binding proteins of chick oviduct. VI. Interaction of purified progesterone-receptor components with nuclear constituents. J. Biol. Chem. 247, 2401-2407.

6. HIGGINS, S.J., ROUSSEAU, G.G., BAXTER, J.D., and TOMKINS, J., (1973). Early events in glucocorticoid action. Activation of the steroid receptor and its subsequent specific nuclear binding studied in a cell-free system. J. Biol. Chem. 248, 5866-5872.

7. KALIMI, M., BEATO, M., and FEIGELSON, P. (1973). Interactions of glucocorticoids with rat liver nuclei. I. Role of the cytosol proteins. Biochemistry 12, 3365-3371.

8. LIPPMAN, M.E. and THOMPSON, E.B. (1974). Steroid receptors and the mechanism of the specificity of glucocorticoid responsiveness of somatic cell hybrids between hepatoma tissue culture cells and mouse fibroblasts. J. Biol. Chem. 249, 2483-2488.

9. CHAMNESS, G.C., JENNINGS, A.W., and McGUIRE, W.L. (1974). Estrogen receptor binding to isolated nuclei. A nonsaturable process. Biochemistry 13, 237.

10. MILGROM, E. and ATGER, M. (1975). Receptor translocation inhibitor and apparent saturability of the nuclear acceptor. Biochemistry 6, 487-492.

11. SIMONS, S.S., MARTINEZ, H.M., GARCEA, R.L., BAXTER, J.D., and TOMKINS, G.M. (1976). Interactions of gulcocorticoid receptor-steroid complexes with acceptor sites. J. Biol. Chem. 251, 334-343.

12. BUGANY, H. and BEATO, M. (1977). Binding of the partially purified glucocorticoid receptor of rat liver to chromatin and DNA. Mol. Cell. Endocrinol. 7, 49-66.

13. BEATO, M., KALIMI, M., KONSTAM, M., and FEIGELSON, P. (1973). Interaction of glucocorticoids with rat liver nuclei: II. Studies on the nature of the cytosol transfer factor and the nuclear acceptor site. Biochemistry 12, 3372-3379.

14. MILGROM, E., ATGER, M., and BAULIEU, E.E. (1973). Acidophilic activation of steroid hormone receptors. Biochemistry 12, 5198.

15. YAMAMOTO, K.R. and ALBERTS, R. (1975). Steroid receptors: elements for modulation of eukaryotic transcription. Cell 4, 301-310.

16. BEATO, M. and DOENECKE, D. (1980). Metabolic effects and modes of action of glucocorticoids. In: "General, comparative and clinical endocrinology of the adrenal cortex," Vol. 3 (I. Chester Jones and I.W. Henderson, eds.) pp. 117-181. Academic Press, New York.

17. YAMAMOTO, K.R. and ALBERTS, B. (1974). On the specificity of the binding of estradiol receptor protein to DNA. J. Biol. Chem. 249, 7076-7086.

18. ROUSSEAU, G.G., HIGGINS, S.J., BAXTER, J.D., GELFAND, D., and TOMKINS, G.M. (1975). Binding of glucocorticoid receptors to DNA. J. Biol. Chem. 250, 6015-6021.

19. KALLOS, J. and HOLLANDER, V.P. (1978). Assessment of specificity of estrogen receptor-DNA interaction by a competitive assay. Nature 272, 177-179.

20. MULVIHILL, E.R., LEPENNEC, J.P., and CHAMBON, P. (1982). Chicken oviduct progesterone receptor: location of specific regions of high affinity binding in cloned DNA fragments of hormone responsive chicken genes. Cell 28, 621-632.

21. SCHEIDEREIT, C., GEISSE, S., WESTPHAL, H.M., ana BEATO, M. (1983). The glucocorticoid receptor binds to defined nucleotide sequences near the promoter of mouse mammary tumour virus. Nature 304, 749-752.

22. RIGGS, A.D., SUZUKI, H., and BOURGEOIS, S. (1970). Lac repressor-operator interaction. I. Equilibrium studies. J. Mol. Biol. 48, 67-83.

23. WRANGE, O., CARLSTEDT-DUKE, D.C., and GUSTAFSSON, J.A. (1979). Purification of the glucocorticoid-receptor from rat liver cytosol. J. Biol. Chem. 254, 9284-9290.

24. OKRET, S., CARLSTEDT-DUKE, J., WRANGE, O., CARLSTROM, K., and GUSTAFSSON, J.A. (1981). Characterization of an antiserum against the glucocorticoid receptor. Biochim Biophys. Acta 677, 205-219.

25. CARLSTEDT-DUKE, J., OKRET, S., WRANGE, O., and GUSTAFSSON, J.A. (1982). Immunochemical analysis of the glucocorticoid receptor: identification of a third domain separated from the steroid-binding and DNA-binding domains. Proc. Natl. Acad. Sci. USA 79, 4260-4264.

26. WESTPHAL, H.M., MOLDENHAUER, G., and BEATO, M. (1982). Monoclonal antibodies to the rat liver glucocorticoid receptor. EMBO J. 1, 1467-1471.

27. WESTPHAL, H.M. and BEATO, M. (1980). The activated glucocorticoid receptor of rat liver. Purification and physical characterization. Eur. J. Biochem. 106, 395-403.

28. WEIGEL, N.L., TASH, J.S., MEANS, A.R., SCHRADER, W.T., and O'MALLEY, B.W. (1981). Phosphorylation of hen progesterone receptor by cAMP-dependent protein kinase. Biochem. Biophys. Res. Commun. 102, 513-519.

29. BAILLEY, A., ATGER, M., ATGER, P., CERBON, M.A., ALIZON, M., HAI, M.T.V., LOGEAT, F., and MILGROM, E. (1983). The rabbit uteroglobin gene. Structure and interaction with the progesterone receptor. J. Biol. Chem. 258, 10384-10389.

30. RINGOLD, G. (1979). Glucocorticoid regulation of mouse mammary tumor virus gene expression. Biochim. Biophys. Acta 560, 487-508.

31. BUETTI, E. and DIGGELMANN, H. (1981). Cloned MMTV virus DNA is biologically active in transferred mouse cells and its expression is stimulated by glucocorticoid hormones. Cell 23, 335-345.

32. HUANG, A.L., OSTROWSKI, M.C., BERARD, D., and HAGER, G.L. (1982). Glucocorticoid regulation of the Ha-MaSVp21 gene conferred by sequences from mousr mammary tumor virus. Cell 27, 245-255.

33. HYNES, N.E., KENNEDY, N., RAHMSDORF, U., and GRONER, B. (1981). Hormone responsive expression of an endogenous proviral gene of MMTV after molecular cloning and gene transfer into cultured cells. Proc. Natl. Acad. Sci. USA 78, 2038-2042.

34. LEE, F., MULLIGAN, R., BERG, P., and RINGOLD, G. (1981). Glucocorticoid regulated expression of dehydrofolate reductase cDNA in mouse mammary tumor virus chimeric plasmids. Nature 294, 228-232.

35. BUETTI, E. and DIGGELMANN, H. (1983). Glucocorticoid regulation of MMTV: identification of a short essential region. EMBO J. 2, 1423-1429.

36. GEISSE, S., SCHEIDEREIT, C., WESTPHAL, H.M., HYNES, N.E., GRONER, B., and BEATO, M. (1982). Glucocorticoid receptors recognize DNA sequence in and around murine mammary tumour virus DNA. EMBO J. 1, 1613-1619.

37. GOVINDAN, M.V., SPIESS, E., and MAJORS, J. (1982). Purified glucocorticoid receptor-hormone complex from rat liver cytosol binds specifically to cloned mouse mammary tumor virus long terminal repeats in vitro. Proc. Natl. Acad. Sci. USA 79, 5157-5161.

38. PAYVAR, F., FIRESTONE, G., ROSS, S.R., CHANDLER, V.L., WRANGE, O., CARLSTEDT-DUKE, J., GUSTAFSSON, J.A., and YAMAMOTO, K.R. (1982). Multiple specific binding sites for purified glucocorticoid receptors on mammary tumor virus DNA. J. Cell. Biochem. 19, 241-247.

39. PFAHL, M. (1982). Specific binding of the glucocorticoid-receptor complex to the mouse mammary tumor proviral promoter region. Cell 31, 475-482.

40. GRONER, B., PONTA, H., BEATO, M., and HYNES, N.E. (1983). The proviral DNA of mouse mammary tumor virus: its use in the study of the molecular details of steroid hormone action. Mol. Cell. Endocrinol. 32, 101-116.

41. PAYVAR, F., DE FRANCO, D., FIRESTONE, G.L., EDGAR, G.L., WRANGE, O., OKRET, S., GUSTAFSSON, J.A., and YAMAMOTO, K.R. (1983). Sequence-specific binding of glucocorticoid receptor to MMTV DNA at sites within and upstream of the transcribed region. Cell 35, 381-392.

42. PAYVAR, F., WRANGE, O., CARLSTEDT-DUKE, J., OKRET, S., GUSTAFSSON, J.A., and YAMAMOTO, K.R. (1981). Purified glucocorticoid receptors bind selectively in vitro to a cloned DNA fragment whose transcription is regulated by glucocorticoids in vivo. Proc. Natl. Acad. Sci. USA 78, 6628-6632.

43. PAYVAR, F. and WRANGE, O. (1983). Relative selectivities and efficiencies of DNA binding by purified intact and protease-cleaved glucocorticoid receptor. In: "Steroid Hormone Receptors: Structure and Function," Nobel Symp. 57. (H. Eriksson, J.A. Gustafsson and B. Hogberg, eds.) Elsevier/North Holland, Biomed. Press, Amsterdam.

44. DELLWEG, H.G., HOTZ, A., MUGELE, K., and GEHRING, U. (1982). Active domains in wild-type and mutant gluco-corticoid receptors. EMBO J. 1, 285-289.

45. HYNES, N.E., VAN OOYEN, A.J.J., KENNEDY, N., HERRLICH, P., PONTA, H., and GRONER, B. (1983). Subfragments of the large terminal repeat cause glucocorticoid responsive ex-pression of mouse mammary tumor virus and of an adjacent gene. Proc. Natl. Acad. Sci. USA 80, 3637-3641.

46. GALAS, D.J. and SCHMITZ, A. (1978). DNase footprinting: a simple method for the detection of protein-DNA binding specificity. Nucleic Acids Res. 5, 3157-3170.

47. SCHEIDEREIT, C., KRAUTER, P., WESTPHAL, H.M., and BEATO, M. (1984). Interaction of glucocorticoid hormone receptors with defined DNA sequences of inducible genes. In: "Hor-mones and Cancer," Vol. 2 (F. Bresciani, et al., eds.) pp. 89-96. Raven Press, New York.

48. SCHEIDEREIT, C. and BEATO, M. (1984). Contacts between receptor and DNA double helix within a glucocorticoid regu-latory element of mouse mammary tumor virus. Proc. Natl. Acad. Sci. USA 81, 3029-3033.

49. SIEBENLIST, U., SIMPSON, R.B., and GILBERT, W. (1980). E. coli RNA polymerase interacts homologously with two dif-ferent promoters. Cell 20, 269-281.

50. OGATA, R.T. and GILBERT, W. (1978). An amino-terminal fragment of the lac repressor binds specifically to lac operator. Proc. Natl. Acad. Sci. USA 75, 5851-5854.

51. WEIHER, H., KONIG, M., and GRUSS, P. (1983). Multiple point mutations affecting the SV40 enhancer. Science 219, 626-631.

52. BANERJI, J., RUSCONI, S., and SCHAFFNER, W. (1981). Ex-pression of a β-globin gene is enhanced by remote SV40 DNA sequences. Cell 27, 299-308.

53. RUSS, P., DAHR, R., and KHOURY, G. (1981). Simian virus 40 tandem repeated sequences as an element of early promoter. Proc. Natl. Acad. Sci. USA 78, 943-947.

54. BANERJI, J., OLSON, L., and SCHAFFNER, W. (1983). A lymphocyte-specific cellular enhancer is located downstream of the joining region in IgG heavy chain genes. Cell 33, 729-740.

55. GILLES, S.D., MORRISON, S.L., OI, V.T., and TONEGAWA, S. (1983). A tissue-specific transcription enhancer element is located in the major intron of a rearranged IgG heavy chain gene. Cell 33, 717-728.

56. QUEEN, C. and BALTIMORE, D. (1983). Immunoglobin gene transcription is activated by downstream sequence elements. Cell 33, 741-748.

57. CHANDLER, V.L., MALER, B.A., and YAMAMOTO, K.R. (1983). DNA sequences bound specifically by glucocorticoid receptor in vitro render a heterologous promoter hormone responsive in vivo. Cell 33, 489-499.

58. KARIN, M. and RICHARDS, R. (1982). Human metallothionein genes. Primary structure of the metallothionein II gene and a related processed gene. Nature 299, 797-802.

59. KARIN, M., HASLINER, A., HOLTGREVE, H., RICHARDS, R.I., KRAUTER, P., WESTPHAL, H.M., and BEATO, M. (1984). Characterization of DNA sequences through which cadmium and glucocorticoid hormones induce human metallothionein II$_A$-gene. Nature 308, 513-519.

60. GLANVILLE, N., DURNAM, D.M., and PALMITER, R.D. (1981). Structure of mouse metallothionein-I gene and its mRNA. Nature 292, 267-269.

61. HUGHES, M.R., COMPTON, J.G., SCHRADER, W.T., and O'MALLEY, B.W. (1981). Interaction of the chick oviduct progesterone-receptor with DNA. Biochemistry 20, 248-291.

62. COMPTON, J.G., SCHRADER, W.T., and O'MALLEY, B.W. (1982). Selective binding of chicken progesterone receptor. A subunit to a DNA fragment containing ovalbumin gene sequences. Biochem. Biophys. Res. Commun. 105, 96-104.

63. COMPTON, J.G., SCHRADER, W.T., and O'MALLEY, B.W. (1983). DNA sequence preference of the progesterone receptor. Proc. Natl. Acad. Sci. USA 80, 16-20.

64. RENKAWITZ, R., BEUG, H., GRAF, T., MATTHIAS, P., GREZ, M., and SCHUTZ, G. (1982). Expression of a chicken lysozyme recombinant gene is regulated by progesterone and dexamethasone after microinjection in oviduct cells. Cell 31, 167-176.

64a. RENKAWITZ, R., SCHUTZ, G., VON DER AHE, D., and BEATO, M. (1984). Identification of hormone regulatory elements in the promoter region of the chicken lysozyme gene. Cell 37, 503-510.

65. GREZ, M., LAND, H., GIESECKE, K., SCHUTZ, M., JUNG, A., and SIPPEL, A.E. (1981). Multiple mRNA's are generated from the chicken lysozyme gene. Cell 25, 743-752.

66. VON DER AHE, D., JANICH, S., SCHEIDEREIT, C., RENKAWITZ, R., SCHUTZ, G., and BEATO, M. (1985). Glucocorticoid and progesterone receptors bind to the same sites in two hormonally-regulated promoters. Nature 313, 706-709.

67. TAKEDA, Y., OHLENDORF, D.H., ANDERSON, V.F., and MATTHEWS, B.W. (1983). DNA binding proteins. Science 221, 1020-1026.

68. AIBA, H. (1983). Autoregulation of the E. coli crp gene: DRP is a transcriptional repressor of its own gene. Cell 32, 141-149.

68a. GUERTIN, M., BARIL, P., BARTKOWIAK, J., ANDERSON, A., and BELANGER, L. (1983). Rapid suppression of x_1-fetoprotein gene transcription by dexamethasone in developing rat liver. Biochemistry 22, 4296-4302.

69. SEEMAN, N.C., ROSENBERG, J.M., and RICH, A. (1976). Sequence-specific recognition of double helical nucleic acids by proteins. Proc. Natl. Acad. Sci. USA 73, 804-808.

70. SADLER, J.R., WATERMAN, M.S., and SMITH, T.F. (1983). Regulatory pattern identification in nucleic acid sequences. Nucleic Acids Res. 11, 2221-2231.

71. DICKERSON, R.E. and DREW, H.R. (1981). Structure of a B-DNA dodecamer. II. Influence of base sequence on helix structure. J. Mol. Biol. 149, 761-786.

STRUCTURE AND REGULATION OF THE HUMAN METALLOTHIONEIN GENE

FAMILY

Adriana Heguy and Michael Karin

Department of Microbiology
University of Southern California
School of Medicine
2011 Zonal Avenue
Los Angeles, California 90033

INTRODUCTION

Metallothioneins (MTs) are a group of low-molecular-weight, heavy-metal binding proteins that are unique in their high cysteine content. MTs specifically bind heavy metals such as Zn, Cd, Cu, Hg, Au, and Ag in their ionic forms (1). These proteins have been isolated and characterized from a large number of animal and plant species, including lower eukaryotes. MTs exist in several molecular forms which are distinguishable by their electrophoretic behavior and are designated MT-I and MT-II. Recently, several other variants have been described in mammals (2).

FUNCTION OF METALLOTHIONEINS

Although the exact function of MTs is not clear, it is thought that they play a central role in the regulation of trace metal metabolism and in the storage of these metal ions in the liver. The induction of MTs after administration of various heavy metal ions has been demonstrated in many different animal species and in cultured cells (1). These results suggest that MTs may have a protective role against heavy metal toxicity. Karin and Herschman (3) demonstrated that the biosynthesis of MTs in cultured cells is induced by glucocorticoids as well as by heavy metal ions. This finding supports the hypothesis that MTs play an important role in the regulation of trace element metabolism and that glucocorticoids, by regulating the rate of MT synthesis, control Cu and Zn homeostasis. Recently, it has been suggested that MTs might be part of the SOS system, which functions to protect cells against damage by free radicals (4).

REGULATION OF METALLOTHIONEIN EXPRESSION

Both heavy metal ions and glucocorticoid hormones induce MT synthesis by elevating the rate of MT gene transcription (5, 6). In addition to metals and glucocorticoids, several other stressful stimuli -- including exposure to heat and cold, strenuous exercise, and tissue injury resulting from the injection of turpentine, carbon tetrachloride, or a bacterial endotoxin (lipopolysaccharide) -- also induce MTs (7). Durnam et al. (8) have shown that the MT response to inflammatory agents is independent of metal ions and glucocorticoid hormones, suggesting a third mechanism of MT gene induction. Recently, it has been reported that interferon causes a transient increase in MT gene transcription (9). Karin et al. also have shown that MT genes respond to interleukin I, a polypeptide hormone thought to mediate the stress response (submitted). Due to the variety of inducers acting on the system, the MT gene family constitutes an attractive system for the study of regulation of gene expression. In this paper, we will review the structure and expression of the human MT genes.

THE HUMAN METALLOTHIONEIN GENE FAMILY

Several of the human MT genes have been isolated and sequenced (10, 11). They comprise a multigene family of at least 12 members that are coordinately induced to different levels by metal ions (11). In man, only one functional gene, $hMT-II_A$, encodes form II of the protein. However, a processed pseudogene, $hMT-II_B$, is present. This pseudogene lacks promoter elements and introns, terminates in a poly(A) tail and is flanked by two direct repeats (10). The $hMT-II_A$ gene encodes a functional protein that confers cadmium resistance to rodent fibroblasts transfected with an autonomously replicating shuttle vector, bovine papilloma virus (BPV) containing the hMT-II$_A$ gene (12).

In contrast, several genes code for MT-I-like variants and at least four of them, hMT-I$_A$, hMT-I$_B$, hMT-I$_C$, and hMT-I$_D$, are tandemly arranged in the genome. Two of these, hMT-I$_C$ and hMT-I$_D$, are pseudogenes and have in-phase termination codons and/or single base pair deletions affecting the reading frame. The hMT-I$_A$ gene is functional on the basis of transformation, using the BPV vector, of rodent fibroblasts to a cadmium resistant phenotype. However, the cells transformed with this vector express only half as much MT mRNA as the cells harboring a BPV-hMT-II$_A$ recombinant plasmid. Consequently, while the hMT-I$_A$ transformed cells are significantly more resistant to Cd than are the parental cells, they are only half as resistant as the cells which express the hMT-II$_A$ gene (11).

STRUCTURE AND REGULATION OF hMT-II$_A$ AND hMT-I$_A$ GENES

The hMT-II$_A$ gene has been studied extensively in our laboratory. Like the other hMT genes, hMT-II$_A$ possesses two introns located at exactly the same position within the coding region as the introns in the mouse MT-I gene (13, 14). The overall organization of the regulatory regions of this gene resembles that of several other eukaryotic genes (15-18 and chapter by Wasylyk, this volume). In addition to the TATA box, there are at least three other functional elements. The sequences that mediate heavy metal and glucocorticoid responsiveness are present in the 5' flanking region and confer the corresponding induction phenotype upon a heterologous promoter (19). We have mapped these elements by generating progressive deletions from both the 5' and 3' ends of the 5' flanking region of the gene and fusing the resulting fragments to the herpes simplex virus (HSV) thymidine kinase (tk) gene (20). The analysis of these in vitro constructed deletion mutants by transfection into rat fibroblasts identified three upstream regulatory elements: the glucocorticoid regulatory element (GRE), located between nucleotides -268 and -237, the metal regulatory elements (MREs), present in at least two copies, situated between -38 to -50 and -138 to -150, and a third element required for maintaining a basal level of expression (Figure 1). The latter element is necessary for transient expression of the hMT-II$_A$-tk fusion gene (21). Comparison of the nucleotide sequences in this region with those present in the mouse MT-I and MT-II genes and in the hMT-II$_A$ and hMT-I$_A$ genes reveals considerable sequence similarity (Figure 2). This suggests that they may serve as a recognition site for a transcription stimulatory factor, responsible for expression of these genes in the absence of inducers. Furthermore, conserved sequences in the hMT-II$_A$ gene are part of an enhancer-like element formed by two direct repeats between positions -68 and -138 and positions -142 and -215. These sequences are similar in their organization to the enhancer element of SV40 (Haslinger and Karin, submitted).

Using assays for protection against DNase I digestion and DMS methylation ("footprinting"), the GRE has been identified as the glucocorticoid receptor binding site. The GRE is located between nucleotides -245 to -265. Comparison of this sequence to the mouse mammary tumor virus long terminal repeat (LTR)

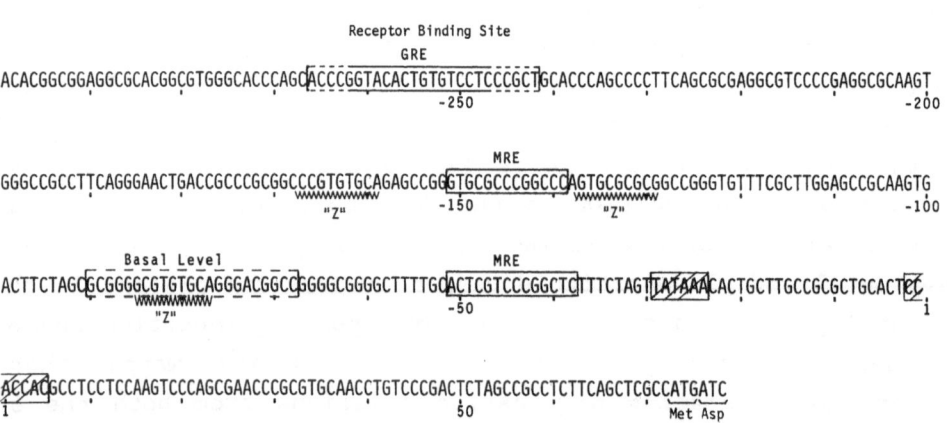

Figure 1. Molecular anatomy of the hMT-II$_A$ promoter. Shown is the sequence of the sense strand of DNA from position -300 to +80. Regions are indicated that are important for hormonal induction, heavy metal induction, and basal level of expression. The TATA box and cap site are hatched. Regions of potential Z-DNA conformation are indicated, as are the first 2 amino acid codons of the translated region. Adapted with permission from Nature, Vol. 308, p. 518, copyright (c) 1984 Macmillan Journals Limited.

```
hMT-I_A      G G C G G G G G C G G A C T C G G C T G G G C
                     ‾‾‾‾‾‾‾‾‾
                        -110                -100

mMT-I        A C G C G G G G C G C G T G A C T A T G C G T
                     ‾‾‾‾‾‾‾‾
                        -100                -90

mMT-II       G G C G G G G G C G C G T G C A T G G T G C C
                     ‾‾‾‾‾‾‾‾‾
                        -90                 -80
```

Figure 2.   Comparison of the sequences important to intrinsic promoter activity in the hMT-II$_A$, hMT-I$_A$, mMT-I, and mMT-II genes.

sequences located between -113 to -132 and -170 to -191 (which also are strongly protected against DNase I digestion upon receptor binding (22)), reveals a high degree of sequence similarity. The conserved elements in these binding sites most likely comprise a consensus nucleotide sequence that is required for glucocorticoid hormone receptor binding to DNA and which allows for hormonal induction. Interestingly, this consensus sequence contains a region of dyad symmetry resembling the repressor binding sites found in prokaryotes. This symmetry suggests that two receptor molecules bind to the DNA at this site (20).

The MRE is a dodecameric sequence that is highly conserved among all functional MT genes whose sequences are known: mouse MT-I (14), mouse MT-II (23), rat MT-I (R. Andersen, personal communication), hMT-II$_A$ (20) and hMT-I$_A$ (11). The MRE is present at least twice in the 5' flanking region of the hMT-II$_A$ gene, although a single copy is sufficient to confer heavy metal ion responsiveness upon various proximal promoter elements (Figure 3). Since the level of metal induction of MT genes increases upon inhibition of protein synthesis (24, 25), we postulated that heavy metal induction is mediated through the removal of a repressor of MT gene expression (27). The simplest hypothesis assumes that the repressor is apothionein itself and that upon metal ion induction, the conversion to metallothionein results in the formation of an inactive form of the repressor. This working hypothesis explains the specificity of the response, since only metals that bind to the apoprotein are effective inducers; it also agrees with the transient nature of the metal induction response (27). However, nickel does not bind to apothionein and Durnam and Palmiter (28) showed that it can induce the mMT-I gene. In our laboratory we have also found that the hMT genes are induced by Ni, but that this response is not mediated by 5' flanking regulatory sequences (Heguy and Karin, unpublished results). Therefore, MT induction by Ni does not contradict the repressor hypothesis.

We also have studied the regulation of expression of the hMT-I$_A$ gene (11). This gene is expressed at a lower level <u>in vivo</u> than the hMT-II$_A$ gene and is not responsive to glucocorticoids. While the hMT-II$_A$ gene is highly responsive to both Cd and Zn, the hMT-I$_A$ gene responds primarily to Cd and is induced only at very high concentrations of Zn. The differential expression of the two characterized hMT genes is due to inherent differences in their promoter and regulatory elements. By fusing the promoter/regulatory region of each of these genes to the HSV-tk gene, we have shown that the induction phenotype of the gene fusions is largely determined by the 5' flanking re-

```
hMT-II_A   1      ACTCGTCCCGG·CTCTTT
                  -50          -40

hMT-II_A   2      GTGCG·CCCGG·CCCAGT
                  -150         -140

hMT-I_A           TTGCG·TCCGG·CCCTCT
                            -50

mMT-I             TTGCG·CCCGGACTCGTC
                  -50          -40

                    TG        T
CONSENSUS:        --CG·CCCGG·C-C
                    CT        C
```

Figure 3. Consensus sequence of the heavy metal ion responsive element (MRE) of the hMT-II$_A$ (2 MREs), at the hMT-I$_A$, and the mMT-I genes. Reprinted by permission from Nature, Vol. 308, pp. 516, copyright (c) 1984 Macmillan Journals Limited.

gions. The absence of a consensus sequence known to be involved in glucocorticoid receptor binding in the regulatory region of hMT-I$_A$ explains that lack of response to glucocorticoids.

We believe that the differences in expression of the two human MT genes are related to their different physiological roles. The hMT-II$_A$ gene may be mostly involved in the regulation of Zn metabolism, whereas the hMT-I class of genes may function primarily as a protective system that is activated following exposure to toxic heavy metal ions.

CHROMOSOMAL LOCATION OF THE hMT GENE FAMILY

Most of the hMT genes, including the hMT-II$_A$, hMT-I$_A$, hMT-I$_B$, and five other MT genes, are clustered in band q22 of chromosome 16 (29, 30). Other MT genes, are dispersed at a minimum of four other autosomal sites on chromosomes 1, 4, 18, and 20 (29). These dispersed genes are not expressed in human-mouse cell hybrids and are likely to be transcriptionally silent (30). The localization of the functional MT genes to 16q22 is particularly interesting because there is an inversion of chromosome 16 -- inv (16) (p13q22) -- which is present in 25% of patients with acute myelomonocytic leukemia (31). In situ hybridization of a probe which hybridizes to all of the MT genes to metaphase spreads from malignant cells carrying the inversion shows labeled sites on both arms of the inverted chromosome. This indicates that the breakpoint at 16q22 splits the MT gene cluster (32). The same breakpoint also was found in patients whose malignant cells have a 16:16 translocation. These findings suggest that MT genes may be integrally involved in the pathogenesis of certain types of leukemia, perhaps by altering granulocyte and monocyte differentiation or by controlling cell proliferation. It also is feasible that some of the MT transcriptional control elements described in this review are involved in the activation of an as yet unidentified cellular gene located at 16p13 whose function is analogous to that of a cellular oncogene. The isolation, by cloning procedures, of the novel junctions present on chromosome 16 should help to determine which of these two possibilities is correct.

CONCLUSIONS

We believe that the metallothionein system can serve as a useful model for studying the regulation of gene expression in mammals (and probably in most other high eukaryotes). While the basic principles of gene regulation by protein-DNA interactions are common to prokaryotes and eukaryotes, there are some special features for regulation in eukaryotes that are clearly different from the situation in bacteria. The most important difference is the ability to control transcription from a given promoter over a large distance (33, 34, 20). We hope that the study of the regulation of MT gene expression will lead to an explanation of how the binding of a regulatory protein (such as the glucocorticoid receptor or the heavy-metal-ion regulatory protein) several hundred base pairs away from the start site of transcription, leads to an increasing frequency of initiation from a proximal promoter element.

ACKNOWLEDGMENTS

We wish to thank members of our laboratory for fruitful discussions and suggestions. Work in our laboratory was supported by grants from the National Institute of Environmental Health Sciences (NIEHS), the Environmental Protection Agency (EPA), and the Department of Energy (DOE).

REFERENCES

1. KAGI, J.H.R. and NORDBERG, M., eds. (1979). "Metallothionein." Birkhauser Verlag, Basel.

2. KLAUSER, S., KAGI, J.H.R., and WILSON, K.J. (1983). Characterization of isoprotein patterns in tissue extracts and isolated samples on metallothionein by reverse-phase high-pressure liquid chromatography. Biochem. J. 209, 71-80.

3. KARIN, M. and HERSCHMAN, H.R. (1979). Dexamethasone stimulation of metallothionein synthesis in HeLa cell cultures. Science 204, 176-177.

4. KARIN, M. (1985). Metallothionein: proteins in search of a function. Cell 41, 9-10.

5. DURNAM, D.M. and PALMITER, R.D. (1981). Transcriptional regulation of the mouse metallothionein-I gene by heavy metals. J. Biol. Chem. 256, 2268-2272.

6. HAGER, L.J. and PALMITER, R.D. (1981). Transcriptional regulation of mouse liver metallothionein-I gene by glucocorticoids. Nature 291, 340-342.

7. OH, S., DEAGEN, J., WHANGER, P., and WESWIG, P. (1978). Biological function of metallothionein. Its induction in rats by various stresses. Am. J. Physiol. 234, E282-E285.

8. DURNAM, D.M., HOFFMAN, J.S., QUAIFE, C.J., BENDITT, E.P., CHEN, H.T., BRINSTER, R.L., and PALMITER, R.D. (1984). Induction of mouse metallothionein-I mRNA by bacterial endotoxin is independent of metals and glucocorticoid hormones. Proc. Natl. Acad. Sci. USA 81, 1053-1056.

9. FRIEDMAN, R.L., MANLY, S.P., McMAHON, M., KERR, I.M., and STARK, G.R. (1984). Transcriptional and post-transcriptional regulation of interferon-induced gene expression in HeLa cells. Cell 38, 745-755.

10. KARIN, M. and RICHARDS, R.I. (1982). Human metallothionein genes: primary structure of the metallothionein-II gene and a related processed gene. Nature 299, 797-802.

11. RICHARDS, R.I., HEGUY, A., and KARIN, M. (1984). Structural and functional analysis of the human metallothionein-I$_A$ gene: differential induction by metal ions and glucocorticoids. Cell 37, 263-272.

12. HARIN, M., CATHALA, G., and NGUYEN-HUU, M.C. (1983). Expression and regulation of a human metallothionein gene carried on an autonomously replicating shuttle vector. Proc. Natl. Acad. Sci. USA 80, 4040-4044.

13. DURNAM, D.M., PERRIN, F., GANNON, F., and PALMITER, R.D. (1980). Isolation and characterization of the mouse metallothionein-I gene. Proc. Natl. Acad. Sci. USA 77, 6511-6515.

14. GLANVILLE, N., DURNAM, D.M., and PALMITER, R.D. (1981). Structure of mouse metallothionein-I gene and its mRNA. Nature 292, 267-269.

15. McKNIGHT, S.L. and KINGSBURY, R. (1982). Transcriptional control signals of a eukaryotic protein-coding gene. Science 217, 316-324.

16. DIERKS, P., VAN OOYEN, A., COCHRAN, M., DOBKIN, D., REISER, J., and WEISSMAN, C. (1983). Three regions upstream from the cap site are required for efficient and accurate transcription of the rabbit β-globin gene in mouse 3T6 cells. Cell 32, 695-706.

17. GROSVELD, G.C., De BOER, E., SHEWMAKER, C.K., and FLAVELL, R.A. (1982). DNA sequences mecessary for transcription of the rabbit β-globin gene in vivo. Nature 295, 120-126.

18. BENOIST, C. and CHAMBON, P. (1981). In vivo sequence requirements of the SV40 early promoter region. Nature 290, 304-310.

19. KARIN, M., HASLINGER, A., HOLTGREVE, H., CATHALA, G., SLATER, E., and BAXTER, J.D. (1984a). Activation of a heterologous promoter in response to dexamethasone and cadmium by metallothionein gene 5' flanking DNA. Cell 36, 371-379.

20. KARIN, M., HASLINGER, A., HOLTGREVE, H., RICHARDS, R.I., KRAUTER, P., WESTPHAL, H.M., and BEATO, M. (1984b). Char-

acterization of DNA sequences through which cadmium and glucocorticoid hormones induce mouse metallothionein-II$_A$ gene. Nature 308, 513-519.

21. KARIN, M. and HOLTGREVE, H. (1984). Nucleotide sequence requirements for transient expression of human metallothionein-II$_A$-thymidine kinase fusion genes. DNA 3, 319-326.

22. SCHEREDEIT, C., GEISSE, S., WESTPHAL, H.M., and BEATO, M. (1983). The glucocorticoid receptor binds to defined nucleotide sequences near the promoter of mouse mammary tumor virus. Nature 304, 749-752.

23. SEARLE, P.F., DAVISON, B.L., STUART, G.W., WILKEI, T.M., NORDSTEDT, G., and PALMITER, R.D. (1984). Regulation, linkage, and sequence of mouse metallothionein I and II genes. Mol. Cell. Biol. 4, 1221-1230.

24. KARIN, M., ANDERSEN, R.D., SLATER, E., SMITH, K., and HERSCHMAN, H.R. (1980). Metallothionein mRNA induction in HeLa cells in response to zinc or dexamethasone is a primary induction response. Nature 286, 295-297.

25. MAYO, K.E. and PALMITER, R.D. (1981). Glucocorticoid regulation of metallothionein-I mRNA synthesis in cultured mouse cells. J. Biol. Chem. 256, 2651-2654.

26. KARIN, M. and RICHARDS, R.I. (1984). The human metallothionein family: structure and expression. Env. Health Persp. 54, 111-115.

27. KARIN, M., SLATER, E.P., and HERSCHMAN, H.R. (1981). Regulation of metallothionein synthesis in HeLa cells by heavy metals and glucocorticoids. J. Cell. Physiol. 106, 63-74.

28. DURNAM, D.M. and PALMITER, R.D. (1984). Induction of metallothionein-I mRNA in cultured cells by heavy metals and iodoacetate: evidence for gratuitous inducers. Mol. Cell. Biol. 4, 484-491.

29. SCHMIDT, C.J., HAMER, D.H., and McBRIDE, O.W. (1984). Chromosomal location of human metallothionein genes: implications for Menkes' disease. Science 224, 1104-1106.

30. KARIN, M., EDDY, R.L., HENRY, W.M., HALEY, L.L., BYERS, M.A., and SHOWS, T.B. (1984c). Human metallothionein genes are clustered on chromosome 16. Proc. Natl. Acad. Sci. USA 81, 5494-5498.

31. LaBEAU, M.M., LARSON, R.A., BITTER, M.A., WARDIMAN, J.W., GOLOMB, H.M., and ROWLEY, J.D. (1983). Association of an inversion of chromosome 16 with abnormal marrow eosinophils in acute myelomonocytic leukemia. N. Engl. J. Med. 309, 630-636.

32. LeBEAU, M.M., DIAZ, M.O., KARIN, M., and ROWLEY, J.D. (1985). Metallothionein gene cluster is split by the inv(16) and t(16,16) in myelomonocytic leukemia. Nature 313, 709-711.

33. BANERJI, J., RUSCONI, S., and SCHAFFNER, W. (1981). Expression of a β-globin gene is enhanced by remote SV40 DNA sequences. Cell 27, 299-308.

34. WASYLYK, B., WASYLYK, C., AUGEREAU, P., and CHAMBON, P. (1983). The SV40 72 bp repeat preferentially potentiates transcription starting from proximal natural or substitute promoter elements. Cell 32, 503-514.

EXPRESSION AND CHROMATIN STRUCTURE OF THE ALPHA-FETOPROTEIN AND ALBUMIN GENES DURING NORMAL DEVELOPMENT AND NEOPLASIA

Jean-Louis Nahon and José M. Sala-Trepat

Laboratoire d'Enzymologie du C.N.R.S.
91190 Gif-sur-Yvette, France

INTRODUCTION

It is generally accepted that embryonic development and cell specialization is achieved via the sequential and selective expression of many genes. However, our knowledge concerning the molecular mechanisms responsible for changes in gene expression during developmental processes is still very limited. Within the last 10 years numerous investigations have aimed to elucidate the regulatory mechanisms that underlie the expression of specific genes associated with the formation of a particular cell type. Most of these studies have made use of terminally differentiated tissues or cells in culture, which in some cases can respond to environmental factors such as hormones (for reviews see references 1 and 2). At the level of cell culture, perhaps myogenesis and erythroid differentiation have received the greatest attention. Recently the use of teratomas has allowed analysis of the expression of individual, identified genes during early embryogenesis (3).

In our laboratory we have focused our attention on the terminal differentiation of the hepatocyte, a system that is particularly suitable for examining the specific molecular events associated with the differentiative process. Indeed, the acquisition of the final hepatocyte phenotype is characterized by the appearance and disappearance of a number of cell-specific proteins in a time dependent manner. Among these proteins, albumin and alpha-fetoprotein (AFP) are particularly interesting since their synthesis is under oncodevelopmental control. In addition, AFP is also a useful marker for the analysis of differentiation in early embryogenesis (4).

In this chapter, we first review some information to explain the interest in the albumin and AFP genes as a model system to study mechanisms underlying differentiation and control of gene activity during developmental and oncogenic processes. We then describe experiments indicating that these two genes are regulated at the transcriptional level, and summarize data showing that modifications at the genomic DNA level (such as amplification, deletion, rearrangements, or changes in methylation pattern) do not appear to be involved in the transcriptional modulation of these genes. Last, we present our investigations on chromatin structure of the albumin and AFP genes in transcriptionally active and inactive rat tissues and cell lines. These studies were designed to provide insight into the control mechanisms that modulate the expression of these two genes.

ALBUMIN AND ALPHA-FETOPROTEIN PRODUCTION: A SYSTEM TO STUDY DIFFERENTIATION AND NEOPLASIA

Albumin and AFP are two major plasma proteins synthesized in mammalian liver. Albumin consists of a single polypeptide chain with a molecular weight of about 66,000 (5). AFP is a single- chain glycoprotein containing about 4% carbohydrate. It has a molecular weight slightly larger than that of albumin (6).

The plasma levels of these two proteins show a reciprocal relationship during mammalian development (7). AFP is the major plasma protein during most of fetal life, reaching concentrations as high as 5 mg/ml in the rat (6). Fetal AFP is produced mainly by the developing liver and the yolk sac (8, 9). The concentration of this protein is drastically decreased in the serum of the normal adult, to less than 50 ng/ml (6, 7, 10). On the other hand, albumin is the dominant plasma protein in adult animals; its concentration increases from low levels early in fetal development to high, approximately constant levels (about 40 mg/ml) in postnatal life (5, 11). The parenchymal cells of the liver are the main site of synthesis of this protein during both embryonic and adult life (12). In addition, albumin is produced by the yolk sac of certain species (human, mouse, and chick) but not of rat (8, 9, 13, 14).

It should be noted that the rate of decline of AFP in the serum of neonatal animals can be manipulated by hormone treatment. Glucocorticoids given to newborn rats markedly reduce the serum AFP level but increase slightly the albumin serum levels (15). It should also be pointed out that AFP synthesis can already be detected in mouse embryos on the 7th day of gestation. At this early stage of development AFP production seems to be confined to the visceral endoderm cells around the embryonic region of the egg cylinder (4).

AFP synthesis by the adult liver can be resumed under certain physiopathological conditions leading to restitutive cell proliferation, such as regeneration of the liver following partial hepatectomy and chemically induced liver necrosis (7, 10, 15). Increased levels of AFP in the serum can also be observed during the preneoplastic stages of liver carcinogenesis (16-18). In all these cases elevations of serum AFP are, however, relatively small. Highly elevated plasma levels of AFP in the adult are generally associated with the appearance of tumors arising from liver cells and yolk sac elements, in analogy with the embryonal sites of AFP synthesis (6, 7, 19, 20). In most of the processes which lead to AFP re-expression, albumin synthesis is diminished in relation to that of normal liver. For instance, most transplantable rat hepatomas which produce high levels of AFP show much reduced rates of albumin synthesis (21-23).

The reciprocal relationship of plasma levels of AFP and albumin during development, their similar physicochemical properties (6), the immunological cross-reactivity of the two proteins in their denatured states (24), the significant similarity in amino-acid sequence (25, 26) and mRNA sequences (27, 28) all suggest that AFP may be the fetal analog of albumin. Moreover, recent studies on the structure of the rat and mouse albumin and AFP genes suggest that both genes arose from a common sequence which underwent successive amplification and divergence (29-31). In addition, it has been shown that in the mouse the albumin and AFP genes are closely linked in tandem on chromosome 5, with the 3' terminus of the albumin preceding the 5' side of the AFP gene (32). This close linkage of the AFP and albumin genes could indicate a coordinated regulation during development comparable to the "switch" of the embryonic-fetal-adult globin genes that occurs in the developing erythroid cells (33, 34).

The albumin and AFP genes provide a useful system then to examine the following questions. 1) What molecular mechanisms are implicated in the early activation of specific genes during embryonic development? 2) Are these mechanisms similar to or different from the molecular events controlling gene activity during terminal differentiation? 3) Do the same molecular control elements operate during oncogenic processes and malignant transformation?

TRANSCRIPTIONAL LEVEL REGULATION OF ALBUMIN AND ALPHA-FETO-PROTEIN GENE EXPRESSION

The molecular analysis of the albumin and AFP system has been undertaken by several laboratories following preparation of specific hybridization probes. First, the isolation and purification of the AFP and albumin mRNA molecules from rat and mouse made possible the preparation of single-stranded cDNA probes (35-39). The application of recombinant DNA technology allowed investigators to obtain rat and mouse albumin and AFP cDNA clones and the corresponding albumin and AFP genomic sequences (27, 30, 40-43). More recently, cloning of human albumin and AFP cDNAs has also been accomplished (28, 44, 45, and Frain and Sala-Trepat, in preparation). Characterization of these cDNA and genomic clones has revealed the information reviewed above on the structure of these genes and their molecular evolution.

Developmental changes in AFP and albumin gene expression were first studied by solution hybridization with specific cDNA probes. Determination of the steady-state levels of AFP and albumin mRNAs in polysomal RNA preparations from rat liver at different stages of development and from different rat hepatomas indicated a close correlation between the concentration of mRNA sequences and the specific albumin and AFP protein synthesis activities (18, 22, 23, 36, 46, 47). These results clearly established that translational control is not a major factor in the control of gene expression of albumin and AFP in rat liver and rat hepatomas.

We have carried out a detailed analysis of the subcellular distribution of albumin and AFP mRNA sequences in developing rat liver and in the Morris hepatoma 7777. As shown in Figure 1, in all stages of liver development and in the hepatoma tissue, less than 2% of the total albumin and AFP mRNA sequences are present in the nuclear compartment. Most of the cellular albumin and AFP mRNA sequences were found to be associated with the polysomes as mature mRNA molecules with no evidence for storage of inactive mRNA sequences in the nuclear or cytoplasmic extrapolysomal compartments (48). These data thus provide no indication that post-transcriptional mechanisms might play an important role in the developmental regulation of the albumin and AFP genes. They rather suggest that the regulation of these genes during normal liver development and neoplasia operates at the transcriptional level. Similar studies carried out with

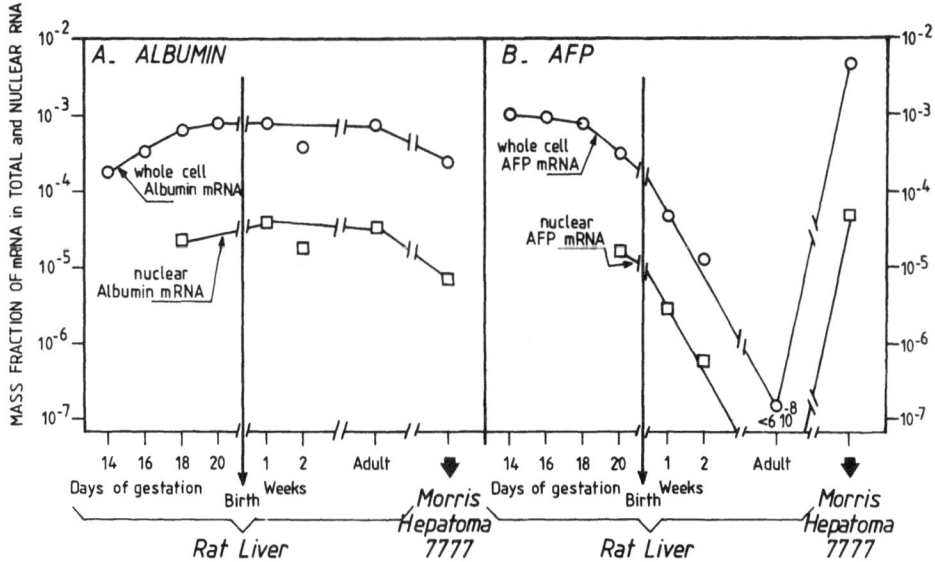

Figure 1.   Changes in albumin (A) and AFP (B) mRNA levels in total cell (o——o) and nuclear RNA preparations (□——□) from developing rat liver and Morris hepatoma 7777. The mass fraction values corresponding to the albumin and AFP mRNA molecules were determined by Rot analysis of hybridization data as shown elsewhere (36, 48).

newborn rats under dexamethasone treatment have also indicated that the glucocorticoid suppression of AFP synthesis in developing liver is probably exerted primarily at the level of transcription (49).

Direct evidence showing that the decrease in AFP mRNA molecules after birth is a result of a much reduced transcription of the AFP gene has been obtained by cell-free nuclear transcription assays. Tilghman and Belayew (50) have thus obtained conclusive results supporting a transcriptional control of the AFP and albumin genes during mouse liver development. Recent nuclear "run-on" experiments from our laboratory have also shown that transcriptional modulation is the primary molecular mechanism underlying the changes in expression of the AFP gene in developing rat liver (Nahon, Danan, and Sala-Trepat, in preparation). Using this more direct method, Guertin et al. (51) have confirmed that the administration of dexamethasone to newborn rats causes a rapid suppression of AFP gene transcription.

All these studies clearly indicate that the major control for AFP and albumin gene expression during rodent liver development and in hepatomas must operate at the level of transcription. The molecular mechanisms responsible for the changes in transcriptional activity could involve modifications at the gene level, changes in chromatin structure, or both.

ABSENCE OF MODIFICATIONS AT THE GENOMIC DNA LEVEL FOR RAT ALPHA-FETOPROTEIN AND ALBUMIN GENES

In our laboratory we have compared, by solution hybridization and Southern blot techniques, the reiteration frequency and the organization of the AFP and albumin genes in chromosomal DNA from different fetal and adult rat tissues (yolk sac, liver, kidney, spleen) and two rat hepatomas. It has been found that these genes are present at a single copy per rat haploid genome in all tissues analyzed, and that the gross organization of the albumin and AFP genes is the same in adult and fetal hepatocytes, as well as in other fetal and adult tissues (reference 39 and Gal et al., submitted for publication). These results indicate that the AFP and albumin genes are not grossly rearranged during development or neoplastic transformation. They appear to remain invariant throughout the regulatory processes involved in their tissue- and time-specific transcription. A similar conclusion has been drawn by Andrews et al. (52) from analysis of the mouse AFP gene in embryonic, adult, and neoplastic tissues.

The modification of specific bases (e.g., DNA methylation) has been invoked as a potential mechanism for altering the transcriptional activity of genes (see references 53 and 54 for reviews). In many higher eukaryotic systems an inverse correlation has been found between the transcriptional expression of a gene and the level of DNA methylation of that gene and particularly the 5' flanking sequences. Studies from our own and other laboratories (55, 56) have shown that specific changes in the level of methylation of the AFP and albumin genes take place during rat liver development. However, the relationship between these changes in methylation pattern and gene activity were found to be complex. For instance, the albumin gene is highly methylated in 18-day fetal hepatocytes and hypomethylated in adult hepatocytes, though no important differences in the transcriptional state of this gene are observed at these two stages of development (see Figures 1 and 2). In contrast, no drastic changes in the methylation state of the AFP gene were found in fetal and adult hepatocytes, but the adult hepatocytes show a much reduced transcription of the AFP gene. These results then indicate that alterations in the methylation patterns of the albumin and AFP intragenic sequences do not seem to play a major role in modulating the transcriptional activity of these genes during rat liver development. However, these studies do not exclude the possible existence of critical methylation sites in the 5' flanking sequences of the albumin and AFP genes that could determine the selective expression of these genes. Investigations directed to search for such hypothetical sites are now underway in our laboratory. In this context, it should be noted that Ott et al. (57) have identified an MspI site at the 5' end of the rat albumin gene whose undermethylation appears to be necessary but not sufficient for expression of the gene in rat hepatoma cells.

It should also be pointed out that Andrews et al. (52) have reported a good correlation between hypomethylation of the AFP

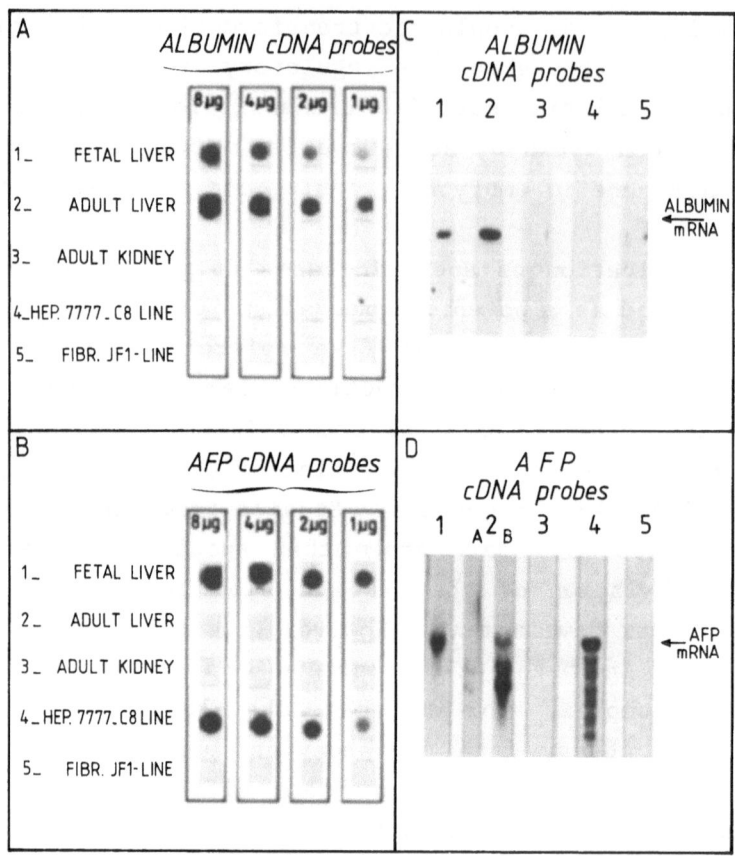

Figure 2. Dot blot (A and B) and Northern blot (C and D) analysis of albumin and AFP mRNA levels in fetal and
adult rat liver, in adult kidney and in two rat cell
lines. Serial dilutions of whole cell RNA were
dotted onto DBM paper and hybridized to cloned ^{32}P-
labeled albumin (A) or AFP (B) cDNA probes as reported elsewhere (64). RNA samples from the different
tissues and cell lines were fractionated on agarose
gels containing 10 mM methylmercury hydroxide, transferred to DBM paper and hybridized with cloned
albumin (C) or AFP (D) cDNA probes as previously
described (48). 1C and 1D, 2 μg of fetal liver total
RNA; 2C, 2 μg of adult liver total RNA; 2D, 25 μg of
total (A) or polyadenylated (B) RNA from adult liver;
3C and 3D, 25 μg of adult kidney total RNA; 4C and
4D, 25 μg and 2 μg of total RNA from the hepatoma
cell line, respectively; 5C and 5D, 25 μg of total
RNA from the JF1 cell line.

gene and its expression in different embryonic, adult, and neoplastic tissues. In particular, the extent of methylation of
the AFP gene was found to be higher in adult than in fetal mouse
liver. The reason for the discrepancy between these results and
those obtained by Vedel et al. (55) and Locker et al. (56) in

the rat is not known, but it might be due to a species-specific difference.

CHROMATIN STRUCTURE OF THE RAT ALPHA-FETOPROTEIN AND ALBUMIN GENES

The fact that eukaryotic DNA is packaged as chromatin has led to attractive models of gene regulation in which chromatin structure plays a critical role. The results of early studies indicated that actively transcribed DNA sequences are complexed with proteins to form nucleosome structures similar to those in the bulk of the genome (see references 58 and 59 for reviews), but it has become clear since the work of Weintraub and Groudine (60) that active genes have altered chromatin structures which render them preferentially sensitive to digestion by the endo-nuclease DNase I (reviewed in references 59 and 61). Several studies suggested that this preferential sensitivity may reflect the potentiality of genes to be expressed in a cell rather than their actual transcriptional activity (59, 61). It has also been found that in addition to the high degree of DNase I sensi-tivity characteristic of active genes there exists small regions of nuclease hypersensitivity usually located 5' to the coding region (references 62, 63, and the chapter by Thomas et al., this volume). These specific sites are more sensitive by an order of magnitude than are active or potentially active gene regions and about two orders of magnitude more accessible than inactive chromatin regions.

We have examined the question of whether alterations in DNase I sensitivity of the albumin and AFP genes are associated with the changes in expression of these genes during development and neoplasia.

Overall DNase I Sensitivity Studies

Nuclei were isolated from newborn and adult rat liver and from culture of a cloned cell line (C8) derived from the Morris hepatoma 7777. The C8 hepatoma cells actively transcribe the AFP gene but albumin transcripts were not detected in RNA pre-parations from these cells (Figure 2). Nuclei were also isolated from adult rat kidney and from a fibroblast cell line, JF-1, to provide information on the DNase I sensitivity of the albumin and AFP genes in non-hepatic cells. Nuclei were incubated with increasing concentrations of DNase I. Isolated DNA was digested with Hind III, and the fragments separated by electrophoresis and transferred onto diazobenzyloxymethyl paper. The blotted DNA was hybridized with cloned probes for rat albumin (the cDNA plasmids pRSA 13 and pRSA 57, Figure 3C; and the genomic probes sub JB, sub C, sub, B, sub A, and sub D, Figure 3A) and for rat AFP (the cDNA plasmids pRAFP 65 and pRAFP 87, Figures 3B and 3D). This procedure allowed visualization of the complete albumin and AFP transcription units. Following hybridization and autoradiographic exposure, the blots were washed and rehybridized to a cDNA probe for tyrosine hydroxylase to provide an internal control for DNase I sensitivity of a gene not expressed in the liver tissue (not shown).

As can be seen from Figures 3A and 3B, the chromatin region containing the albumin and AFP genes is much more sensitive to the nucleolytic action of DNase I in adult liver than in adult kidney. Both albumin and AFP genes appear to be very sensitive to DNase I in adult liver. A quantitative comparison of the accessibility to DNase I digestion of the albumin and AFP gene regions, relative to the reference "inactive" tyrosine hydroxylase gene, has shown that in adult liver the albumin and AFP genes are 6 to 12 times more sensitive to DNase I than is the tyrosine hydroxylase gene (64). In newborn liver the level of sensitivity of the albumin and AFP genes is not significantly different from that found in adult liver (not shown). In contrast, in adult kidney the albumin, AFP, and tyrosine hydroxylase genes are equally resistant to the DNase I action (64). The DNA fragments containing these three genes are also resistant to DNase I in nuclei from the fibroblast JF-1 cell line (Figures 3C and 3D, and reference 64).

In the hepatoma cells the AFP gene is rapidly digested by the nuclease while the albumin gene is only mildly sensitive (see Figures 3C and 3D). Quantitative analysis of the densitometric data indicates that in the hepatoma cells the AFP gene is 4 to 5 times more sensitive to DNase I than the albumin gene (64). These data support a close relationship between DNase I sensitivity and gene expression in this hepatoma cell line. The lower level of DNase I sensitivity of the albumin gene in the hepatoma cells than in adult liver suggests that alterations in

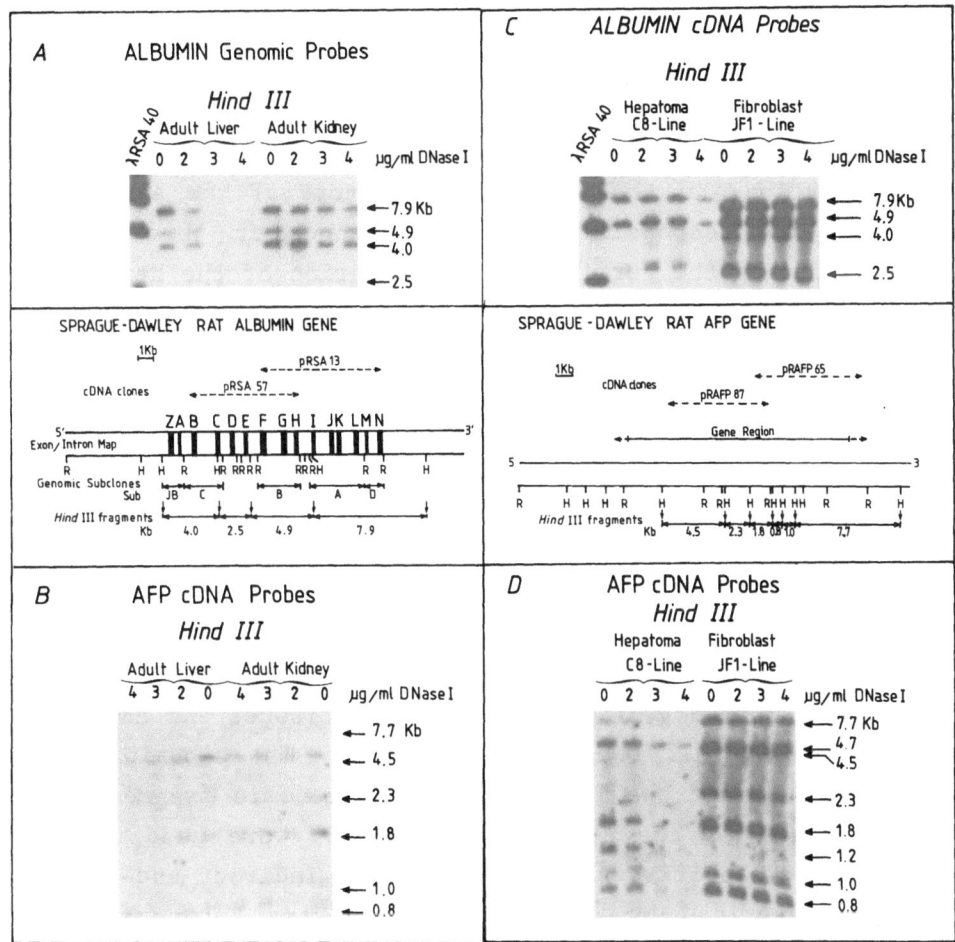

Figure 3. DNase I sensitivity of the albumin and AFP genes in adult rat tissues and cell lines. Nuclei from adult rat liver and kidney and from hepatoma C8 and fibroblast JF1 cells were incubated with increasing amounts of DNase I as indicated elsewhere (64). The DNA was extracted, digested with HindIII and analysed by the method of Southern with albumin (C) or AFP cDNA (B, D) probes or with the albumin genomic probes subJB, subA, subB, subC, and subD (27, 29, 40). The size of the genomic DNA fragments containing specific albumin or AFP gene sequences is given on the right of the figure. The restriction maps of the Sprague-Dawley albumin (29, 40) and AFP genes (43) are shown in the center of the figure. The difference in the HindIII restriction patterns of these genes in the C8 hepatoma cells and in other cells is due to allelic polymorphism. The Morris hepatoma 7777 was originally derived from a Buffalo rat.

the chromatin structure of this gene might occur during the process of carcinogenesis.

The high degree of DNase I sensitivity of the AFP gene in adult liver, despite its extremely low level of transcription, may reflect its potential for re-expression in oncogenic processes. Since the levels of DNase I sensitivity of the albumin and AFP genes were found to be similar in newborn and adult liver, these studies provide no evidence to support the idea that alterations in the chromatin structure of the albumin and AFP genes are responsible for the changes in gene activity occurring during the terminal differentiation of the hepatocyte. However, the drastic differences in DNase I sensitivity of these genes between hepatic and non-hepatic tissues suggest that such alterations in chromatin structure might be involved in the early establishment of the tissue-specific potential of overt gene expression.

DNase I Hypersensitive Sites

We have recently searched for the presence of DNase I-hypersensitive cleavage sites in the chromatin regions flanking the AFP and albumin genes in the same rat tissues and cell lines analyzed for overall DNase I sensitivity. The albumin genomic subclone sub JB and the Hind III - EcoRI genomic fragment close to the 5' end of the AFP gene (Figure 3) were used to probe nuclease-hypersensitive sites by using the indirect end-labeling technique described by Wu (62). Three DNase I hypersensitive sites could be mapped in the 5' flanking region of the albumin gene in chromatin from newborn and adult liver but not in the chromatin of the C8 hepatoma cell line and on non-hepatic cells (adult kidney, JF-1 fibroblasts). Two of these sites are located within 0.5 kb of the 5' end and the third site is found about 2.5 to 3 kb upstream of the albumin gene (65). The presence of these sites appears, then, to be directly correlated with the actual state of transcription of the albumin gene in the tissues and cell lines analyzed.

We have also detected at least three nuclease hypersensitive sites in the 5' end and flanking regions of the AFP gene in chromatin from the actively transcribing newborn rat hepatocytes. All these sites are also present in the C8 hepatoma cell line but could not be detected in adult kidney or the JF-1 fibroblast cells. Interestingly, only the more distal site located at about 2 to 3 kb from the 5' end of the AFP gene is detected in adult liver (Nahon and Sala-Trepat, in preparation). The presence of this site could then be related to the differentiated state of the cell, while the other two sites near the 5' end of the AFP gene would be directly correlated with active transcription. The fact that these two sites are absent in adult liver provides the first indication for the occurrence of alterations in chromatin structure reflecting the transition from the active to the inactive state of the AFP gene during the terminal differentiation of the hepatocyte.

CONCLUSIONS

The albumin and AFP genes provide a powerful model system to investigate the molecular mechanisms responsible for changes in gene expression during developmental and oncogenic process. Several studies have indicated that the expression of these two genes during rodent liver development and in different hepatomas is regulated mainly at the transcriptional level. Changes in the transcriptional template capacity of these genes during development and neoplasia do not appear to result from alterations in gene number or gross rearrangements within the genome. Furthermore, no evidence has implicated changes in methylation of specific gene sequences in the transcriptional modulation of these genes during rat liver development.

The structural organization of chromatin is thought to determine the state of differentiation and activity of eukaryotic genes. Our investigations on the chromatin structure of the AFP and albumin genes in developing rat liver and other tissues and cell lines have shown important differences in the conformation of these genes in hepatic and non-hepatic cells. Further, different sets of DNase I hypersensitive sites have been found upstream from the albumin and AFP genes depending on the state of differentiation of the cells and on the transcriptional state of these genes in the tissues and cell lines analyzed. The disappearance of two DNase I hypersensitive sites in the 5' flanking region of the AFP gene during late liver development indicated that changes in chromatin structure of this gene occur during the terminal differentiation of the hepatocyte in strict correlation with differential gene activity. It is likely that the special chromatin structures, that we have detected with the nuclease probe at the 5' end of the AFP and albumin genes, play an important role in the early establishment of the tissue-specific potential of overt gene expression and in the transcriptional regulation of these genes during the terminal differentiation. The short stretches of DNA amidst the 5' hypersensitive sites might mark sequences onto which regulatory proteins can be bound. It is now of obvious interest to search for specific proteins which will preferentially bind to these regions of the genome.

ACKNOWLEDGMENTS

We wish to thank Drs. A. Poliard, J.L. Danan, and T. Erdos for careful reading of the manuscript. This work was supported by an "Action de Soutien Programmée" from the Centre National de le Recherche Scientifique (Biologie de la Reproduction et du Développement) and by grants from Institut National de la Santé et de la Recherche Médicale and the Association pour le Développement de la Recherche sur le Cancer.

REFERENCES

1. AHMAD, F., SCHULTZ, J., RUSSEL, T.R., and WERNER, R., eds. (1978). "Differentiation and Development," Miami Winter Symposia, Vol. 15. Academic Press, New York.

2. BUCKINGHAM, M.E., ed. (1981). "Development and Differentiation." CRC Press, Boca Roton, Florida.

3. SILVER, L.M., MARTIN, G.R., and STRICKLAND, S., eds. (1983). "Teratocarcinoma Stem Cells," Cold Spring Harbor Conferences on Cell Proliferation, Vol. 10. Cold Spring Harbor Laboratory, New York.

4. DZIADEK, M. and ADAMSON, E. (1978). Localization and synthesis of alphafetoprotein in post implantation mouse embryos. J. Embryol. Exp. Morph. 43, 289-313.

5. PETERS, T., Jr. (1975). In: "The Plasma Proteins," Vol. 1 (F.W. Putman, ed.) pp. 133-181. Academic Press, New York.

6. RUOSLAHTI, E. and SEPPALA, M. (1979). α-Fetoprotein in cancer and fetal development. Adv. Cancer Res. 29, 275-346.

7. ABELEV, G.I. (1971). Alpha-fetoprotein in oncogenesis and its association with malignant tumors. Adv. Cancer Res. 14, 295-358.

8. GITLIN, D. and BOESMAN, M. (1967). Sites of serum α-fetoprotein synthesis in the human and in the rat. J. Clin. Invest. 46, 1010-1016.

9. GITLIN, D. and PERRICELLI, A. (1970). Synthesis of serum albumin, prealbumin, α-fetoprotein, 1α-antitrypsin, and transferrin by the human yolk sac. Nature 228, 995-997.

10. DE NECHAUD, B. and URIEL, J. (1971). Antigenes cellulaires transitoires du foie de rat. I. Secretion et synthese des proteines seriques foetospecifiques au cours du developpement et de la regeneration hepatiques. Int. J. Cancer 8, 71-80.

11. VAN FURTH, R. and ADINOLFI, M. (1969). In vitro synthesis of the fetal α_1-globulin in man. Nature 222, 1296-1299.

12. GITLIN, D. and GITLIN, J.D. (1975). In: "The Plasma Proteins," Vol. 2 (F.W. Putnam, ed.) pp. 264-237. Academic Press, New York.

13. GITLIN, D. and KITZES, J. (1967). Synthesis of serum albumin, embryo specific α-globulin and conalbumin by the chick yolk sac. Biochim. Biophys. Acta 147, 334-340.

14. SELLEM, C., FRAIN, M., ERDOS, T., and SALA-TREPAT, J.M. (1984). Differential expression of albumin and α-fetoprotein genes in fetal tissues of mouse and rat. Dev. Biol. 102, 51-60.

15. BELANGER, L., HAMEL, D., LACHANCE, L., DUFOUR, D., TREMBLAY, M., and GAGNON, P.M. (1975). Hormonal regulation of α_1-fetoprotein. Nature 256, 657-659.

16. SELL, S., BECKER, F., LEFFERT, H., and WATABE, H. (1976). Expression of an oncodevelopmental gene product (α-fetoprotein) during fetal development and adult oncogenesis. Cancer Res. 36, 4239-4249.

17. WATABE, H. (1971). Early appearance of embryonic α-globulin in rat serum during carcinogenesis with 4-dimethylamino azobenzene. Cancer Res. 31, 1192-1194.

18. SELL, S., SALA-TREPAT, J.M., SARGENT, T., THOMAS, K., NAHON, J.L., GOODMAN, T.A., and BONNER, J. (1980). Molecular mechanisms of control of albumin and alpha-fetoprotein production: a system to study the early effects of chemical hepatocarcinogens. Cell Biol. Intern. Rep. 4, 235-254.

19. ABELEV, G.I. (1974). α-Fetoprotein as a marker of embryo-specific differentiations in normal and tumor tissue. Transplant. Rev. 20, 3-37.

20. SELL, S. and MORRIS, H.P. (1974). Relationship of rat α_1-fetoprotein to growth rate and chromosome composition of Morris hepatomas. Cancer Res. 34, 1413-1417.

21. SCHREIBER, G., ROTERMUND, H.M., MAENO, H., WEIGAND, K., and LESHS, R. (1969). The proportion of the incorporation of leucine into albumin to that into total protein in rat

liver and hepatoma Morris 5123TC. <u>Eur</u>. <u>J</u>. <u>Biochem</u>. <u>10</u>, 355-361.

22. TSE, T.P.H., MORRIS, H.P., and TAYLOR, J.M. (1978). Molecular basis of reduced albumin synthesis in Morris Hepatoma 7777. <u>Biochemistry</u> <u>17</u>, 21121-2128.

23. SELL, S., THOMAS, K., MICHAELSON, M., SALA-TREPAT, J.M., and BONNER, J. (1979). Control of albumin and α-fetoprotein expression in rat liver and in some transplantable hepatocellular carcinomas. <u>Biochim</u>. <u>Biophys</u>. <u>Acta</u> <u>564</u>, 173-178.

24. RUOSLAHTI, E. and ENGVALL, E. (1976). Immunological cross-reaction between alpha-fetoprotein and albumin. <u>Proc</u>. <u>Natl</u>. <u>Acad</u>. <u>Sci</u>. <u>USA</u> <u>73</u>, 4641-4644.

25. RUOSLAHTI, E. and TERRY, W.D. (1976). Alpha-foetoprotein and serum albumin show sequence homology. <u>Nature</u> <u>260</u>, 804-805.

26. LIAO, W.S.L., HAMILTON, R.W., and TAYLOR, J.M. (1980). Amino acid sequence homology between rat α-fetoprotein and albumin at the COOH-terminal regions. <u>J</u>. <u>Biol</u>. <u>Chem</u>. <u>255</u>, 8046-8049.

27. JAGODZINSKI, L., SARGENT, T.D., YANG, M., GLACKIN, C., and BONNER, J. (1981). Sequence homology between RNAs encoding rat α-fetoprotein and rat serum albumin. <u>Proc</u>. <u>Natl</u>. <u>Acad</u>. <u>Sci</u>. <u>USA</u> <u>78</u>, 3521-3525.

28. MORINAGA, T., SAKAI, M., WEGMANN, T.G., and TAMAOKI, T. (1983). Primary structures of human α-fetoprotein and its mRNA. <u>Proc</u>. <u>Natl</u>. <u>Acad</u>. <u>Sci</u>. <u>USA</u> <u>80</u>, 4604-4608.

29. SARGENT, T., JAGODZINSKI, L., YANG, M., and BONNER, J. (1981). Fine structure and evolution of the rat serum albumin gene. <u>Mol</u>. <u>Cell</u>. <u>Biol</u>. <u>1</u>, 871-883.

30. KIOUSSIS, D., EIFERMAN, F., VAN DE RIJN, P., GORIN, M.B., INGRAM, R.S., and TILGHMAN, S.H. (1981). The evolution of α-fetoprotein and albumin. II. The structures of the α-fetoprotein and albumin genes in the mouse. <u>J</u>. <u>Biol</u>. <u>Chem</u>. <u>256</u>, 1960-1967.

31. ALEXANDER, F., YOUNG, P.R., and TILGHMAN, S.H. (1984). Evolution of the albumin-α-fetoprotein ancestral gene from the amplification of a 27 nucleotide sequence. <u>J</u>. <u>Mol</u>. <u>Biol</u>. <u>173</u>, 159-176.

32. INGRAM, R.S., SCOTT, R.W., and TILGHMAN, S.M. (1981). α-Fetoprotein and albumin genes are in tandem in the mouse genome. <u>Proc</u>. <u>Natl</u>. <u>Acad</u>. <u>Sci</u>. <u>USA</u> <u>78</u>, 4694-4698.

33. GROUDINE, M. and WEINTRAUB, H. (1981). Activation of globin genes during chicken development. <u>Cell</u> <u>24</u>, 353-401.

34. MANIATIS, T., FRISTSCH, E.F., LAUER, J., and LAWN, R.M. (1980). The molecular genetics of human hemoglobins. <u>Ann</u>. <u>Rev</u>. <u>Genet</u>. <u>14</u>, 145-178.

35. INNIS, M.A. and MILLER, D. (1977). Quantitation of rat α-fetoprotein messenger RNA with a complementary DNA probe. <u>J</u>. <u>Biol</u>. <u>Chem</u>. <u>252</u>, 8469-8475.

36. SALA-TREPAT, J.M., DEVER, J., SARGENT, T.D., THOMAS, K., SELL, S., and BONNER, J. (1979). Changes in expression of albumin and α-fetoprotein genes during rat liver development and neoplasia. <u>Biochemistry</u> <u>18</u>, 2167-2178.

37. MIURA, K., LAW, S.W., NISHI, S., and TAMAOKI, T. (1979). Isolation of α-fetoprotein messenger RNA from mouse yolk sac. J. Biol. Chem. 254, 5515-5521.

38. BROWN, R.D. and PAPACONSTANTINOU, J. (1979). Mouse albumin mRNA in liver and a hepatoma cell line. J. Biol. Chem. 254, 5177-5183.

39. SALA-TREPAT, J.M., SARGENT, T.D., SELL, S., and BONNER, J. (1979). α-Fetoprotein and albumin genes of rats: no evidence for amplification-deletion or rearrangement in rat liver carcinogenesis. Proc. Natl. Acad. Sci. USA 76, 695-699.

40. SARGENT, T.D., WU, J.R., SALA-TREPAT, J.M., WALLACE, R.B., REYES, A.A., and BONNER, J. (1979). The rat serum albumin gene: analysis of cloned sequences. Proc. Natl. Acad. Sci. USA 18, 3256-3260.

41. TILGHMAN, S.M., KIOUSSIS, D., GORIN, M.B., GARCIA RUIZ, J.P., and INGRAM, R.S. (1979). The presence of intervening sequences in the α-fetoprotein gene of the mouse. J. Biol. Chem. 254, 7393-7399.

42. LAW, S., TAMAOKI, T., KREUZALER, M., and DUGAICZYK, A. (1980). Molecular cloning of DNA complementary to a mouse α-fetoprotein mRNA sequence. Gene 10, 53-61.

43. GAL, A., NAHON, J.L., LUCOTTE, G., and SALA-TREPAT, J.M. (1984). Structural variants of the α-fetoprotein gene in different inbred strains of rat. Mol. Gen. Genet. 195, 153-158.

44. DUGAICZYK, A., LAW, S.W., and DENNISON, D.E. (1982). Nucleotide sequence and the encoded amino-acids of human serum albumin mRNA. Proc. Natl. Acad. Sci. USA 79, 71-75.

45. FRAIN, M. (1984). Structure et expression des genes codant pour deux proteines marqueurs de la differenciation hepatique chez l'homme: l'albumine et l'alpha-foetoproteine. These d'Etat, University of Paris.

46. LIAO, W.S.L., CONN, A.R., and TAYLOR, J.M. (1980). Changes in rat α1-fetoprotein and albumin mRNA levels during fetal and neonatal development. J. Biol. Chem. 255, 10036-10039.

47. CASSIO, D., WEISS, M.C., OTT, M.O., SALA-TREPAT, J.M., FRIES, J., and ERDOS, T. (1981). Expression of the albumin gene in rat hepatoma cells and their dedifferentiated variants. Cell 27, 351-358.

48. NAHON, J.L., GAL, A., FRAIN, M., SELL, S., and SALA-TREPAT, J.M. (1982). No evidence for post-transcriptional control of albumin and α-fetoprotein gene expression in developing rat liver and neoplasia. Nucleic Acids Res. 10, 1895-1911.

49. BELANGER, L., FRAIN, M., BARIL, P., GINGRAS, M.C., BARTKOWIAK, J., and SALA-TREPAT, J.M. (1981). Glucocorticoid supression of α1-fetoprotein synthesis in developing rat liver. Evidence for selective gene repression at the transcriptional level. Biochemistry 20, 6665-6671.

50. TILGHMAN, S.H. and BELAYEW, A. (1982). Transcriptional control of the murine albumin/α-fetoprotein locus during development. Proc. Natl. Acad. Sci. USA 79, 5254-5257.

51. GUERTIN, M., BARIL, P., BARTKOWIAK, J., ANDERSON, A., and BELANGER, L. (1983). Rapid supression of α1-fetoprotein

gene transcription by dexamethasone in developing rat liver. Biochemistry 22, 4296-4302.

52. ANDREWS, G.K., DZIADEK, M., and TAMAOKI, T. (1982). Expression and methylation of the mouse α-fetoprotein gene in embryonic, adult, and neoplastic tissues. J. Biol. Chem. 257, 5148-5153.

53. RAZIN, A. and RIGGS, A.D. (1980). DNA methylation and gene function. Science 210, 604-610.

54. DOERFLER, W. (1983). DNA methylation and gene activity. Ann. Rev. Biochem. 52, 93-124.

55. VEDEL, M., GOMEZ-GARCIA, M., SALA, M., and SALA-TREPAT, J.M. (1983). Changes in methylation pattern of albumin and α-fetoprotein genes in developing rat liver and neoplasia. Nucleic Acids Res. 11, 4335-4354.

56. KUNNATH, L. and LOCKER, J. (1983). Developmental changes in the methylation of the rat albumin and α-fetoprotein genes. EMBO J. 2, 317-324.

57. OTT, M.O., SPERLING, L., CASSIO, D., LEVILLIERS, J., SALA-TREPAT, J.M., and WEISS, M.C. (1982). Undermethylation at the 5'-end of the albumin gene is necessary but not sufficient for albumin production by rat hepatoma cells in culture. Cell 30, 825-833.

58. FELSENFELD, G. (1978). Chromatin. Nature 271, 115-122.

59. MATHIS, D., OUDET, P., and CHAMBON, P. (1980). Structure of transcribing chromatin. Progr. Nucleic Acid Res. Mol. Biol. 24, 1-54.

60. WEINTRAUB, H. and GROUDINE, M. (1976). Chromosomal subunits in active genes have an altered conformation: globin genes are digested by deoxyribonuclease I in red blood cell nuclei but not in fibroblast nuclei. Science 193, 848-856.

61. WEISBROD, S. (1982). Active chromatin. Nature 297, 289-295.

62. WU, C. (1980). The 5' end of Drosophila heat shock genes in chromatin are hypersensitive to DNase I. Nature 286, 854-860.

63. ELGIN, S.C.R. (1981). DNase I-hypersensitive sites on chromatin. Cell 27, 413-415.

64. NAHON, J.L., GAL, A., ERDOS, T., and SALA-TREPAT, J.M. (1984). Differential DNase I sensitivity of the albumin and α-fetoprotein genes in chromatin from rat tissues and cell lines. Proc. Natl. Acad. Sci. USA 81, 5031-5035.

65. NAHON, J.L. and SALA-TREPAT, J.M. (1984). Tissue-specific DNase I hypersensitive sites in rat chromatin are present upstream from the 5' ends of the albumin and α-fetoprotein genes. J. Cell Biol. 99, 139a.

TRANSCRIPTIONAL AND POST-TRANSCRIPTIONAL CONTROL OF HISTONE GENE
EXPRESSION

Daniel Schümperli

Institut für Molekularbiologie II
Universität Zürich
ETHZ-Hönggerberg
8093 Zürich, Switzerland

INTRODUCTION

The structure and developmental regulation of histone genes
have been the subject of recent reviews (1, 2). This chapter
will, therefore, deal only with experimental studies designed to
identify transcriptional and post-transcriptional regulation
signals and factors interacting with them.

THE ANATOMY OF HISTONE GENE PROMOTERS

The promoter of the early embryonic H2A gene of the sea
urchin Psammechinus miliaris has been analysed in great detail
by surrogate genetics and microinjection into X. laevis oocytes
(3-5). The picture emerging from these experiments is that the
H2A gene promoter contains at least three distinct functional
elements: the modulator, selector, and initiator elements (3).
The modulator is located far upstream of the gene, between positions -165 and -111, and is responsible for determining the rate
of transcription of the gene (3-5). It is similar in many
respects to viral enhancer elements. For instance, it functions
in an orientation-independent fashion. It may, therefore, be
significant that functionally important sequences within the H2A
modulator element are similar to the Moloney murine sarcoma
virus enhancer and to topologically related 5' long terminal

repeat (LTR) sequences from other viruses (5). The selector element containing the TATA box primarily specifies the correct 5' end of the mRNA, whereas the main function of the initiator segment is to facilitate transcription initiation (3).

Similar surrogate genetics were also applied to the study of transcription of a cloned Xenopus laevis H4 gene (6). This gene contains a duplicated promoter and, accordingly, two mRNAs differing from each other by 115 nucleotides at their 5' ends are produced. Deletion experiments showed that a segment between position -65 and -35 of the stronger downstream transcription start site was important for maximal expression of the H4 gene in microinjected X. laevis oocytes. DNA sequence comparisons revealed that this important area of the H4 promoter contains two highly conserved DNA motifs near positions -51 to -46 in all H4 genes analysed (6).

The X. laevis oocyte expression system has also been used to isolate factors that will complement or stimulate the faulty or weak expression of early embryonic sea urchin histone genes (7, 8). For instance, a protein factor isolated and partly purified from chromatin of sea urchin embryos strongly stimulates expression of the H2B gene in frog oocytes. The factor has an apparent molecular weight of 40,000 on sucrose density gradients and, as expected for a nucleic acid binding protein, it binds to heparin sepharose columns (8). Numerous mutations in the 5' flanking region of the H2B gene failed to reveal any effect on stimulation by the factor (8), although some of them drastically reduced the overall rates of H2B transcription. In particular, these studies revealed the importance of the sequence CAAT (-100) and of a H2B-specific conserved sequence motif (-60) for maximal H2B transcription (Mächler, Mous, Stunnenberg, and Birnstiel, manuscript in preparation). However, deletions and substitution mutations of the H2B gene have clearly established that the interaction site(s) for the factor is located downstream from the transcription start site (8).

Thus, the oocyte injection system has revealed a great variety of sequences important for histone gene transcription. Transcriptional control elements have been found upstream and downstream of the transcription start site. However, even within the limited set of histone gene promoters studied to date, one can also begin to find some commonalities. For instance, the TATA box and the initiator element are important components, although to different extents, of both the H2A and H2B promoters. (They were not analysed in the case of the X. laevis H4 gene.) The CAAT sequence as well as subtype-specific conserved sequence motifs around position -50 were found to be functionally important for H2B and H4, but not for H2A. Furthermore, the H2A gene modulator and the H2B stimulatory factor interaction site could both belong to yet another class of promoter elements which may be located either upstream or downstream of the transcription start site and which can function independent of their orientation.

3' EDITING OF HISTONE mRNAs

Histone mRNAs, unlike other eukaryotic mRNAs, carry at their 3' terminus a RNA hairpin structure rather than a poly(A) tail. The hairpin is part of a 23 bp sequence motif which is highly conserved among histone genes throughout evolution (9). About 6 bp downstream from this sequence there is another, somewhat less well conserved sequence motif, CAAGAAAGA, which is absent from the mature mRNA (10).

The sequences required for the generation of correct 3' ends of the sea urchin H2A gene were mapped by mutational analyses and oocyte injections to reside in the hairpin itself but also within the first 100 bp of spacer DNA (11, 12). Point mutations in the hairpin structure indicated that the dyad symmetry and not the sequence itself is important for the generation of 3' ends (12). Injections of DNA heteroduplexes between the point mutants and the wild-type sequence revealed that the hairpin was required at the RNA rather than at the DNA level (12). Taken together, these experiments suggested that RNA processing rather than accurate termination might be the mechanism by which the authentic 3' ends of histone mRNAs are formed. 3' processing of histone mRNA was then directly demonstrated by oocyte injection of precursor RNA (13, 14) synthesized in vitro and, more recently, also in partly purified in vitro systems (reference 15, and D. Schümperli, unpublished observation).

173

Further DNA injection experiments of processing-deficient mutants of the H2A gene revealed that transcription terminates heterogeneously in the post-H2A spacer and that the transcripts are then processed to the mature 3' ends (13).

As is the case for transcription initiation, 3' editing of sea urchin histone mRNAs can also be complemented in the X. laevis oocyte. For unknown reasons, H3 transcripts are processed very inefficiently in the oocyte. The elongated transcripts are apparently quite stable, since they are present in amounts comparable to those of H2A and H2B mRNAs (16). This defect can be complemented by a factor extracted from the chromatin of sea urchin embryos (7). This factor behaves during the purification like a nucleic acid binding protein with an apparent molecular weight of 200-250,000 (7). Complementation of 3' editing of H3 in vivo transcripts can also be achieved with a small poly(A)-RNA of approximately 60 nucleotides (60N-RNA) from sea urchin embryos (17). A functional RNA of this size could also be extracted from the active factor preparation (17). Correct 3' editing was also observed with H3 in vitro precursor RNA injected into oocytes that had been pre-injected with the 60 nucleotide RNA (13). This provided final proof that this novel snRNP particle is involved in the processing reaction. Several small deletions, insertions or linker scanning mutations of the two conserved 3' motifs almost invariably abolish the 60N-RNA-dependent 3' editing of H3 mRNA (18). Several cDNA clones for the 60N-RNA, now called U7 RNA, have recently been isolated and sequenced (19). They contain regions of extensive sequence similarity to the two conserved 3' motifs. It is therefore likely that this RNA interacts with histone mRNA precursors by base pairing. This interaction would result in unfolding of the mRNA hairpin structure and exposure of the processing site which are presumably crucial steps for processing to occur.

CONCLUSIONS

Functional analysis of gene expression has revealed a rather complex organization of histone promoter elements. It can be inferred from these findings that transcription initiation is not merely the product of an interaction of RNA polymerase with DNA and nucleotides, but rather must be controlled by several additional factors. Post-transcriptionally, histone gene expression is subject to additional levels of control, i.e., 3' processing involving U7 snRNPs, nucleo-cytoplasmic transport, degradation or stabilization of precursor transcripts or processed mRNAs in the nuclear or cytoplasmic compartment. Little is known about the cellular components involved in these processes and about their targets on histone DNA or RNA. It will also be important to know which of these processes are used to regulate histone gene expression during embryonic development, tissue differentiation, or during the cell division cycle. For instance, it is now clearly established that both transcriptional and post-transcriptional processes must be involved in changing histone mRNA levels during the animal cell cycle. Moreover, histone gene switches have been well documented, the prime examples occuring during early embryonic development and spermatogenesis in sea urchins and during the differentiation of erythroid cells in birds (1, 2). Further functional analysis of histone genes in a variety of experimental systems will be required to understand the molecular basis for these regulatory phenomena.

REFERENCES

1. HENTSCHEL, C.C. and BIRNSTIEL, M.L. (1981). The organization and expression of histone gene families. Cell 25, 301-313.

2. MAXSON, R., MOHUN, T., COHN, R., and KEDES, L. (1983). Expression and organization of histone genes. Ann. Rev. Genetics 17, 239-277.

3. GROSSCHEDL, R. and BIRNSTIEL, M.L. (1980). Identification of regulatory sequences in the prelude sequences of an H2A histone gene by the study of specific deletion mutants in vivo. Proc. Natl. Acad. Sci. USA 77, 1432-1436.

4. GROSSCHEDL, R. and BIRNSTIEL, M.L. (1980). Spacer DNA sequences upstream of the TATAAATA sequence are essential for promotion of H2A histone gene transcription. Proc. Natl. Acad. Sci. USA 77, 7102-7106.

5. GROSSCHEDL, R., MACHLER, M., ROHRER, U., and BIRNSTIEL, M.L. (1983). A functional component of the sea urchin H2A gene modulator contains an extended sequence homology to a viral enhancer. Nucleic Acids Res. 11, 8123-8136.

6. CLERC, R.G., BUCHER, P., STRUB, K., and BIRNSTIEL, M.L. (1983). Transcription of a cloned Xenopus laevis H4 histone gene in the homologous frog oocyte system depends on

an evolutionary conserved sequence motif in the -50 region. <u>Nucleic</u> <u>Acids</u> <u>Res.</u> <u>11</u>, 8641-8657.

7. STUNNENBERG, H.G. and BIRNSTIEL. M.L. (1982). Bioassay for components regulating eukaryotic gene expression: a chromosomal factor involved in the generation of histone mRNA 3' termini. <u>Proc.</u> <u>Natl.</u> <u>Acad.</u> <u>Sci.</u> <u>USA</u>, <u>79</u>, 6201- 6204.

8. MOUS, J, STUNNENBERG, H., GEORGIEV, O., and BIRNSTIEL, M.L. (1985). Stimulation of sea urchin H2B histone gene transcription by a chromatin-associated protein fraction depends on gene sequences downstream of the transcription start-site (submitted).

9. BUSSLINGER, M., PORTMANN, R., and BIRNSTIEL, M.L. (1979). A regulatory sequence near the 3' end of sea urchin histone genes. <u>Nucleic</u> <u>Acids</u> <u>Res.</u> <u>6</u>, 2997-3008.

10. HENTSCHEL, C., IRMINGER, J.C., BUCHER, P., and BIRNSTIEL, M.L. (1980). Sea urchin histone mRNA termini are located in gene regions downstream from putative regulatory sequences. <u>Nature</u>, <u>285</u>, 147-151.

11. BIRCHMEIER, C., GROSSCHEDL, R., and BIRNSTIEL, M. L. (1982). Generation of authentic 3' termini of an H2A mRNA in vivo is dependent on a short inverted DNA repeat and spacer sequences. <u>Cell</u> <u>28</u>, 739-745.

12. BIRCHMEIER, C., FOLK, W., and BIRNSTIEL, M.L. (1983). The terminal stem-loop structure and 80 bp of spacer DNA are required for the formation of 3' termini of sea urchin H2A mRNA. <u>Cell</u> <u>35</u>, 433-440.

13. BIRCHMEIER, C., SCHUMPERLI, D., SCONZO, G., and BIRNSTIEL, M.L. (1984). 3' editing of mRNA's: sequence requirements and involvement of a 60-nucleotide RNA in maturation of histone mRNA precursors. <u>Proc.</u> <u>Natl.</u> <u>Acad.</u> <u>Sci.</u> <u>USA</u> <u>81</u>, 1057-1061.

14. KRIEG, P.A. and MELTON, D.A. (1984). Formation of the 3' end of histone mRNA by post-transcriptional processing. <u>Nature</u> <u>308</u>, 203-206.

15. PRICE, D.H. and PARKER, C.S. (1984). The 3' end of Drosophila histone H3 mRNA is produced by a processing activity in vitro. <u>Cell</u> <u>38</u>, 423-429.

16. HENTSCHEL, C., PROBST, E., and BIRNSTIEL, M. L. (1980). Transcriptional fidelity of histone genes injected into Xenopus oocyte nuclei. <u>Nature</u> <u>288</u>, 100-102.

17. GALLI, G., HOFSTETTER, H., STUNNENBERG, H.G., and BIRNSTIEL, M.L. (1983). Biochemical complementation with RNA in the Xenopus oocyte: a small RNA is required for the generation of 3' histone mRNA termini. <u>Cell</u> <u>34</u>, 823-828.

18. GEORIEV, O. and BIRNSTIEL, M.L. (1985). The conserved CAAGAAAGA spacer sequence is an essential element for the formation of 3' termini of the sea urchin H3 histone mRNA by RNA processing. <u>EMBO</u> <u>J.</u> <u>4</u>, 481-489.

19. STRUB, K., GALLI, G., BUSSLINGER, M., and BIRNSTIEL, M.L. (1984). The cDNA sequences of the sea urchin U7 small nuclear RNA suggest specific contacts between histone mRNA precursor and U7 RNA during RNA processing. <u>EMBO</u> <u>J.</u> <u>3</u>, 2801-2807.

GENE EXPRESSION AND INTERFERON

Bruce D. Korant

Central Research & Development Department
Experimental Station
E. I. du Pont de Nemours & Company, Inc.
Wilmington, Delaware 19898

INTRODUCTION

The regulation of gene expression in the interferon system falls into two distinct sub-topics. First is the induction of the interferon structural genes in cells infected with a virus or treated with some other inducer, such as double-stranded RNA or lectins. The second involves the mechanism of interferon action, in which several distinct genes are induced in interferon-treated cells to be transcribed and translated as the cells become virus resistant or cease their usual division cycle. In neither the induction of interferon nor its action is much known about specific effects on chromatin structure, but it is presumed that perturbations do occur as in other examples of regulated gene expression. Besides providing yet another "model" system, the interferons offer the following features: very rapid alterations in gene expression initiated at the cell surface by a relatively few inducing molecules (only one molecule in some cell types), leading to profound biological sequelae (refractility to viruses and blockade of cell cycle movements). The interferons are particularly interesting because cell division often stops in treated cells, in comparison to other hormonally-regulated systems where division rates are enhanced. The property of interferons to make cell cultures quiescent thus avoids the problem of studying cascades of gene expression which occur in growth-stimulated cells.

THE INTERFERON PROTEINS

The interferons are a family of polypeptides with extremely potent biological activities, whether measured in cultured cells (primary or continuous) or in animals. They also have potential clinical applications as anti-infectives or cytostatic agents, although their efficacy still remains to be clearly established in treatment of human disease.

All the interferons are single polypeptides, with approximately 166 amino acid residues. Some of the protein species, notably fibroblast (β) and immune (γ), contain carbohydrate when produced in animal cells, although the role of the carbohydrate in their activity seems to be minimal. The present number of distinct species is fourteen for the α (leucocyte) type and one each for the β and γ types. The nomenclature is derived from the cell type which produces the interferon, rather than the responder.

All the interferons induce viral resistance in cell cultures. In such an assay, the cells are treated briefly (30 or more minutes) with a solution containing a dilution of the interferon and then are challenged with a virus. Usually, the cells are visually examined at daily intervals thereafter, until the untreated cells have been completely destroyed (one to three days). The interferon-treated cells will usually show protection against the viral cytopathic effect, with one unit of interferon activity representing the dilution causing fifty percent protection. Variability in the assay is two-fold or less. Homogeneous interferons having become available, it is now possible from such an assay to define a specific biological activity. For most interferons, this is close to 2×10^8 units per milligram of interferon protein, although higher specific activities have been reported (for a review, see reference 1). This translates into one unit (50% protection) representing picomole amounts of interferon. Cell growth inhibition is observed at similar levels, but may be higher for some cells with particular interferons. The old concept of species specificity for interferons has too many exceptions to be considered valid any longer.

It is also clear from studies with homogeneous natural interferons and individual species from recombinant DNA sources that interferons are multipotent, and can stimulate cells to become virus resistant, stop dividing, or become activated in various cellular immune reactions (e.g., killer cell activation). Whether these multiple responses are due to more than one class of interferon receptor, distinct domains on each interferon polypeptide triggering different cellular responses, or the manner in which interferon signals are transmitted from the cell surface to the nucleus is far from clear.

INDUCTION OF INTERFERON STRUCTURAL GENES

The classical inducers of the interferon structural genes are viruses. Most viruses will induce some interferon in cell cultures, but the most effective are RNA viruses, especially those which are relatively ineffective at blocking cellular protein synthesis. Enhancement of the induction may be accomplished by irradiation of the inducing virus with low doses of ultraviolet light, which prevents a full-blown infection but leaves sufficient (unknown) viral genes intact to act as an inducing agent. In studies reviewed by Marcus (2) it is hypothesized that one or more molecules of double-stranded RNA within the infected cell are sufficient to trigger the induction of interferon structural genes. This induction is largely at the transcriptional level, although there are reports of inducers also stabilizing existing interferon messenger RNAs (3). Synthetic double-stranded RNA molecules, such as polyribo-inosine-polyribocytosine, poly(IC), are also able to induce interferon expression, although effective induction requires many copies of RNA polymer per cell. The poly(IC) is thought to act at the cell surface, from studies using particulate supports for the inducer, but it is difficult to know absolutely that none of the RNA is released by cleavage and may subsequently penetrate the cells. The induction event is fairly rapid, and occurs within 30-60 minutes of exposure. The effect of the inducer may be potentiated by careful, timed addition of inhibitors of RNA or protein synthesis (super-induction), although the exact mechanism is unclear. Prior addition to the cells of minimal quantities of interferon also helps to potentiate the response to an inducer. For certain cell types (T-cells), induction of interferon is best done with antigens or lectins, rather than double-stranded RNA. In general, small molecules such as halogenated pyrimidines are useful inducers of interferon only in whole animals (4).

Following addition of the inducer to the cells, there is a burst of transcriptional activity of the interferon structural genes. However, the relative level of transcription is low, with at most only a few thousand interferon mRNAs produced. The mRNA may be isolated in an active form, and translated in cell-free systems or oocytes to produce biologically-active interferon. Most of the mRNA produced is specific for the cell type induced, so that fibroblast cells produce greater than 90% β interferon message, and a minority of α species. The situation is reversed in leucocytes. The reason for this is unclear, but may involve tissue-specific enhancer sequences (5). The high specific activity of the interferons permitted assay of cloned interferon cDNAs by hybrid selection and were taken advantage of in isolating the rare clones containing interferon sequences (5).

The cDNA sequences of the interferons were consistent with published amino-terminal protein sequences of purified interferons (6, 7) and were used to complete the primary protein sequence analyses. Moreover, cDNAs were introduced into expression plasmids and cloned into bacteria, yeast, or mammalian cells in order to produce very large quantities of interferon, relative to what had been available prior to genetic engineering. It is premature to suggest that interferon supply is entirely solved, but for most laboratories interested in using interferon in biochemical research experiments, there is no longer a problem in obtaining sufficient amounts of biologically active interferon (8).

The induction of interferons has been addressed most effectively by using cloned cDNAs or genomic DNAs of α or β interferons. One of the first problems addressed was careful examination of the chromosomal locations of interferon structural genes using cDNAs as probes. The work of Trent et al. (9) clearly showed that the human α and β interferons were located near one terminus of chromosome 9. More recent studies (10-14) analyzed in detail the DNA sequences upstream from the α and β coding regions, and deletions of segments of the non-coding portions of the interferon genes have established the regions required for inducibility. The results are reminiscent of other inducible eukaryotic genes (see chapter by Wasylyk, this volume) and may be summarized as follows:

Fiers and colleagues (12) inserted the human fibroblast interferon gene into SV40 viral vectors and transfected monkey cells. There was specific induction (10-30 fold) observed upon addition of poly(IC) to the cells, and no effect of the inducer on the interferon gene lacking 5' non-coding sequences. Deletion mapping of the inducible region showed that nucleotides between position -186 and -144 upstream from the mRNA initiation (cap) site comprised the responsive sequences. It was found by sequence analysis that a block of 18 residues (-174 to -157) was similar to sequences in unrelated genes which are induced by glucocorticoid hormones (see chapter by Beato et al., this volume). The TATA box is present at position -35 from the cap site for β interferon. The situation is slightly different for the α interferons. Weissman and colleagues (11, 13, 14), using a similar approach but different plasmids and host cells, found that not more than 117 upstream nucleotides were required for induction by a virus of α interferon mRNA synthesis at the correct initiation site. In the absence of inducer, RNA was transcribed from the interferon gene, but of disperse and in appropriate size and start site. Sequencing showed a purine-rich 5' region conserved in all α interferon genes studied. It was also found that removal of sequences upstream of position -117 enhanced expression of the interferon gene, suggesting these sequences were involved in suppression of transcription. Revel (5), in reviewing the subject of regulation of interferon transcription, concluded that the α and β genes are homologous in 5' elements. The genes are also located in the same macroscopic region of human chromosome 9, although there have been no descriptions of the linear distance between the two classes of interferon. Therefore, it still is unclear why α and β genes are regulated differently in cells of distinct lineage (fibroblasts vs. lymphocytes) and respond quite specifically to different inducers. The next major advances will likely come from in vitro reconstitution experiments in which correct transcription initiation can be studied on defined genes in the absence of complications encountered with intact cells. At present, the argument can still be offered that the primary effect of the inducer is on some segment of DNA not contiguous with the interferon genes, to produce a DNA, RNA, or protein which subsequently regulates interferon gene transcription.

INDUCTION OF CELLULAR GENES BY INTERFERON

Mechanism studies have indicated that interferon-treated cells became virus resistant following induction of cellular genes and synthesis of one or more proteins. Although the assignment of one of the induced proteins to a role in the antiviral state has not yet been achieved, it is clear that a number of interesting enzymatic activities are induced in interferon-treated cells, including protein kinases, nucleases, and oligo-adenylate synthase, as well as structural proteins such as HLA surface antigens (1).

It is believed that interferon action is initiated by binding of one or more interferon molecules to specific cellular receptors. The number of interferon receptors on cells in culture is rather low, with a typical figure of 10^2 to 10^3 (15). The receptors appear to be (glyco) proteins, with a subunit size of greater than 150,000, perhaps in association with ganglio-sides (16). It is likely that the receptors are available only on the surface of sensitive cells, rather than within the cells, but it may be that the interferon-receptor complexes require internalization before a specific effect on gene activation can occur, by analogy to other peptide hormones. The nature of the signal from the cell surface to the nucleus following interferon treatment is presently unknown. Interferons were initially thought to be species specific, but there are many interesting exceptions, including the very surprising one that human interferons act to protect plant cells against virus challenge (17). The "receptor site" for animal interferons on plant cells has not been studied.

The details of the interferon proteins' binding to their receptors are not well described. Although several interferon primary amino acid sequences are published, there have been no presentations of detailed models of interferon three-dimensional structure. There are proposed structural similarities between α and β (18) and between β and γ (19) interferons, but biochemical support is so far unavailable. A current view is that interferons α and β use a common receptor, but the one for interferon γ is distinct. This view is largely based on biological experiments and is subject to future revision.

Efforts addressed at preparing active fragments of interferons have been mostly fruitless. Synthetic peptides of interferon β were prepared and examined for interferon mimicking activity or blocking activity, without success (unpublished results), although some of the segments induced antibodies to intact β interferon (20). Genetic engineers have expressed truncated versions of interferon, but removal of more than a few residues from either end reduced activity substantially (Wetzel and Petteway, personal communications), as have the products of proteolytic digestion of native interferon. There is a paradox in these results in that interferon α and β activities are resistant to denaturants such as urea, guanidine, or dodecyl sulfate, even at high temperatures. There is, however, instability of interferon in reducing agents (e.g., 2-mercaptoethanol) and there is further biochemical evidence to establish an intrachain disulfide bond which is required for activity. At present it appears that no single linear domain of interferon is sufficient for receptor binding or activity in standard bioassays.

Shortly after interferon binds to a sensitive cell, there is a burst of synthesis of several mRNA species, and new proteins appear in the cytoplasm (21). The level of induction ranges from only a few-fold to more than two orders of magnitude, and after several hours mRNAs for a 15 kDa protein and a 56 kDa protein rise to almost 1% of total poly(A) message (22-24). One or more of the induced proteins are involved in establishment of the antiviral state, and possibly in cell-division inhibition. In human cells, chromosome 21 has a regulatory role, since extra copies of it lead to greatly enhanced sensitivity to interferon. It is presently unclear whether chromosome 21 codes for the interferon receptor, or some key regulatory component leading to enhanced transcription of interferon-sensitive genes.

Several interferon-regulated genes have been identified and isolated as cDNA or genomic clones (25). The response to interferon for at least two of these genes represents a response at the transcriptional level, based on in vitro transcription experiments (24), and induction of the gene activity is responsive to the concentration of interferon applied. The induction occurs within minutes following addition of the interferon to the cell culture.

The interferon-responsive sequences are not yet described. However, genomic clones of oligoadenylate synthase (25) and other interferon-sensitive genes are in hand and it should become clear shortly whether they possess common inducible sequences. Construction of interferon-inducible vectors for mammalian cells should subsequently become practical. There is no evidence yet about alterations in chromatin structure, either at a gross level or at local sites near interferon-induced genes, but one would predict that local perturbations will be detected. With time after removal of interferon, transcription of the induced genes declines and ceases, and the return to a resting state requires protein synthesis (24). It is hoped that detailed study of chromatin alterations in interferon-sensitive regions will shed new light on the current picture of gene activation/deactivation cycles in cell cultures and in animals.

Studies with inhibitors of polyamine biosynthesis show that coincident with reduction of intracellular polyamines, there is potentiation of the biological effects of interferon (26-28) and the potentiation includes the level of transcription of inducible genes (Korant et al., manuscript in preparation). One explanation for the effect is that polyamines (e.g., spermidine or putrescine) supress transcription of interferon-inducible genes by binding to DNA directly. However, it is equally arguable that polyamines alter binding of repressing or activating proteins to DNA, or participate in supressing the signal from the interferon-treated cell surface to the nucleus or the avail-

TABLE 1. Features of interferon-regulated gene expression

Rapid, specific induction of primary transcription.

Initiated from the cell surface by a few molecules of bound interferon.

Potentiated by additional copies of human chromosome 21, and by artificial reduction of intracellular polyamine concentrations.

Several distinct genes are induced.

New mRNAs and proteins are synthesized before/during the establishment of the antiviral state and cessation of cell division.

Several of the proteins reach intracellular levels of greater than 0.2% of total (more than 5×10^5 copies per cell).

The mRNA for a 15 kDa protein and a second for a 56 kDa protein become transiently abundant following induction, then decay as their transcription ceases. Decay of induced mRNAs is slowed by inhibition of protein synthesis.

Cell mutants are isolable which vary their responses to different interferons, and in the induction of specific genes.

---

ability of surface receptors. Some of these questions are directly testable with current methods. One also may inquire whether polyamine fluctuations may alter chromatin structure generally, and affect other regulated gene activities.

SUMMARY

The main points of this brief review on interferon-inducible genes are provided in Table 1.

While interferons continue to show promise as therapeutics, their use, especially in cancer therapy, is complicated by the spectrum of responses encountered. It may be possible in the future to explain (or predict) variability in individual responses to these drugs on the basis of the responses of multiple genes to the interferon signal. For example, at least one gene responsive to α or β interferons is virtually silent when γ interferon is applied (24). Derivatives of cell lines which show a characteristic genetic response to interferon are sometimes altered selectively and display qualitatively distinct response patterns (1), including the expression of oncogenes (29). Therefore, the intelligent medical application of hormone like growth promotors or inhibitors, such as interferons or lymphokines, must be based on a primary understanding of the multiple genetic responses possible in a given cell. The benefits of such studies will be at the same time practical and relevant to a clarified view of control of gene expression.

REFERENCES

1. LEBLEU, B. and CONTENT, J. (1982). Mechanisms of interferon action: biochemical and genetic approaches. In: "Interferon 4," (I. Gresser, ed.) p. 47. Academic Press, New York.

2. MARCUS, P. (1983). Interferon induction by viruses: one molecule of dsRNA as the threshold for induction. In: "Interferon 5," (I. Gresser, ed.) p. 115. Academic Press, New York.

3. RAJ, N. and PITHA, P. (1983). Two levels of regulation of β-interferon gene expression in human cells. Proc. Natl. Acad. Sci. USA 80, 3923.

4. BARON, S. (1984). Overview of progress in interferon research, 1979-1983. In: "Interferon: Research, Clinical Application, and Regulatory Consideration," (K. Zoon, P. Noguchi, and T. Liu, eds.) p. 3. Elsevier, New York.

5. REVEL, M. (1983). Genetic and functional diversity of interferons in man. In: "Interferon 5," (I. Gresser, ed.) p. 206. Academic Press, New York.

6. KNIGHT, E., HUNKAPILLER, M., KORANT, B., HARDY, R., and HOOD, L. (1980). Human fibroblast interferon: amino acid analysis and amino terminal amino acid sequence. Science 207, 525.

7. ZOON, K., SMITH, M., BRIGDEN, P., ANFINSEN, C., HUNKAPILLAR, M., and HOOD, L. (1980). Amino terminal sequence of the major component of human lymphoblastoid interferon. Science 207, 527.

8. DERYNCK, R. (1983). More about interferon cloning. In: "Interferon 5," (I. Gresser, ed.) p. 181. Academic Press, New York.

9. TRENT, J., OLSON, S., and LAWN, R. (1982). Chromosomal localization of human leucocyte, fibroblast, and immune interferon genes by means of in situ hybridization. Proc. Natl. Acad. Sci. USA 79, 7809.

10. ULLRICH, A., GRAY, A., GOEDDEL, D., and DULL, T. (1982). Nucleotide sequence of human chromosome 9 containing a leucocyte interferon gene cluster. J. Mol. Biol. 156, 467.

11. MANTEI, N. and WEISSMAN, C. (1982). Controlled expression of a human interferon α gene introduced into mouse L cells. Nature 297, 128.

12. TRAVERNIER, J., GHEYSEN, D., DUERINCK, F., VAN DER HEYDEN, J., and FIERS, W. (1983). Deletion mapping of the inducible promoter of human interferon β gene. Nature 301, 634.

13. RAGG, H. and WEISSMAN, C. (1983). Not more than 117 base pairs of 5'-flanking sequence are required for inducible expression of a human interferon α gene. Nature 303, 439.

14. WEIDLE, U. and WEISSMAN, C. (1983). The 5'-flanking region of a human interferon α gene mediates viral induction of transcription. Nature 303, 442.

15. AGUET, M. and MOGENSEN, K. (1983). Interferon receptors. In: "Interferon 5," (I. Gresser, ed.) p. 1. Academic Press, New York.

16. JOSHI, A., SARKAR, F., and GUPTA, S. (1982). Crosslinking of human leukocyte interferon α-2 to its receptor on human cells. J. Biol. Chem. 257, 3884.

17. ORCHANSKY, P., RUBENSTERN, M., and SELA, I. (1982). Human interferons protect plants from virus infection. Proc. Natl. Acad. Sci. USA 79, 2278.

18. HAYES, T. (1980). Chou-Fasman analysis of the secondary structure of F and Le interferons. Biochem. Biophys. Res. Commun. 95, 852.

19. DeGRADO, W., WASSERMAN, Z., and CHOWDHRY, V. (1982). Sequence and structural homologies among type I and type II interferons. Nature 300, 379.

20. U.S. patents 4,311,639; 4,341,761; and 4,438,030.

21. KNIGHT, E. and KORANT, B. (1979). Fibroblast interferon induces synthesis of four proteins in human fibroblast cells. Proc. Natl. Acad. Sci. USA 76, 1924.

22. KORANT, B., BLOMSTROM, D., JONAK, G., and KNIGHT, E. (1984). Purification and characterization of a 15,000 molecular weight protein from human and bovine cells, induced by interferon. J. Biol. Chem. 259, 14835.

23. CHEBATH, J., MERLIN, G., METZ, R., BENECK, P., and REVEL, M. (1983). Interferon-induced 56,000 M_r protein and its mRNA in human cells: molecular cloning and partial sequence of the cDNA. Nucleic Acids Res. 11, 1213.

24. LARNER, A., JONAK, G., CHENG, Y., KORANT, B., KNIGHT, E., and DARNELL, J. (1984). Transcriptional induction of two genes by interferon-β in human cells. Proc. Natl. Acad. Sci. USA 81, 6733.

25. MERLIN, G., CHEBATH, J., BENECH, P., METZ, R., and REVEL, M. (1983). Molecular cloning and sequence of partial cDNA for interferon induced oligo A synthetase mRNA from human cells. Proc. Natl. Acad. Sci. USA 80, 4904.

26. LEE, E. and SREEVALSAN, T. (1981). Interferon as an inhibitor of polyamine enzymes. In: "Advances in Polyamine Research," (C. Caldera, ed.) p. 175. Raven Press, New York.

27. SUNKARA, P., PRAKASH, N., MAYER, G., and SJOERDSMA, A. (1983). Tumor suppression with a combination of α-difluoromethyl ornithine and interferon. Science 219, 851.

28. SEKAR, V., ATMAR, V., JOSHI, A., KRIM, M., and KUEHN, G. (1983). Inhibition of ornithine decarboxylase in human fibroblast cells by type I and type II interferons. Biochem. Biophys. Res. Commun. 114, 950.

29. JONAK, G. and KNIGHT, E. (1984). Selective reduction of c-myc mRNA in Daudi cells by human β-interferon. Proc. Natl. Acad. Sci. USA 81, 1747.

Z-DNA AND ITS BINDING PROTEINS

Fernando Azorin and Alexander Rich

Department of Biology
Massachusetts Institute of Technology
Cambridge, Massachusetts 02139

INTRODUCTION

Since the recent discovery of the left-handed Z conformation of DNA, many efforts have been directed toward uncovering
its biological relevance. The Z-DNA conformation was first
described in an atomic resolution X-ray diffraction crystal
structure of the hexanucleotide d(CpGpCpGpCpG) (1). As shown in
Figure 1, there are important conformational differences between
Z-DNA and B-DNA:.

1. In Z-DNA the double helix is left-handed, while B-DNA
is a right-handed double helix.

2. Z-DNA has only one groove instead of the two grooves
present in B-DNA. The position corresponding to the major
groove in B-DNA forms the outer convex surface in Z-DNA. The
bases have an external position in the Z conformation, while in
the B conformation they are more in the center of the molecule.
Thus, the bases are more accessible to the solvent in Z-DNA.
This increased accessibility of the bases in the Z conformation
might be an important factor in the specific recognition of Z-
DNA sequences by proteins.

3. The sugar-phosphate backbone has a characteristic zig-
zag organization in Z-DNA in contrast with the smooth helical
organization shown in the right-handed B-conformation.

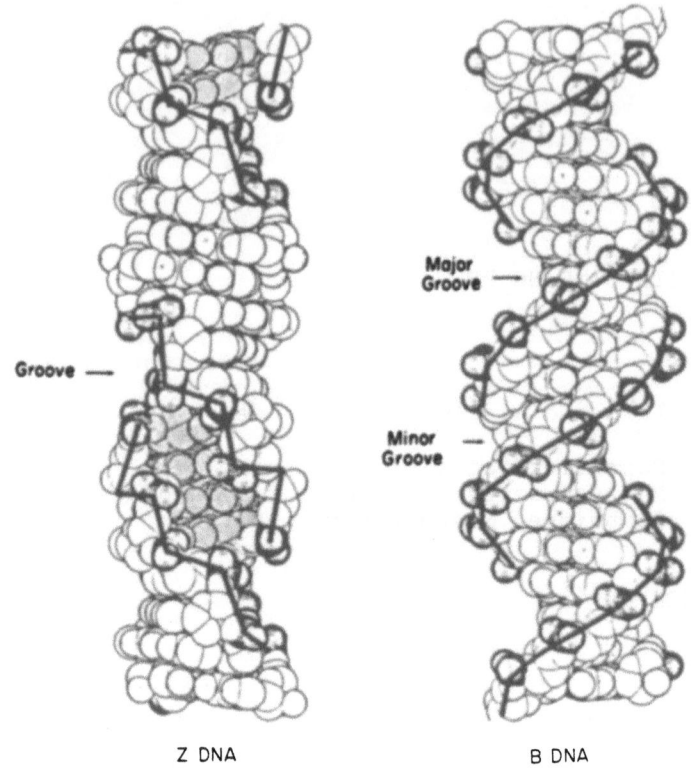

Figure 1. Van der Waals models of Z-DNA and B-DNA. The Z-DNA diagram shows the molecule as it appears in the d(CpGpCpGpCpG) crystal.

All of these structural differences are the result of internal conformational changes. Simple rotation of the double helix in the opposite sense is not sufficient to convert the right-handed B conformation into the left-handed Z conformation. In addition, the DNA molecule has to undergo several internal conformational changes that include a flipping over of the bases, so that the upper surface of the bases in the B conformation corresponds to the lower surface in the Z conformation and vice versa. This flipping over of the bases is associated with rotation of the guanine residues around the glycosydic bond so that they adopt the <u>syn</u> conformation instead of the more common <u>anti</u> conformation present in the B form. Z-DNA is characterized by the alternation of the <u>syn</u> and <u>anti</u> conformation of the bases along the polynucleotide chain. Every other nucleotide in Z-DNA adopts the <u>syn</u> conformation, while in B-DNA all residues adopt the <u>anti</u> conformation.

Purines can adopt the <u>syn</u> conformation more easily than can pyrimidines (2). Thus, adoption of the left-handed Z-conformation is favored in sequences that have alternations of purine and pyrimidine residues. Formation of Z-DNA has been observed in poly(dG-dC) (3-6) and in poly(dC-dA)/poly(dG-dT) (7-9) sequences. It has never been detected for poly(dA-dT) sequences, even though it has been shown that Z-DNA can be crystallized in sequences containing AT base pairs, such as in the hexanucleotide d(CpGpTpApCpG) (10). The crystal structure of this hexanucleotide shows significant variation in the ordering of the water molecules in the deep groove. Around the TA base pairs they are disordered in contrast to the CG base pairs, where the water molecules are well-ordered. Maintenance of an ordered interaction with solvent molecules is probably an important factor in the stabilization of the Z-DNA conformation.

Sequences that are not strictly alternating purine-pyrimidine can also adopt the left-handed Z conformation. For example, the hexanucleotide d(CpGpApTpCpGp) crystallizes in the Z conformation when the C5 position of the cytosine residues is either methylated or brominated (11). Note that the two internal nucleotides (AT) are out of purine-pyrimidine alteration in relation to the CG nucleotides of either end of the molecule. Thus, although one third of its nucleotides are out of purine-pyrimidine alternation, this oligomer can still form left handed Z-DNA. The energetic cost of forming Z-DNA in such sequences is greater than that of forming Z-DNA in a strictly alternating sequence. Stabilization of the Z conformation in this case requires either methylation or bromination of the cytosines, modifications that are known to facilitate the B to Z transition (4, 12-14), while the strictly alternating d(CpGpCpGpCpGp) requires no modifications to crystallize as Z-DNA.

Z-DNA STABILIZATION BY IONS AND BY DNA MODIFICATION

Right-handed B-DNA and left-handed Z-DNA form an equilibrium system, with left-handed Z-DNA being the conformation of higher free energy. Several factors are known to affect the B to Z equilibrium. For example, Z-DNA can be stabilized by ions, chemical modification of the bases, negative supercoiling of DNA, or interaction with specific proteins.

It is known that high concentrations of monovalent cations induce the transition to the Z conformation in poly(dG-dC) (3). The midpoint of the B-Z transition occurs at approximately 2.5 - 2.7 M NaCl. Divalent cations are more effective in inducing the B to Z transition; 700 mM $MgCl_2$ is enough to stabilize the Z conformation in poly(dG-dC) (4). Of all the cations studied so far, cobalt hexamine shows the strongest effect upon the stabilization of Z-DNA in poly(dG-dC); the transition occurs at a concentration of approximately 20 μM (4). X-ray crystallographic studies have shown that cobalt hexamine binds in an ordered fashion to the left-handed form of poly(dG-dC). The ion interacts specifically through hydrogen bonding with the guanine residue on one strand and a neighboring phosphate group (15). This specific interaction results in considerable stabilization of the left-handed Z conformation. In contrast, sodium ions and monovalent cations in general show a somewhat territorial binding to Z-DNA instead of the specific binding shown by cobalt hexamine.

One of the most common modifications in eukaryotic DNA is methylation at the C5 position of cytosine in CG dinucleotides. Behe and Felsenfeld (4) showed that methylation at this position strongly facilitates the B to Z transition in poly(dG-dC). The methylated version of poly(dG-dC) undergoes the transition at approximately 0.7 M NaCl, instead of 2.5 - 2.7 M for the unmethylated poly(dG-dC). Only millimolar concentrations of divalent cations are necessary to stabilize Z conformation of poly(dG-m^{5}dC). It is interesting to note that the polyamines, spermine and spermidine, have a very strong stabilizing effect upon the formation of Z-DNA in poly(dG-m^{5}dC). This effect is detected at concentrations that are about ten times lower than those measured in vivo (4). The biological relevance of this observation remains unknown. However, CG sequences are highly under-represented in eukaryotic genomes (16), and a considerable amount of evidence suggests that methylation of CG residues is associated with gene inactivation (17).

Other chemical modifications of the bases are known to have strong effects on the B to Z transition (see reference 18 for a review). Of all of them, bromination of either the C5 position of cytosine or the C8 position of guanine shows the strongest effect. When poly(dG-dC) is held in the Z conformation in 4 M NaCl and then brominated, the resulting Br-poly(dG-dC) polymer is stable in the Z conformation at NaCl concentrations as low as 50 mM (19).

EFFECT OF SUPERCOILING ON Z-DNA STABILITY

Neither ions (with the exception of polyamines) nor chemical modification of the bases (with the exception of methylation) are likely to play an important role in Z-DNA stabilization in vivo. Other factors (supercoiling and Z-DNA binding proteins) seem more likely to be physiologically important.

In biological systems DNA is generally found in a topologically constrained state, either in a circular form (bacterial chromosomes, plasmids, and many viruses) or in topologically closed domains as in eukaryotic chromatin. Supercoiling is generated in DNA whenever the number of turns of the double helix is not equal to the number of helical turns the molecule would have in a linear or relaxed form. The degree of supercoiling can be expressed in terms either of the number of supercoils or the superhelical density, which is the number of supercoils per turn of the double helix. In prokaryotes, DNA is maintained in a negatively supercoiled state. The actual degree of supercoiling is controlled very precisely by a complex group of enzymes, the DNA topoisomerases. Most of the DNA in eukaryotic chromatin appears to be in a relaxed state, since nucleosomes take up most of the superhelical tension. However, DNA topoisomerases have been isolated from eukaryotic systems, indicating that eukaryotic DNA must be under topological stress either transiently or locally.

Negatively supercoiled DNA has a higher free energy than relaxed DNA. The free energy difference is proportional to the square of the number of negative supercoils or superhelical

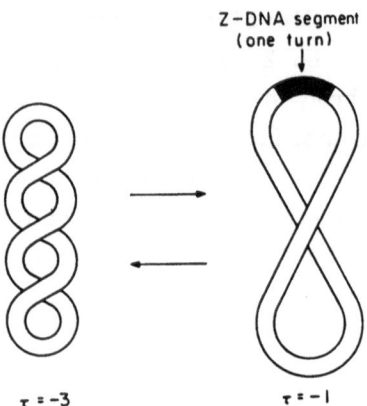

Figure 2. A schematic diagram illustrating the effect on super-coiling of forming a helical turn of Z-DNA in a negatively supercoiled closed circular DNA. In the diagram, a 12-bp segment of B-DNA has converted to Z-DNA, resulting in the loss of two negative super-coils.

density. With negatively supercoiled DNA, any process is fa-vored that would relieve the torsional strain. Formation of Z-DNA reduces the average negative superhelical density. Con-version of one helical turn of B-DNA into the Z conformation reduces the number of negative superhelical turns by two, as illustrated schematically in Figure 2.

Induction of Z-DNA by increasing negative supercoiling can be monitored by a number of different methods. Formation of Z-DNA results in partial relaxation of a plasmid. This reduction in the average superhelical density associated with the forma-tion of Z-DNA can be visualized by one or two-dimensional gel electrophoresis (6, 9, 10). Using this technique, formation of Z-DNA has been detected in plasmids containing inserts of either $d(CG)_n$ (6, 20) or $d(CA/GT)_n$ (9).

Another way to detect formation of Z-DNA in a supercoiled molecule is through the use of anti-Z-DNA specific antibodies (8, 26, 27). In contrast to B-DNA, Z-DNA is highly immunogenic, and both polyclonal and monoclonal antibodies against the left-handed Z conformation have been obtained by several groups of workers (21-25). Figure 3A illustrates how anti-Z-DNA anti-bodies can be used to detect formation of Z-DNA in negatively

194

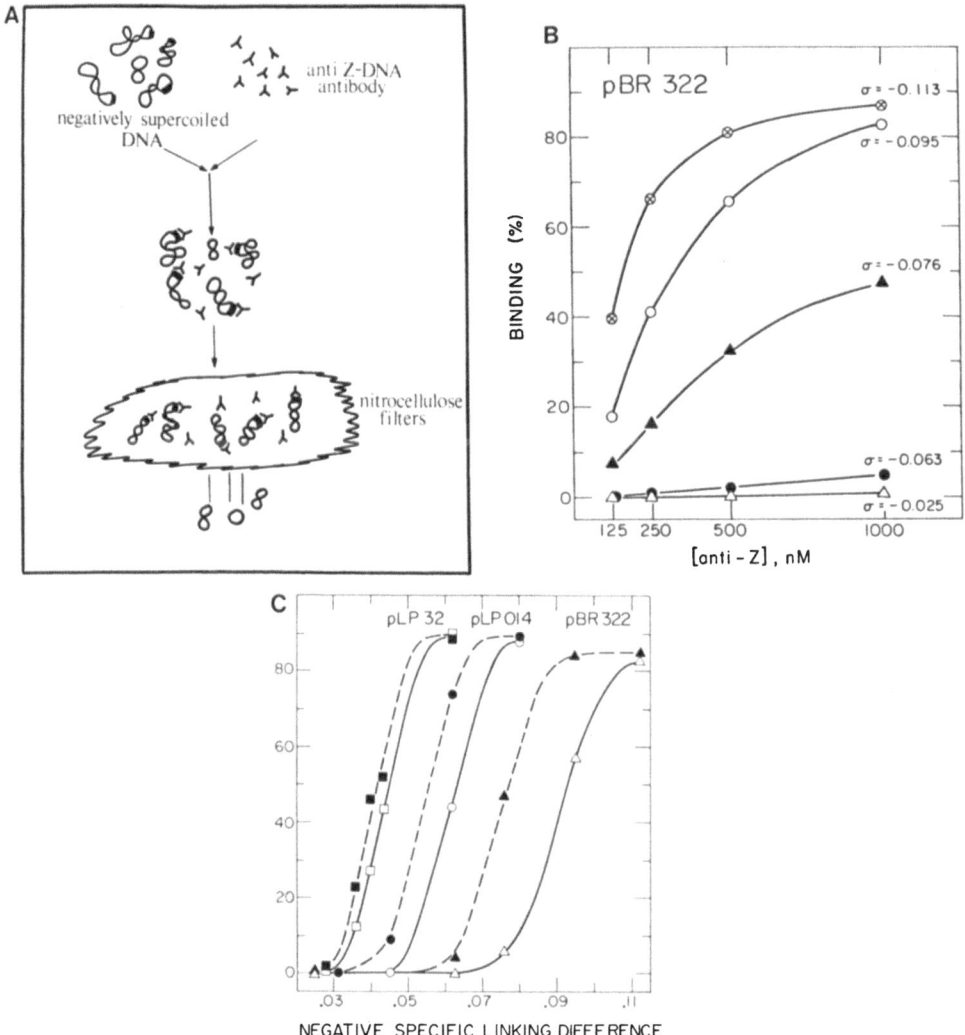

Figure 3. (A) Schematic diagram illustrating the use of anti-Z-DNA antibodies to detect Z-DNA (▬) in negatively supercoiled plasmids. (B) The binding of negatively supercoiled pBR322 to increasing concentrations of anti-Z-DNA antibodies is detected by nitrocellulose filter binding. σ indicates the superhelical density of each one of the samples used. (C) The effect of increasing NaCl concentrations upon formation of B-DNA in negatively supercoiled plasmids. Percent retention on nitrocellulose filters is plotted as a function of the negative superhelical density for three different plasmids. pLP32 and pLP014 are pBR322 derivatives that contain an alternating (CG) insert 32 base pairs and 14 base pairs long respectively. Formation of Z-DNA at the (CG) insert in pLP32 and pLP014 is detected at lower negative superhelical densities than formation of Z-DNA in pBR322 itself. The results obtained at 250 mM NaCl are shown by the solid line, while the dotted line corresponds to the retention observed at 150 mM NaCl.

supercoiled DNA molecules. If a negatively supercoiled molecule contains sequences in the Z conformation, they will be specifically recognized by the antibodies, so that an antibody-DNA complex will be formed. Subsequent filtration through nitrocellulose filters will result in retention on the filters of those DNA molecules containing Z-DNA, since nitrocellulose filters are known to retain protein molecules as well as protein-DNA complexes, but not naked double-stranded DNA. The percentage of negatively supercoiled DNA molecules retained on the filters will thus be a measure of the amount of Z-DNA present in the initial sample. Figure 3B shows the effect of adding anti-Z-DNA antibodies to different plasmid (pBR322) samples that differ in the degree of negative supercoiling. The antibodies do not retain relaxed pBR322, but an increasing retention is observed as the negative superhelical density of the plasmid increases (26, 27).

The Z-DNA sequences in pBR322 can be identified by finding the antibody binding sites. A modification of the experiment in Figure 3A provides a way to localize the DNA sequences that are bound to the antibodies. After formation of the antibody-DNA complexes, the antibody molecules are covalently cross-linked to the DNA with glutaraldehyde, and the complexes are then cleaved with restriction endonucleases. Subsequent filtration through nitrocellulose filters results in the retention of only those restriction fragments that contain an antibody molecule covalently attached to them. These fragments can be identified by gel electrophoresis. In this way, the main antibody binding site in pBR322 has been identified (Figure 4) (26, 27). The sequence consists of 14 bp of alternating purine and pyrimidine residues with one base pair out of alternation. Binding of the anti-Z-DNA antibodies to this sequence results in blockage of the three restriction sites shown in Figure 4.

The midpoint of the B to Z transition in pBR322 occurs at a negative superhelical density of approximately 0.09 when the assay is carried out in 250 mM NaCl (Figure 3C). However, as shown by the dotted line in Figure 3C, when incubation is carried out in 150 mM NaCl, Z-DNA formation occurs at much lower superhelical densities (27): the midpoint of the transition is now 0.075. Thus, Z-DNA forms more readily at lower salt concentrations. The inhibitory effect that NaCl shows on the B to Z transition in negatively supercoiled plasmids (27, 28) occurs in the concentration range 10 mM - 300 mM. This contrasts with the stabilizing effect seen at higher NaCl concentrations (3). The inhibitory effect is restricted to closed circular DNA molecules, since increasing NaCl concentrations in this range does not show any effect on the formation of Z-DNA in linear molecules (27).

In contrast to linear DNA, formation of Z-DNA in a negatively supercoiled molecule involves the stabilization of two B-Z junctions. Formation of a B-Z junction is associated with a relatively unfavorable free energy change (5 kcal/junction) (20, 26, 27). The inhibitory effect described above may be due to salt effects on the formation of B-Z junctions, since at higher salt concentrations, a higher energy is required to stabilize a B-Z junction (27). The dimensions of a B-Z junction have been estimated to be between 4 and 8 base pairs through the use of restriction endonucleases (29).

PROTEINS THAT STABILIZE Z-DNA

Negative supercoiling of the DNA molecule might be an important factor regulating the formation of Z-DNA in vivo. Induction of the Z conformation in naturally occurring DNA sequences occurs at negative superhelical densities that can exist inside the cells (26, 27, 30). However, whatever the

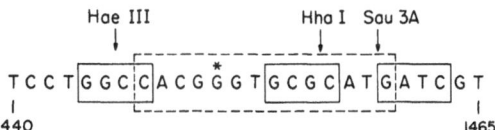

Figure 4.   The nucleotide sequence of plasmid pBR322 from residues 1440 to 1465. The solid boxes correspond to recognition sites for the Hae III, Hha I, and Sau 3A restriction endonucleases, which are blocked by the binding of the anti-Z-DNA antibody to the negatively supercoiled plasmid. Enclosed in the dashed box is a 14 bp sequence of alternating purine and pyrimidine residues with one base pair (G) out of alternation. This box correspond to the proposed binding site of the antibody.

biological role of Z-DNA, it is likely to involve interaction with specific proteins that will stabilize the left-handed Z conformation upon binding. Affinity chromatography has been used to isolate specific Z-DNA binding proteins (31-33). In these experiments, total nuclear proteins are fractionated by means of their relative affinity for Br-poly(dG-dC) (Z-DNA) versus regular B-DNA. This technique has been used successfully to isolate Z-DNA binding proteins from Drosophila cells (31), wheat germ (33) and, in particular, from SV40 minichromosomes (32).

Nordheim and Rich (30) have shown that increasing the negative superhelical density of the SV40 DNA results in the formation of Z-DNA as judged by antibody binding. All of the antibody binding sites are localized within the SV40 control region. In particular, binding of the antibodies results in the blocking of the two Sph I restriction sites localized in the 72 bp repeats of the SV40 enhancer, suggesting that sequences at or around these restriction sites are forming left-handed Z-DNA upon negative supercoiling. The Z-DNA binding proteins isolated from the SV40 minichromosome have been shown to affect the accessibility of the Sph I sites in naked SV40 DNA as well as in the SV40 minichromosomes (32). Although no functional assays have been carried out, the fact that the Z-DNA binding proteins isolated from SV40 minichromosomes interact with the SV40 control region suggests that they may have a regulatory function.

We do not yet have direct evidence for the biological role(s) that Z-DNA might play in vivo. However, further characterization of the Z-DNA binding proteins will undoubtedly help in determining the biological role(s) for Z-DNA.

Z-DNA is only one among several conformations that DNA can adopt. Increasing evidence suggests that the cell takes advantage of this conformational flexibility. Thus, in addition to the sequence information, the DNA molecule also contains conformational information. Different DNA conformations might be used in biological systems as a code for the binding of proteins that control the functions of particular regions of DNA.

ACKNOWLEDGMENTS

This research was supported by grants from the National Institutes of Health, the American Cancer Society, and the Office of Naval Research. F.A. was a fellow of the Charles King Trust, Boston, Massachusettes USA.

REFERENCES

1. WANG, A.H.-J., QUIGLEY, G.J., KOLPAK, F.J., CRAWFORD, J.L., VAN BOOM, J.H., VAN DER MAREL, G., and RICH, A. (1979). Molecular structure of a left-handed double helical DNA fragment at atomic resolution. Nature 282, 680-686.

2. HASCHEMEYER, A.E.V. and RICH, A. (1969). J. Mol. Biol. 27, 369-384.

3. POHL, F.M. and JOVIN, T.M. (1972). Salt-induced co-operative conformational change of a synthetic DNA: equilibrium and kinetics studies with poly(dG-dC). J. Mol. Biol. 67, 375-396.

4. BEHE, M. and FELSENFELD, G. (1981). Effects of methylation on a synthetic polynucleotide: the B-Z transition in poly(dG-m5dC)-poly(dG-m5dC). Proc. Natl. Acad. Sci. USA 78, 4801-4804.

5. SINGLETON, C.K., KLYSIK, J., STIRDIVANT, S.M., and WELLS, R.D. (1982). Left-handed Z-DNA is induced by supercoiling in physiological conditions. Nature 299, 312-316.

6. PECK, L.J., NORDHEIM, A., RICH, A., and WANG, J.C. (1982). Flipping of cleaved d(pC-G)$_n$ DNA sequences from right to left-handed helical structure by salt Co(III) or negative supercoiling. Proc. Natl. Acad. Sci. USA 79, 4560-4564.

7. ARNOTT, S., CHANDRASEKARAN, R., BIRDSALL, D.L., LESLIE, A.G.W., and RATLIFF, R.L. (1980). Left-handed DNA helices. Nature 283, 743-745.

8. NORDHEIM, A. and RICH, A. (1983). The sequence (dC-dA)-(dG-dT) forms left-handed Z-DNA in negatively supercoiled plasmids. Proc. Natl. Acad. Sci. USA 80, 1821-1825.

9. HANIFORD, D.B. and PULLEYBLANK, D.E. (1983). Facile transition of poly[(dTG)-(dCA)] into a left-handed helix in physiological conditions. Nature 302, 632-634.

10. WANG, A.H.-H., HAKOSHIMA, T., VAN DER MAREL, G., VAN BOOM, J.H., and RICH, A. (1984). AT base pairs are less stable than GC base pairs in Z-DNA: the crystal structure of d(m5CGTAm5CG). Cell 37, 321-331.

11. WANG, A.H.-J., GESSNER, R.V., VAN DER MAREL, G., VAN BOOM, J.H., and RICH, A. (1985). Submitted.

12. FUJII, S., WANG, A.H.-J., VAN DER MAREL, G., VAN BOOM, J.H., and RICH, A. (1982). Molecular structure of (m5dC-dG): the role of methyl group on 5-methyl cytosine in stabilizing Z-DNA. Nucleic Acids Res. 10, 7879-7892.

13. KLYSIK, J., STIRDIVANT, S.M., SINGLETON, C.K., ZACHARIAS, W., and WELLS, R.D. (1983). Effects of 5 cytosine methylation on the B-Z transition in DNA fragments and recombinant plasmids. J. Mol. Biol. 168, 51-71.

14. McINTOSH, L.P., GREIGER, J., ECKSTEIN, F., ZARLING, D.A., VAN DE SANDE, J.M., and JOVIN, T.M. (1983). Left-handed helical conformation of poly [d(A-m^{5}C)-d(G-T)]. Nature 304, 83-86.

15. GESSNER, R.V., QUIGLEY, G.J., WANG, A.H.-J., VAN DER MAREL, G., VAN BOOM, J.H., and RICH, A. (1985). Structural bonds for stabilization of Z-DNA by cobalt hexamine and magnesium cations. Biochemistry 24, 237-240.

16. BIRD, A.P. (1980). DNA methylation and the frequency of CpG in animal DNA. Nucleic Acids Res. 8, 1499-1504.

17. DOERFLER, W. (1983). DNA methylation and gene activity. Ann. Rev. Biochem. 52, 93-124.

18. RICH, A., NORDHEIM, A., and WANG, A.H.-J. (1984). The chemistry and biology of left-handed Z-DNA. Ann. Rev. Biochem. 53, 791-846.

19. MOLLER, A., NORDHEIM, A., KOZLOWSKI, S.A., PATEL, D., and RICH, A. (1984). Bromination stabilizes poly(dG-dC) in the Z-DNA form under low salt conditions. Biochemistry 23, 14-62.

20. PECK, L.J. and WANG, A. (1983). Energetics of B-to-Z transition in DNA. Proc. Natl. Acad. Sci. USA 80, 6206-6210.

21. LAFER, E.M., MOLLER, A., NORDHEIM, A., STOLLAR, B.D., and RICH, A. (1981). Antibodies specific for left-handed Z-DNA. Proc. Natl. Acad. USA 78, 3546-3550.

22. MALFOY, B. and LENG, H. (1981). Antiserum to Z-DNA. FEBS Lett. 132, 45-48.

23. ZARLING, D.A., McINTOSH, L.P., ARNDT-JOVIN, D.J., ROBERT-NICOUD, M., and JOVIN, T.M. (1984). Interactions of anti-poly[d(G-br^{5}C)] with synthetic, viral, and cellular Z-DNAs. J. Biomol. Struct. Dynam. 1, 1081-1107.

24. MOLLER, A., GABRIELS, J.E., LAFER, E.M., NORDHEIM, A., RICH, A., and STOLLAR, B.D. (1982). Monoclonal antibodies recognize different parts of Z-DNA. J. Biol. Chem. 257, 12081-12085.

25. THOMAE, R., BECK, S., and POHL, F.M. (1983). Isolation of Z-DNA containing plasmids. Proc. Natl. Acad. Sci. USA 80, 5550-5553.

26. NORDHEIM, A., LAFER, E.M., PECK, L.J., WANG, J., STOLLAR, B.D., and RICH, A. (1982). Negatively supercoiled plasmids contain left-handed Z-DNA segments as detected by antibody binding. Cell 31, 309-318.

27. AZORIN, R., NORDHEIM, A., and RICH, A. (1983). Formation of Z-DNA in negatively supercoiled plasmids is sensitive to small changes in salt concentration within the physiological range. EMBO J. 2, 649-658.

28. SINGLETON, C.K., KLYSIK, J., STIRDIVANT, S.M., and WELLS, R.D. (1982). Left-handed Z-DNA is induced by supercoiling in physiological conditions. Nature 299, 312-316.

29. AZORIN, R., HAHN, R., and RICH, A. (1984). Restriction endonucleases can be used to study B-Z junctions in supercoiled DNA. Proc. Natl. Acad. Sci. USA 81, 5714-5718.

30. NORDHEIM, A. and RICH, A. (1983). Negatively supercoiled simian virus 40 DNA contains Z-DNA segments within transcriptional enhancer sequences. Nature 303, 674-679.

31. NORDHEIM, A., TESSER, P., AZORIN, R., KWON, Y., MOLLER, A., and RICH, A. (1982). Isolation of Drosophila proteins that bind selectively to left-handed Z-DNA. Proc. Natl. Acad. Sci. USA 79, 7729-7733.

32. AZORIN, F. and RICH, A. (1985). Isolation of Z-DNA binding proteins from SV40 minichromosomes: evidence for binding to the viral control region. Cell (in press).

33. LAFER, E.M., SOUSA, R., ROSEN, B., HSU, A., and RICH, A. (1985). Submitted.

BIOCHEMISTRY AND MOLECULAR BIOLOGY OF DNA REPLICATION IN YEAST

Josef Arendes

Institut für Physiologische Chemie
Johannes-Gutenberg-Universität
D-6500 Mainz, West Germany

INTRODUCTION

For the past two decades, the study of the mechanism of DNA replication has been focused mainly on the chromosomes of the simple prokaryotes and their viruses (1). The complexity of the eukaryotic genome and multiple levels of control during the replication of eukaryotic chromosomes have until recently prevented similar studies. In recent years, a lower eukaryote, the yeast Saccharomyces cerevisiae, has become a major focus of efforts in molecular biology. In this chapter, I will briefly review accomplishments in this area. Yeast is an ideal model system for studies on the structure and replication of the eukaryotic chromosome. Yeast cells are easy to grow and study biochemically. Genetic analysis of S. cerevisiae has reached a more advanced stage of sophistication than in other eukaryotic systems. The availability in yeast of defined mutants defective in progression through the cell division cycle is a particular advantage for studying detailed replication mechanism (for comprehensive treatments see references 2 and 3). Application of recombinant DNA methodology and yeast DNA transformation techniques have provided important new tools for analyzing DNA structure and function in yeast cells (for reviews see references 4 and 5).

CHROMOSOMAL DNA AND CHROMATIN STRUCTURE

The haploid genome of $\underline{S}$. $\underline{cerevisiae}$ consists of 1.4×10^7 bp or 9×10^9 daltons (6), which is only fourfold the complexity of the $\underline{E}$. $\underline{coli}$ genome. Yeast DNA is distributed in 17 chromosomes. Size measurements of yeast chromosomal DNA suggest that each chromosome contains a single DNA molecule. The range in sizes of these molecules is from 150 kb to about 2500 kb (6-8).

In yeast, chromosomal elements such as centromeres and telomeres, which play essential roles in the stable inheritance of eukaryotic chromosomes, have been identified and their function can be studied (reviewed in reference 9). Functional centromere DNA segments (CEN) have been isolated from yeast genomic libraries and have been cloned (10, 11). These functional centromeres cause replicating molecules to segregate properly through mitosis and meiosis. Telomeric sequences, which occur at the ends of most yeast chromosomes, have also been cloned on a linear DNA vector (9, 12). Studies of the function of the telomeres in yeast suggest that they provide stable and fully replicatable ends of the chromosomes (13, 14).

Certain chromosomal yeast DNA sequences allow hybrid molecules to replicate autonomously in yeast cells and, consequently, to transform yeast at high frequency (15-17). The ability of these autonomously replicating sequences (ARS) to permit replication of hybrid molecules in yeast strongly suggests that they serve as chromosomal origins (5). About 450 ARS sequences are found in the yeast genome and they occur about once in 30-40 kb of chromosomal DNA (18, 19). This number agrees well with the number of replication origins in chromosomal DNA as determined by electron microscopy (20). In addition to yeast DNA sequences, DNA fragments from a wide variety of eukaryotes promote autonomous plasmid replication in yeast (21, 22). These data suggest that possible replication origins from various eukaryotic DNAs are recognized by the yeast replication apparatus and that initiation sequences may be similar or identical in all eukaryotes. However, similarities between different ARS elements of yeast are limited to a small consensus sequence within an extremely AT-rich region (23).

While ARS-containing plasmids replicate efficiently, they do not segregate properly. By inserting yeast centromere sequences, these chimeric plasmids were mitotically and meiotically stabilized (10). Thus, the ARS-CEN plasmids behave in yeast as stable minichromosomes (11, 24).

Until recently, it was believed that DNA methylation is an universal obligatory function in all eukaryotic cells. However, yeast lacks any detectable methylated bases in the nuclear DNA (25, 26). Absence of methylated bases has also been shown for Drosophila melanogaster DNA (27). Thus, DNA methylation does not seem to be a general feature of all eukaryotic cells.

The organization of yeast chromatin is similar to that of higher eukaryotes. Yeast chromatin contains the four core histones H2a, H2b, H3, and H4 (28-31). It is uncertain whether yeast has a histone that is comparable to H1 of higher eukaryotes. Several proteins with similar electrophoretic properties have been isolated, but they differ in amino acid composition from histone H1 (30, 31). HMG-like proteins have been isolated from yeast chromatin (31, 22).

The DNA in yeast chromatin is in typical nucleosomal subunits, which consist of an octamer of two each of the core histones and about 140 bp of DNA wrapped around this core (33-35). However, the spacer between adjacent nucleosome cores is fairly uniform at about 20 bp, which is less than that observed for higher eukaryotes (33-35). Yeast chromosomes do not condense during the mitotic cell cycle, which might be due to the lack of histone H1 (35). Centrifugation studies with lysed yeast cells revealed that the yeast chromatin can form higher order structures, which contain domains of negative supercoils (36, 37).

CONTROL OF DNA REPLICATION IN YEAST

Yeast DNA replication occupies a restricted portion of the cell cycle. As in other eukaryotes, the yeast cell cycle is conventionally divided into: G_1 phase, which precedes the initiation of chromosomal DNA replication; S phase, during which chromosomal DNA is replicated; a subsequent G_2 phase and an M phase, during which mitosis and nuclear division occur (reviewed in reference 38). Late in G_1 phase, the spindle-pole body becomes duplicated and the initiation of chromosome replication occurs. With this entry into S phase, a small bud emerges at the cell surface. The duration of S phase under optimal growth conditions (1.5 - 2 h doubling time) is about 25% of the cell cycle (39). In cells growing on a poor nitrogen source, S phase is considerably longer (40). During DNA synthesis, the bud grows continuously. At the end of S phase, a complete spindle is formed. Somewhat later, the nucleus migrates to the neck connecting mother cell and bud, and nuclear division occurs. Cytokinesis and complete cell separation follow soon after completion of nuclear division (38).

Yeast chromosomes initiate DNA replication from multiple replication origins (presumably the ARS sequences). Multiple origins were demonstrated by electron microscopy (41, 42) as well as DNA fiber autoradiography (43, 44). From the initiation sites, DNA replication proceeds bidirectionally (43, 44). Inter-origin distances were estimated to be about 10-40 nm, by electron microscopy (41, 42), and about 15-60 nm, by DNA fiber autoradiography (42, 43). The average distance between origins, as derived from these studies, is 60-100 kb. For the small chromosomal DNAs isolated from synchronized S phase cells, a value of 36 kb was determined (20). Most adjacent replication origins are activated at the beginning of the S phase (20, 43) but some origins appear to be activated throughout the S phase (44). The rate of replication fork movement at 24°C was estimated to be 2.1 kb/min in a diploid strain (43) and 6.3 kb/min at 30°C in a haploid strain (44).

The ordered sequence of steps during the cell cycle depends on the expression of specific genes. For the analysis of the cell cycle and studies of DNA replication, cycle-specific mutants are very profitable. Many temperature-sensitive mutants that arrest at characteristic points in the cell cycle have been isolated from S. cerevisiae (reviewed in references 38, 45, 46). These mutants have led to identification of over 50 cell division cycle (CDC) genes (38). The gene products of some CDC mutants are involved in the control of DNA replication, because these mutants either fail to enter S phase or stop DNA synthesis. Figure 1 shows the sequence of CDC genes which act sequentially in control of DNA synthesis. The products of CDC 28, CDC 4, and CDC 7 are required, in this order, prior to the initiation of DNA synthesis, whereas the products of CDC 8 and CDC 21 are required continuously during S phase (47, 48). Mutations in CDC 28 and several additional genes, which were later identified (49, 50), arrest the cell at the end of G_1 phase. At the same stage of the cycle, yeast cells are arrested by the mating pheromones a (51) and α (47). It was suggested that at this stage of the cell cycle a unique control point exists at which the cell probes the environment for all essential factors before the normal cellular program proceeds. The control point has been termed "start" and its completion is a prerequisite for the initiation of DNA synthesis (38, 52).

The identity of CDC gene products involved in DNA replication has been established for some genes. The product of CDC 21 was shown to be thymidylate synthetase (53, 54) and the CDC 9 gene codes for a DNA ligase (55). Recently, the CDC 8 protein has been purified using in vitro replication systems (56-58). The purified protein binds to single-stranded DNA and stimulates yeast DNA polymerase I. The gene has been cloned recently (59, 60). Using a different approach, it has now been shown that the

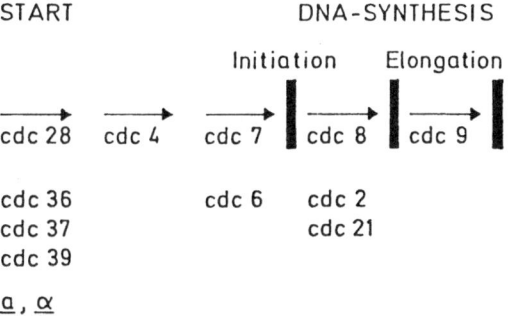

Figure 1. Control of DNA synthesis by CDC genes.

herpes simplex virus thymidine kinase gene complements the CDC 8 defect (61). Genetic and biochemical data were presented that demonstrate that the CDC 8 gene encodes thymidylate kinase (61, 62). How this enzyme might act as part of a multiprotein complex involved in DNA replication remains to be studied.

By analogy with prokaryotic systems, it is expected that many proteins are involved in DNA replication, and more genes with roles in DNA synthesis need to be identified. New selection methods for isolating replication mutants have been used and the mutants are now under study (58, 63).

IN VITRO DNA REPLICATION

Much of the basic information about chromosomal replication in prokaryotes was obtained by studying in vitro replication systems (1). Similarly, most of our understanding of eukaryotic DNA replication mechanisms is derived from studies on viral and plasmid model systems (64). The development of a well-defined in vitro replication system requires a thoroughly characterized template DNA, whose in vitro synthesis can be followed and mimics in vivo synthesis. In yeast, the 2 μm plasmid meets these requirements.

Most laboratory strains of S. cerevisiul harbor a class of double-stranded, closed circular DNA molecules with a contour length of 2 μm (reviewed in references 65-67). The plasmid consists of 6,318 bp and is present at a level of 50-100 copies per cell. The complete nucleotide sequence of this 2 μm DNA has been determined (68). The plasmid contains two identical inverted repeats of 599 bp, which divide the DNA into two unique regions of approximately equal lengths. Plasmid molecules having both orientations of these unique regions with respect to each other (forms A and B) are present in equal numbers in the cell (65-67). The 2 μm DNA can be isolated from yeast cells in supercoiled form, but intracellularly it is associated with histones and occurs in nucleosomes (69, 70). The plasmid appears to be associated with the cell nucleus and it interacts with the folded chromosome (71).

The 2 μm DNA replicates as a theta form from one origin. The major and a minor origins of replication have been mapped (66, 72). In vivo replication of the plasmid is under the same control as that of nuclear DNA, because the plasmid replication requires the products of CDC 28, 4, and 7 (73, 74). The 2 μm plasmid replicates only once early in S phase (75). Thus, the plasmid is particularly useful for the study of mechanisms of chromosomal DNA replication and their control.

Cellular extracts from exponentially growing S. cerevisiae cells have been prepared, which are capable of replicating exogenously added 2 μm DNA (76-78). In such in vitro replication systems, DNA synthesis is initiated at the in vivo origin and proceeds bidirectionally as it does in vivo. Extracts prepared from CDC mutants are defective for 2 μm DNA replication in vitro at restrictive temperature, suggesting that the in vitro system mimics replication in vivo (76-78). The 2 μm DNA replication is sensitive to aphidicolin (77, 78), a specific inhibitor of eukaryotic DNA polymerase α (79), and was shown to be mediated by yeast DNA polymerase I (80). In the in vitro system, plasmids which contain ARS sequences are also replicated, and DNA replication is initiated at the specific origin (58, 78). Using the in vitro system, it was possible to purify the CDC 8 gene product by complementation assay (56, 57).

Replicating activity was isolated in a high molecular weight form (M_r = 2 x 10^6) from yeast extracts (81). When such fractions, containing the putative replication complex, were incubated with DNA, protein knobs were found to be associated with the 2 μm DNA origins (81) as well as with ARS sequences (82). In the high-molecular-weight replicative fraction, several proteins have been identified that are possibly involved in DNA replication. These include DNA polymerase I, DNA ligase, DNA primase, CDC 8 protein, and DNA topoisomerase II (83). The crude extract has been fractionated into several components and in vitro reconstitution of the activity has been achieved (58). During the fractionation, a single-stranded DNA-binding protein, DNA polymerase I, DNA primase, and DNA topoisomerase II have been identified and purified (58). A "preinitiation complex" has been isolated and characterized, which, in addition to the above mentioned components also contained RNase H activity (58).

All the results obtained with these _in vitro_ replication systems provide support for the idea that a multiprotein complex (or replisome) is involved in the replication of both the extrachromosomal 2 μm DNA and chromosomal DNA in yeast.

REPLICATION PROTEINS

Two antigenically distinct nonmitochondrial DNA polymerases have been detected in S. cerevisiae (84-86). DNA polymerase I is the major enzyme. It represents more than 90% of total DNA polymerase activity in the cell and has biochemical properties similar to those of eukaryotic polymerases α. DNA polymerase II, which constitutes less than 10% of the total activity, resembles prokaryotic polymerases in that it has an associated proofreading 3'-exonuclease (86, 87). Yeast polymerases are high molecular weight proteins, and an analog of DNA polymerase β of higher eukaryotes is not present in yeast (88).

The polypeptide structure of DNA polymerase I has been unclear. In early preparations, polypeptides of 70,000 daltons were predominant. However, in recent preparations a high molecular weight form of DNA polymerase I (M_r = 140,000) has always been detected together with additional polypeptides (57, 89). It was proposed that the low molecular weight forms are due to extensive proteolysis during purification. Recently, it has been demonstrated that the DNA polymerase I activity is associated with a polypeptide of 140,000 daltons (90). All fractions contained an additional active form of 110,000 daltons, which was presumably produced by proteolysis. Thus, precautions against proteolysis have to be taken (91). There is evidence that a high molecular weight core enzyme might be conserved during evolution (91). If this is true, it would have important implications for the interaction of DNA polymerase with other replication proteins. At present, nothing is known about the polypeptide structure of DNA polymerase II, because the low level of this protein in the cell has not permitted similar analysis (90).

Both DNA polymerases of yeast were shown to be equally sensitive to inhibition by aphidicolin (92), a specific inhibitor of eukaryotic DNA polymerases α (79). However, the high molecular weight DNA polymerase I was more sensitive to aphidicolin than the enzyme used in the previous study (90). Because aphidicolin also inhibits yeast DNA synthesis in vivo and in cellular extracts (80), it is likely that the high molecular weight enzyme is closely related in structure to the enzyme in the cell. An aphidicolin-resistant DNA polymerase I mutant has been isolated and it was shown that DNA polymerase I appears to be the yeast replicase (80).

None of the DNA polymerases of yeast is capable of initiating DNA chains de novo (86). It has been shown that all three yeast RNA polymerases can synthesize DNA primer which can be used to initiate replication of single-stranded DNA (93, 94). However, by fractionation of an in vitro replication extract, a distinct DNA primase activity has been detected and purified (58). Two additional reports on identification and purification of yeast DNA primase have recently been published (89, 95). The DNA primase is well distinguished from the RNA polymerases of yeast. The enzyme appears to be physically associated with, but not a part of, yeast DNA polymerase I (89, 95). It synthesizes oligoribonucleotides of discrete length (8-12 nucleotides). These are used as primer for DNA synthesis (58, 89, 95). The association of DNA polymerase I with the DNA primase suggests that this complex may function in the synthesis of Okazaki fragments at chromosomal replication forks (89).

A single-stranded DNA binding protein which specifically stimulates DNA polymerase I has been isolated from yeast (96). This protein seems to be similar to the one isolated using a complementation assay in an in vitro replication system (57, 58). The protein increases the processivity of DNA polymerase I, and it is assumed that this protein participates in DNA replication. It presumably destabilizes secondary structures in DNA (57, 96). A somewhat similar stimulatory protein was purified earlier (97), but it was found to bind to both single-stranded and double-stranded DNAs (97).

The ability to stimulate DNA synthesis catalyzed by DNA polymerase I has also been reported recently for a new RNase H activity isolated from yeast (98). This enzyme differs from two previously described RNase H activities from yeast (99). RNase H activity might be involved in the removal of RNA primers from DNA, but the physiological role of the yeast enzymes remains to be determined. DNA-dependent ATPases, which may play a role in unwinding the DNA helix, have also been isolated from yeast (100). However, it has not been shown that these enzymes are involved in DNA replication.

DNA topoisomerases have been subjected to intensive studies in recent years. In S. cerevisiae, as in all other eukaryotic organisms that have been examined, two DNA topoisomerases have been identified (101-104). An ATP-independent DNA topoisomerase I has been demonstrated (101) and partially purified as a 76,000 dalton protein (102). However, in a purification with minimized proteolysis the enzyme was shown to be a monomeric protein of 90,000 daltons (104). The ATP-dependent DNA topoisomerase II has also been purified near to homogeneity and shown to be a single subunit enzyme with a monomer weight of M_r = 150,000 (103, 104). The properties and catalytic characteristics of the yeast topoisomerases are very similar to other eukaryotic topoisomerases, and both enzymes are nuclear proteins (104). It has been suggested that DNA topoisomerases might be involved in initiation of DNA replication and/or in the segregation of daughter molecules at the termination of DNA replication (105). A temperature-sensitive DNA topoisomerase II mutant of S. cerevisiae has been identified (106). It was shown that the topoisomerase II is essential for viability. The mutant is defective in the segregation of daughter chromosomes (106). The structural gene for yeast DNA topoisomerase II has recently been cloned and shown to be a single-copy, essential gene (107). With this clone it is possible to apply new developed methods to mapping the chromosomal position of the gene and to construct mutants for genetic and biological studies. Yeast, therefore, provides an excellent system to define the role of eukaryotic DNA topoisomerases II in DNA replication at the molecular level.

CONCLUSION

The chromosomes of the yeast Saccharomyces cerevisiae are similar in their structure and mechanisms of replication to those of higher eukaryotes. The combination of classical genetics, recombinant DNA methodology, and yeast transformation techniques has allowed identification of functional elements in the genome besides structural genes. Assays have been devised by combining these techniques to study chromosome structure and function in the yeast cell. Thus, yeast is useful for studying many basic questions in eukaryotic molecular biology. Experiments with yeast will continue to increase our understanding of chromosome structure and behaviour. In addition, these studies will be useful in the development of systems to analyze the phenomena in the chromosomes of higher eukaryotic cells which are not seen in yeast.

REFERENCES

1. KORNBERG, A. (1980). "DNA Replication." W.H. Freeman, San Francisco.

2. STRATHERN, J.N., JONES, E.W., and BROACH, J.R., eds. (1981). "The molecular biology of the yeast Saccharomyces. Life cycle and inheritance." Cold Spring Harbor Monograph 11 A, Cold Spring Harbor, New York.

3. STRATHERN, J.N., JONES, E.W., and BROACH, J.R., eds. (1982). "The molecular biology of the yeast Saccharomyces. Metabolism and gene expression." Cold Spring Harbor Monograph 11 B, Cold Spring Harbor, New York.

4. PETES, T.D. (1980). Molecular genetics of yeast. Ann. Rev. Biochem. 49, 845-876.

5. STRUHL, K. (1983). The new yeast genetics. Nature 305, 391-397.

6. LAUER, G.O., ROBERTS, T.J., and KLOTZ, L.C. (1977). Determination of the nuclear DNA content of Saccharomyces cerevisiae and implications for the organization of DNA in yeast chromosomes. J. Mol. Biol. 114, 507-526.

7. PETES, T.D. and FANGMAN, W.O. (1972). Sedimentation properties of yeast chromosomal DNA. Proc. Natl. Acad. Sci. USA 69, 1188-1191.

8. PETES, T.D., NEWLON, C.S., BYERS, B., and FANGMAN, W.L. (1974). Yeast chromosomal DNA: size, structure, and replication. Cold Spring Harbor Symp. Quant. 38, 9-16. Cold Spring Harbor, New York.

9. BLACKBURN, E.H. and SZOSTAK, J.W. (1980). The molecular structure of centromeres and telomeres. Ann. Rev. Biochem. 53, 163-194.

10. CLARKE, L. and CARBON, J. (1980). Isolation of a yeast centromere and construction of functional small circular chromosomes. Nature 287, 504-509.

11. CARBON, J. (1984). Yeast centromeres: structure and function. Cell 37, 351-353.

12. SZOSTAK, J.W. and BLACKBURN, E.H. (1982). Cloning yeast telomeres on linear plasmid vectors. Cell 29, 245-255.

13. SHAMPAY, J., SZOSTAK, J.W., and BLACKBURN, E.H. (1984). DNA sequences of telomeres maintained in yeast. Nature 310, 154-157.

14. WAMSLEY, R.W., CHAN, C.S.M., TYE, B.-K., and PETES, T.D. (1984). Unusual DNA sequences associated with the ends of yeast chromosomes. Nature 310, 157-160.

15. HSIAO, C.L. and CARBON, J. (1979). High-frequency transformation of yeast by plasmids containing the cloned yeast ARG 4 gene. Proc. Natl. Acad. Sci. USA 76, 3829-3833.

16. STINCHCOMB, D.T., STRUHL, K., and DAVIS, R.W. (1979). Isolation and characterization of a yeast chromosomal replicator. Nature 282, 39-43.

17. STRUHL, K., STINCHCOMB, D.T., SCHERER, S., and DAVIS, R.W. (1979). High frequency transformation of yeast: autonomous replication of hybrid DNA molecules. Proc. Natl. Acad. Sci. 76, 1035-1039.

18. CHAN, C.S.M. and TYE, B.-K. (1980). Autonomously replicating sequences in Saccharomyces cerevisiae. Proc. Natl. Acad. Sci. USA 77, 6329-6333.

19. BEACH, D., PIPER, M. and SHALL, S. (1980). Isolation of chromosomal origins of replication in yeast. Nature 284, 185-187.

20. NEWLON, C.S. and BURKE, W. (1980). Replication of small chromosomes in yeast. In: "Mechanistic Studies of DNA Replication and Genetic Recombination," ICN-UCLA Symposia on Molecular and Cellular Biology 19 (B. Alberts and C.C. Fox, eds.) pp. 339-409. Academic Press, New York.

21. STINCHCOMB, D.T., THOMAS, M., KELLY, J., SELKER, E., and DAVIS, R.W. (1980). Eukaryotic DNA segments capable of autonomous replication in yeast. Proc. Natl. Acad. Sci. USA 77, 4559-4563.

22. ZAKIAN, V.A. (1981). Origin of replication from Xenopus laevis mitochondrial DNA promotes high-frequency transformation of yeast. Proc. Natl. Acad. Sci. USA 78, 3128-3132.

23. BROACH, J.R., LI, Y.-Y., FELDMAN, J., JAYARAM, M., ABRAHAM, J., NASMYTH, K.A., and HICKS, J.B. (1983). Localization and sequence analysis of yeast origins of DNA replication. Cold Spring Harbor Symp. Quant. Biol. 47, 1165-1173. Cold Spring Harbor, New York.

24. STINCHCOMB, D.T., MANN, C., and DAVIS, R.W. (1982). Centromeric DNA from Saccharomyces cerevisiae. J. Mol. Biol. 158, 157-179.

25. HATTMAN, S., KENNY, C., BERGER, L., and PRATT, K. (1978). Comparative study of DNA methylation in three unicellular eucaryotes. J. Bacteriol. 135, 1156-1157.

26. PROFFITT, J.H., DAVIE, J.R., SWINTON, D., and HATTMAN, S. (1984). 5-Methylcytosine is not detectable in Saccharomyces cerevisiae DNA. Mol. Cell Biol. 4, 985-988.

27. URIELI-SHOVAL, S., GRUENBAUM, Y., SEDAT, J., and RAZIN, A. (1982). The absence of detectable methylated bases in Drosophila melanogaster. FEBS Lett. 146, 148-152.

28. WINTERSBERGER, U., SMITH, P., and LETNANSKY, K. (1973). Yeast chromatin. Preparation from isolated nuclei, histone composition, and transcription capacity. Eur. J. Biochem. 33, 123-130.

29. BRANDT, W.F. and VON HOLT, C. (1976). The occurrence of histone H3 and H4 in yeast. FEBS Lett. 65, 386-390.

30. THOMAS, J.G. and FURBER, V. (1976). Yeast chromatin structure. FEBS Lett. 66, 274-280.

31. SOMMER, A. (1978). Yeast chromatin: search for histone H1. Mol. Gen. Genet. 161, 323-331.

32. WEBER, S. and ISENBERG, I. (1980). HMG proteins of Saccharomyces cerevisiae. Biochemistry 19, 22236-22240.

33. LOHR, D., CORDEN, J., TATCHELL, K., KOVACIC, R.T., and VAN HOLDE, K.E. (1977). Comparative subunit structure of HeLa, yeast, and chicken erythrocyte chromatin. Proc. Natl. Acad. Sci. USA 74, 79-83.

34. NELSON, D.A., BELTZ, W.R., and RILL, R.L. (1977). Chromatin subunits from baker's yeast: isolation and partial characterization. Proc. Natl. Acad. Sci. USA 74, 1343-1347.

35. FANGMAN, W.L. and ZAKIAN, V.A. (1981). Genome structure and replication. In: "The molecular biology of the yeast Saccharomyces. Life cycle and inheritance," (J.N. Strathern, E.W. Jones, and J.R. Broach, eds.) pp. 27-58. Cold Spring Harbor Monograph 11 A. Cold Spring Harbor, New York.

36. PINON, R. and SALTS, Y. (1977). Isolation of folded chromosomes from the yeast Saccharomyces cerevisiae. Proc. Natl. Acad. Sci. USA 74, 2850-2854.

37. PINON, R. (1979). Folded chromosomes in meiotic yeast cells: analysis of early meiotic events. J. Mol. Biol. 129, 433-437.

38. PRINGLE, J.R. and HARTWELL, L.H. (1981). The Saccharomyces cerevisiae cell cycle. In: "The molecular biology of the yeast Saccharomyces. Life cycle and inheritance," (J.N. Strathern, E.W. Jones, and J.R. Broach, eds.) pp. 97-142. Cold Spring Harbor Monograph 11 A. Cold Spring Harbor, New York.

39. WILLIAMSON, D.H. (1965). The timing of deoxyribonucleic acid synthesis in the cell cycle on Saccharomyces cerevisiae. J. Cell Biol. 25, 511-528.

40. RIVIN, C.J. and FANGMAN, W.L. (1980). Cell cycle phase expansion in nitrogen-limited cultures of Saccharomyces cerevisiae. J. Cell Biol. 85, 96-107.

41. PETES, T.D. and NEWLON, C.S. (1974). Structure of DNA in DNA replication mutants of yeast. Nature 251, 637-639.

42. NEWLON, C.S., PETES, T.D., HEREFORD, L.M., and FANGMAN, W.L. (1974). Replication of yeast chromosomal DNA. Nature 247, 32-35.

43. PETES, T.D. and WILLIAMSON, D.H. (1975). Fiber autoradiography of replicating yeast DNA. Exp. Cell Res. 95, 103-110.

44. RIVIN, C.J. and FANGMAN, W.L. (1980). Replication fork rate and origin activation during S phase of Saccharomyces cerevisiae. J. Cell Biol. 85, 108-115.

45. HARTWELL, L.H. (1974). Saccharomyces cerevisiae cell cycle. Bacteriol. Rev. 38, 164-198.

46. SIMCHEN, G. (1978). Cell cycle mutants. Ann. Rev. Genet. 12, 161-191.

47. HEREFORD, L.M. and HARTWELL, J.H. (1974). Sequential gene function in the initiation of Saccharomyces cerevisiae DNA synthesis. J. Mol. Biol. 84, 445-461.

48. HARTWELL, L.H. (1976). Sequential function of gene products relative to DNA synthesis in the yeast cell cycle. J. Mol. Biol. 104, 803-814.

49. REED, S.I. (1980). The selection of Saccharomyces cerevisiae mutants defective in the start event of cell division. Genetics 95, 561-577.

50. BRETER, H.-J., FERGUSON, J., PETERSON, T.A., and REED, S.I. (1983). Isolation and transcriptional characterization of three genes which function at start, the controlling event of the Saccharomyces cerevisiae cell division cycle: CDC 36, CDC 37, CDC 39. Mol. Cell.Biol. 3, 881-891.

51. WILKINSON, L.E. and PRINGLE, J.R. (1974). Transient G1 arrest of S. cerevisiae cells of mating type α by a factor produced by cells of mating type a. Exp. Cell. Res. 89, 175-188.

52. HARTWELL, L.H., CULOTTI, J., PRINGLE, J.R., and REID, B.J. (1974). Genetic control of the cell division cycle in yeast. Science 183, 46-51.

53. GAME, J.C. (1976). Yeast cell-cycle mutant cdc21 is a temperature-sensitive thymidylate auxotroph. Mol. Gen. Genet. 146, 313-315.

54. BISSON, L. and THORNER, J. (1977). Thymidine 5'-monophosphate-requiring mutants of Saccharomyces cerevisiae are deficient in thymidylate synthetase. J. Bacteriol. 132, 44-50.

55. JOHNSTON, L.H. and NASMYTH, K. (1978). Saccharomyces cerevisiae cell cycle mutant cdc9 is defective in DNA ligase. Nature 274, 891-893.

56. KUO, C.-L. and CAMPBELL, J.L. (1982). Purification of the cdc8 protein of Saccharomyces cerevisiae by complementation in an aphidicolin-sensitive in vitro DNA replication system. Proc. Natl. Acad. Sci. USA 79, 4243-4247.

57. ARENDES, J., KIM, K.C., and SUGINO, A. (1983). Yeast 2-μm plasmid DNA replication in vitro: purification of the CDC8 gene product by complementation assay. Proc. Natl. Acad. Sci. USA 80, 673-677.

58. SUGINO, A., SAKAI, A., WILSON-COLEMAN, F., ARENDES, J., and KIM, K.C. (1983). In vitro reconstitution of yeast 2-μm plasmid DNA replication. In: "Mechanism of DNA replication and recombination," (N.R. Cozzareli, ed.) pp. 527-552. Alan R. Liss, New York.

59. KUO, C.-L. and CAMPBELL, J.L. (1983). Cloning of Saccharomyces cerevisiae DNA replication genes: isolation of the CDC8 gene and two genes that compensate for the cdc8-1 mutation. Mol. Cell.Biol. 3, 1730-1737.

60. BIRKENMEYER, L.G., HILL, J.C., and DUMAS, L.B. (1984). Saccharomyces cerevisiae CDC8 gene and its product. Mol. Cell.Biol. 4, 583-590.

61. SCLAFANI, R.A. and FANGMAN, W.L. (1984). Yeast gene CDC8 encodes thymidylate kinase and is complemented by herpes thymidine kinase gene TK. Proc. Natl. Acad. Sci. USA 81, 5821-5825.

62. JONG, A.Y.S., KUO, C.-L., and CAMPBELL, J.L. (1984). The CDC8 gene of yeast encodes thymidylate kinase. J. Biol. Chem. 259, 11052-11059.

63. KUO, C.-L., HUANG, N.-H., and CAMPBELL, J.L. (1983). Isolation of yeast replication mutants in permeabilized cells. Proc. Natl. Acad. Sci. USA 80, 6455-6459.

64. CHALLBERG, M.D. and KELLY, T.J. (1982). Eukaryotic DNA replication: viral and plasmid model systems. Ann. Rev. Biochem. 51, 901-934.

65. BROACH, J.R. (1981). The yeast plasmid 2 μ circle. In: "The molecular biology of yeast Saccharomyces. Life cycle and inheritance," (J.N. Strathern, E.W. Jones, and J.R. Broach, eds.) pp. 445-470. Cold Spring Harbor Monograph 11 A, Cold Spring Harbor, New York.

66. BROACH, J.R. (1982). The yeast plasmid 2 μ circle. Cell 28, 203-204.

67. GUNGE, N. (1983). Yeast DNA plasmids. Ann. Rev. Microbiol. 37, 253-276.

68. HARTLEY, J.L. and DONELSON, J.G. (1980). Nucleotide sequence of the yeast plasmid. Nature 286, 860-864.

69. LIVINGSTON, D.M. and HAHNE, S. (1979). Isolation of a condensed intracellular form of the 2 μm DNA plasmid of Saccharomyces cerevisiae. Proc. Natl. Acad. Sci. USA 76, 3727-3731.

70. NELSON, R.G. and FANGMAN, W.L. (1979). Nucleosome organization of the yeast 2 μm DNA plasmid. A eucaryotic mini-chromosome. Proc. Natl. Acad. Sci. USA 76, 6515-6519.

71. TAKETO, M., JAZWINSKI, S.M., and EDELMAN, G.M. (1980). Association of the 2 μm DNA plasmid with yeast folded chromosomes. Proc. Natl. Acad. Sci. USA 77, 3144-3148.

72. BROACH, J.R. and HICKS, J.B. (1980). Replication and recombination functions associated with the yeast plasmid 2 μ circle. Cell 21, 501-508.

73. PETES, T.D. and WILLIAMSON, D.H. (1975). Replicating circular DNA molecules in yeast. Cell 4, 249-253.

74. LIVINGSTON, D.M. and KUPFER, D.M. (1977). Control of Saccharomyces cerevisiae 2 μm DNA replication by cell division cycle genes that control nuclear DNA replication. J. Mol. Biol. 116, 249-260.

75. ZAKIAN, V.A., BREWER, B.J., and FANGMAN, W.L. (1979). Replication of each copy of the yeast 2 micron DNA plasmid occurs during the S phase. Cell 17, 923-934.

76. JAZWINSKI, S.M. and EDELMAN, G.M. (1979). Replication in vitro of the 2 μm DNA plasmid of yeast. Proc. Natl. Acad. Sci. USA 76, 1223-1227.

77. KOJO, H., GREENBERG, B.D., and SUGINO, A. (1981). Yeast 2 μm plasmid DNA replication in vitro: origin and direction. Proc. Natl. Acad. Sci. USA 78, 7261-7265.

78. CELNIKER, S.E. and CAMPBELL, J.L. (1982). Yeast DNA replication in vitro: initiation and elongation events mimic in vivo processes. Cell 31, 201-213.

79. HUBERMAN, J.A. (1981). New views of the biochemistry of eucaryotic DNA replication revealed by aphidicolin, an unusual inhibitor of DNA polymerase. Cell 23, 647-648.

80. SUGINO, A., KOJO, H., GREENBERG, B., BROWN, P.O., and KIM, K.C. (1981). In vitro replication of yeast 2 μm plasmid DNA. In: "ICN-UCLA Symposia on Molecular and Cellular Biology," 21 (D.S. Ray and C.F. Fox, eds.) pp. 529-553. Academic Press, New York.

81. JAZWINSKI, S.M. and EDELMAN, G.M. (1982). Protein complexes from active replicative fractions associate in vitro with the replication origins of yeast 2-μm DNA plasmid. Proc. Natl. Acad. Sci. USA 79, 3428-3432.

82. JAZWINSKI, S.M., NIEDZWIECKA, A., and EDELMAN, G.M. (1983). In vitro association of a replication complex with a yeast chromosomal replicator. J. Biol. Chem. 258, 2754-2757.

83. JAZWINSKI, S.M. and EDELMAN, G.M. (1984). Evidence for participation of a multiprotein complex in yeast DNA replication in vitro. J. Biol. Chem. 259, 6852-6857.

84. WINTERSBERGER, U. and WINTERSBERGER, E. (1970). Studies on deoxyribonucleic acid polymerases from yeast. I. Partial purification and properties of two DNA polymerases from mitochondria-free cell extracts. Eur. J. Biochem. 13, 11-19.

85. WINTERSBERGER, U. and WINTERSBERGER, E. (1970). Studies on deoxyribonucleic acid polymerases from yeast. II. Partial purification and characterization of mitochondrial DNA polymerase from wild-type and respiration-deficient yeast cells. Eur. J. Biochem. 13, 20-27.

86. CHANG, L.M.S. (1977). DNA polymerases from baker's yeast. J. Biol. Chem. 252, 1873-1880.

87. WINTERSBERGER, E. (1978). Yeast DNA polymerases: antigenetic relationship, use of RNA primer and associated exonuclease activity. Eur. J. Biochem. 84, 167-172.

88. WINTERSBERGER, U. (1974). Absence of a low molecular weight DNA polymerase from nuclei of yeast Saccharomyces cerevisiae. Eur. J. Biochem. 50, 197-202.

89. SINGH, S. and DUMAS, L.B. (1984). A DNA primase that copurifies with the major DNA polymerase from the yeast Saccharomyces cerevisiae. J. Biol. Chem. 259, 7936-7940.

90. BADARACCO, G., CAPUCCI, L., PLEVANI, P., and CHANG, L.M.S. (1983). Polypeptide structure of DNA polymerase I from Saccharomyces cerevisiae. J. Biol. Chem. 258, 10720-10726.

91. HUBSCHER, U., SPANOS, A., ALBERT, W., GRUMMT, F., and BANDS, G.R. (1981). Evidence that a high molecular weight replicative DNA polymerase is conserved during evolution. Proc. Natl. Acad. Sci. USA 78, 6771-6775.

92. PLEVANI, P., BADARACCO, G., GINELLI, E., and SORA, S. (1980). Effect and mechanism of action of aphidicolin on yeast deoxyribonucleic acid polymerase. Antimicrob. Agents Chemother. 18, 50-57.

93. PLEVANI, P. and CHANG, L.M.S. (1977). Enzymatic initiation of DNA synthesis by yeast RNA polymerases. Proc. Natl. Acad. Sci. USA 74, 1937-1941.

94. PLEVANI, P. and CHANG, L.M.S. (1978). Initiation of enzymatic DNA synthesis by yeast RNA polymerase I. Biochemistry 17, 2530-2536.

95. PLEVANI, P., BADARACCO, G., AUGL, C., and CHANG, L.M.S. (1984). DNA polymerase I and DNA primase complex in yeast. J. Biol. Chem. 259, 7532-7539.

96. LaBONNE, S.G. and DUMAS, L.B. (1983). Isolation of a yeast single-strand deoxyribonucleic acid binding protein that specifically stimulates yeast DNA polymerase I. Biochemistry 22, 3214-3219.

97. CHANG, L.M.S., LURIE, K., and PLEVANI, P. (1978). A stimulatory factor for DNA polymerase. Cold Spring Harbor Symp. Quant. Biol. 43, 587-595.

98. KARWAN, R., BLUTSCH, H., and WINTERSBERGER, U. (1983). Physical association of a DNA polymerase stimulating activity with a ribonuclease H purified from yeast. Biochemistry 22, 5500-5507.

99. WYERS, F., HUET, J., SENTENAC, A., and FROMAGEOT, P. (1976). Role of DNA RNA hybrids in eukaryotes. Characterization of yeast ribonucleases H_1 and H_2. Eur. J. Biochem. 69, 385-395.

100. PLEVANI, P., BADARACCO, G., and CHANG, L.M.S. (1980). Purification and characterization of two forms of DNA dependent ATPase from yeast. J. Biol. Chem. 255, 4957-4963.

101. DURNFORD, J.M. and CHAMPOUX, J.J. (1978). The DNA untwisting enzyme from Saccharomyces cerevisiae. J. Biol. Chem. 253, 1086-1089.

102. BADARACCO, G., PLEVANI, P., RUYECHAN, W.T., and CHANG, L.M.S. (1983). Purification and characterization of yeast topoisomerase I. J. Biol. Chem. 258, 2022-2026.

103. GOTO, T. and WANG, J.C. (1982). Yeast DNA topoisomerase II. An ATP-dependent type II topoisomerase that catalyzes the catenation, decatenation, unknotting, and relaxation of double-stranded DNA rings. J. Biol. Chem. 257, 5866-5872.

104. GOTO, T., LAIPIS, P., and WANG, J.C. (1984). The purification of DNA topoisomerases I and II of the yeast Saccharomyces cerevisiae. J. Biol. Chem. 259, 10422-10429.

105. GELLERT, M. (1981). DNA topoisomerases. Ann. Rev. Biochem. 50, 879-910.

106. DiNARDO, S., VOELKEL, K., and STERNGLANZ, R. (1984). DNA topoisomerase II mutant of Saccharomyces cerevisiae: topoisomerase II is required for segregation of daughter molecules at the termination of DNA replication. Proc. Natl. Acad. Sci. USA 81, 2616-2620.

107. GOTO, T. and WANG, J.C. (1984). Yeast DNA topoisomerase II is encoded by a single-copy, essential gene. Cell 36, 1073-1080.

THE STRUCTURES AND FUNCTIONS OF THE LOW MOLECULAR WEIGHT HMG PROTEINS

G.H. Goodwin, R.H. Nicolas, C.A. Wright, and S. Zavou

Chester Beatty Laboratories
Institute of Cancer Research
Fulham Road
London, SW3 6JB UK

INTRODUCTION

It is a widely held view that during the development of an organism genes that are to be activated in a particular cell type are assembled in a specific chromatin structure prior to the onset of transcription. Considerable effort has been devoted to elucidating the components of such "transcriptionally competent" chromatin. Evidence that active genes have a different chromatin structure from inactive genes has come from nuclease digestion experiments. In such experiments it has been found that all active genes or potentially active genes are very much more sensitive to digestion by DNase I than inactive genes. This sensitivity was primarily attributed to an altered core particle structure of the nucleosomes bound to active genes (1). The discovery that the small HMG proteins[*], HMG-14 and HMG-17, could induce such a conformational change in nucleosomes (2) raised the possibility that the structural alterations that occur in nucleosomes prior to or during transcription could be analysed in detail.

---

[*]The high mobility group (HMG) proteins are operationally defined as being those proteins which are extracted from nuclei with 0.35 M NaCl and which are soluble in 2% trichloroacetic acid or 5% perchloric acid. They have faster electrophoretic mobilities than the bulk of the non-histone proteins in the nucleus, hence their name. For details see reference 1a.

This review sets out to summarize current knowledge on these small HMG proteins, their structure, occurrence, multiplicity, and association with transcribed sequences. The structure and function of the larger HMG proteins, HMG-1 and HMG-2, will not be covered in this chapter since they form a quite distinct class of nuclear proteins (see chapter by Puigdomènech and Jose, this volume).

COMPOSITION AND OCCURRENCE

Table 1 summarizes some of the properties of the low molecular weight HMG proteins. Originally, two small HMG proteins were found in all mammalian and avian cells. These two DNA-binding proteins were termed HMG-14 and HMG-17 (3, 4). Since then, however, it has become apparent they may be two members of a more complex class of proteins. This is exemplified by recent HPLC reverse phase chromatography analyses. Figure 1 shows an elution profile of HMG proteins from rat thymus nuclei and Figure 2 shows the electrophoretic analysis of the fractions. HMG-17 elutes in peak A. HMG-14 elutes in peaks C and D, together with other minor components which could also be HMG variants or modified HMG-14 or HMG-17 proteins. A major component with a mobility slower than HMG-14 elutes in peak F. This

TABLE 1. Properties of low molecular weight HMG proteins

1. Small basic hydrophilic proteins of 9,000-11,000 molecular weight.

2. Constitute approximately 1% of the total chromosomal proteins.

3. Loosely bound within the cell nucleus -- can be extracted with 0.35 M NaCl.

4. Soluble in 5% perchloric acid.

5. Bind to DNA.

6. Widespread occurrence -- found in the nuclei of numerous tissues of higher eukaryotes.

7. Have few hydrophobic and aromatic amino acids, and are random coil polypeptides in solution.

protein has an amino acid composition (see Table 2) similar to
the HMG-I protein described in HeLa cells (5). An unknown
protein, possibly H1^{O}, elutes in fraction G. H1, HMG-1, and
HMG-2 elute at the end of the chromatography. Two very small
proteins elute in fractions B and H and are probably proteins
HMG-18, 19A, or 19B, which were previously isolated from calf
thymus (6) and which resemble H1 and the HMG proteins (Table 3).
Since only very limited sequence data are available for these
proteins, it is not certain whether these proteins should be
included in the HMG-14/HMG-17 class. The amino acid compositions of rat HMG-17, HMG-14, and HMG-I (Table 2) are similar and
show that they form a family of three proteins, with HMG-14
having properties intermediate between HMG-17 and HMG-I. HMG-I
appears to be present only in proliferating cells since it is
found in young rat thymus (4-6 week old animals) and tissue
culture fibroblasts, but not in liver (see Figure 3). In rat
fibroblasts transformed with the avian sarcoma virus, HMG-I is
present together with a slightly slower additional band HMG-I'
(Figures 2 and 3). HMG-I' has not yet been isolated. HeLa
cells blocked in metaphase also have a protein migrating behind
HMG-I called HMG-M (5). HMG-M has the same amino acid composition as HMG-I and has been found to be highly phosphorylated in
metaphase cells. It is probably a highly phosphorylated version

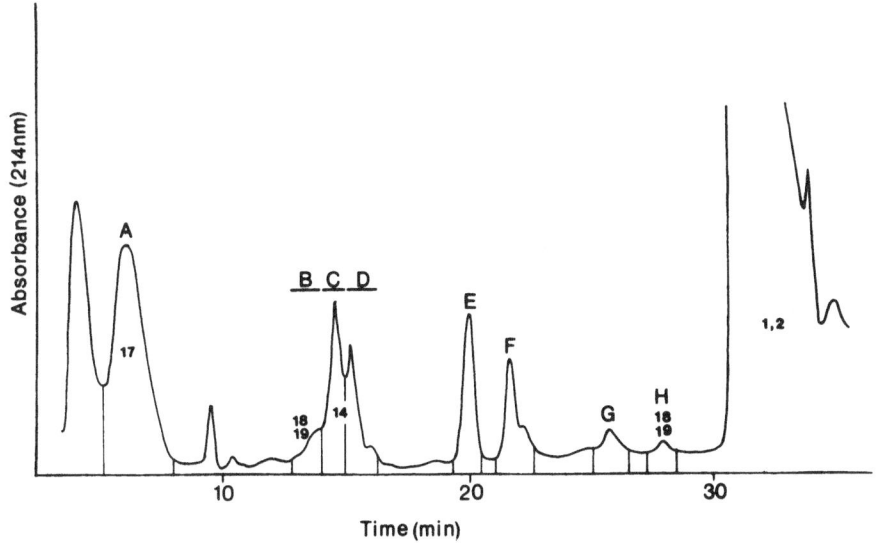

Figure 1.   Reverse phase HPLC   chromatography   of PCA-extracted
proteins from 6-week rat thymus nuclei.

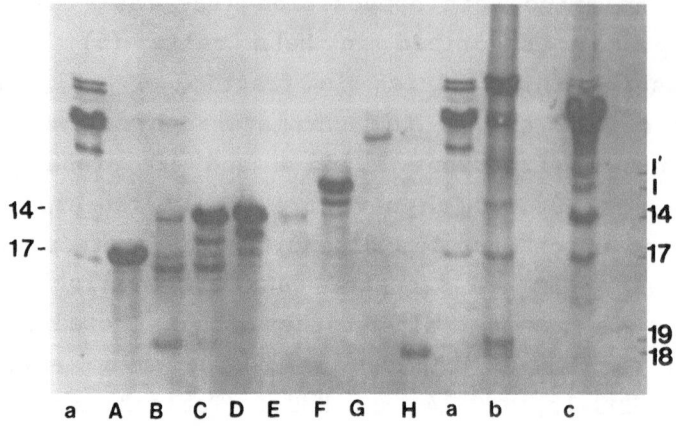

Figure 2. Acid-urea gel electrophoresis of fractions from the HPLC chromatography:

a: rat-thymus nuclei percholoric acid extract
b: calf thymus HMG proteins (calf HMG-14 has a slower mobility than rat HMG-14)
c: PCA extract of rat V1T fibroblast nuclei

A-H: fractions from the HPLC chromatography, the letters referring to the peaks of Figure 1.

of HMG-I, but its relationship to HMG-I' is not yet understood since HMG-I' is not phosphorylated and is present in interphase cells.

Three proteins, termed HMG-14, HMG-17, and HMG-Y, have been isolated from avian cells (Figure 3 and Table 2). Calf and chicken HMG-17 have very similar amino acid compositions, and electrophoretic mobilities and indeed sequence analyses of these two proteins show that the proteins are identical in all but five positions of the 98 residues that make up this protein (7). The avian protein thought to be equivalent to mammalian HMG-14 from its electrophoretic mobility and chromatography does not have such a close resemblance to mammalian HMG-14. The amino acid compositions and sequences (see below), although related, are not highly conserved. Two dimensional electrophoretic analyses of calf and avian HMG-14 reveals that both proteins have four or five subfractions (8, 9), which may relate to some of the bands eluting with HMG-14 on the HPLC chromatography. Chicken oviduct and erythrocytes have a doublet of proteins HMG-Y which migrate ahead of HMG-17 (10). This protein(s) has a composition similar to HMG-14 and HMG-17, but no sequence data

TABLE 2. Amino acid compositions (mol %) and molecular weights of low molecular weight HMG proteins from mammals, birds, and fish.

Amino acid	Rat (thymus)			Chicken (erythrocyte and oviduct)			Trout (liver)		
	HMG-I	HMG-14	HMG-17	HMG-14	HMG-17	HMG-Y	C	D	H6
Asx	4.9	7.8	13.7	9.3	9.1	6.2	6.0	8.0	6.7
Thr	7.1	4.6	1.8	4.6	3.0	5.0	2.6	4.1	1.6
Ser	11.9	8.2	2.8	5.2	4.3	4.8	4.4	4.3	5.6
Glx	17.8	16.4	9.0	15.6	11.7	13.7	23.8	21.9	6.1
Pro	9.0	7.3	10.0	10.5	12.1	10.9	10.9	8.4	12.3
Gly	11.1	8.1	9.3	5.6	10.0	8.2	2.8	3.3	7.4
Ala	5.2	13.2	20.2	18.0	17.2	14.8	16.0	16.6	25.4
Cys	-	-	-	-	-	-	-	-	-
Val	4.7	3.2	2.6	tr	2.2	1.8	4.2	2.5	3.4
Met	-	0.1	-	-	-	-	tr	tr	-
Ile	1.0	1.1	1.1	-	-	0.3	0.6	-	-
Leu	1.6	2.3	-	1.1	1.2	1.8	0.8	1.0	1.2
Tyr	-	-	-	tr	tr	-	tr	-	-
Phe	-	-	-	tr	tr	0.4	tr	-	-
His	0.6	0.2	-	1.1	tr	0.8	tr	0.9	-
Lys	16.5	20.9	25.8	24.0	23.6	26.2	19.6	23.7	23.1
Arg	9.2	7.8	4.1	4.1	4.6	4.2	4.6	4.2	7.2
Mol Wt	-	10,700	9,200	12,800	9,200	-	-	-	7,200

are available. In addition, avian cells have bands migrating in the HMG-I position but these have not been isolated yet.

In fish, the situation is quite different. Although there are clearly three proteins, C, D, and H6 (11, 12), in this class of proteins with similar compositions and sequences to the mammalian and avian proteins (Figure 3 and Table 2), they are not

TABLE 3. Amino acid compositions (mol %) of thymus H1- and HMG-like proteins.

Amino acid	HMG-18	HMG-19A	HMG-19B
Asx	5.0	4.3	9.4
Thr	4.8	2.2	2.5
Ser	3.8	3.8	6.0
Glx	5.2	18.0	18.1
Pro	6.1	5.4	4.2
Gly	8.3	10.5	7.2
Ala	15.6	10.6	10.2
Cys	-	-	-
Val	7.9	1.1	2.0
Met	0.9	1.3	1.8
Ile	-	1.3	1.5
Leu	1.2	5.1	3.9
Tyr	1.3	0.8	0.9
Phe	2.6	2.0	1.7
His	1.2	0.4	0.6
Lys	27.6	29.8	22.4
Arg	8.5	3.4	7.8

sufficiently similar to relate them directly to individual mammalian or avian HMG proteins.

Only four of the above nine HMG proteins have been completely sequenced: calf HMG-17 (13), chicken HMG-17 (7), calf HMG-14 (14), and trout H6 (18). The sequences clearly show similarities throughout the proteins, demonstrating that they belong to the same class of proteins. Partial sequences of some of the other proteins (15) demonstrate that they also belong to this class (Figure 4). For a computer-generated alignment of the low molecular weight HMG proteins see reference 18a.

Although HMG-like proteins have been isolated from <u>Droso-</u><u>phila</u>, plants, and yeast, there are no sequence data to show that HMG-14/HMG-17 equivalents are present in these species, and the amino acid compositions of such proteins do not closely resemble the mammalian, avian, or fish proteins.

POSTSYNTHETIC MODIFICATIONS

The HMG proteins are modified by phosphorylation, acetyla-tion, poly ADP-ribosylation, and glycosylation (see chapter by Laland et al., this volume). Phosphorylation has been exten-sively studied by a number of laboratories and it seems certain that HMG-14 and HMG-I are phosphorylated <u>in</u> <u>vivo</u>. There is some disagreement, however, as to phosphorylation of HMG-17, probably due to misidentification of protein bands or loss of phosphates during protein isolation. Thus, Saffer and Glazer (16) found both HMG-14 and HMG-17 are phosphorylated (mainly at serine residues) in CHO cells with little change though the cell cycle.

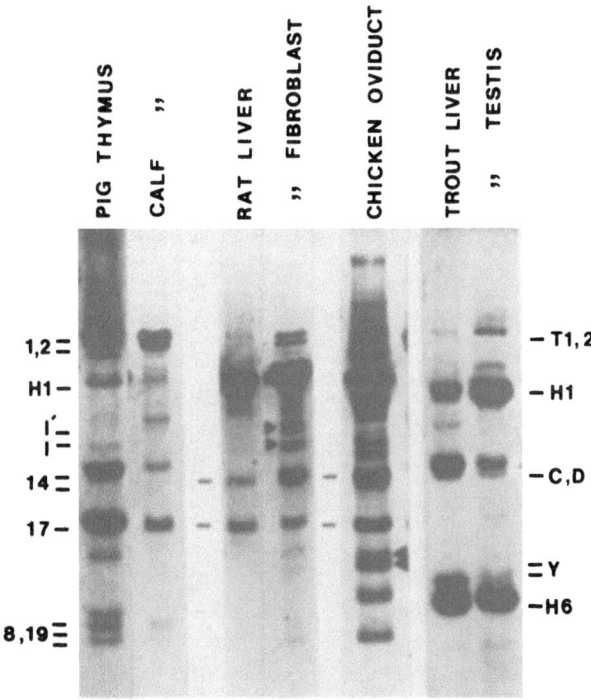

Figure 3. Acid-urea polyacrylamide gel electrophoresis of HMG proteins from mammals, birds, and fish.

```
                                                   10                                                          18
Trout D    Pro-Lys-Arg- o -Lys- o - o -Gln-Gly-Ala-Ala-Gly-Asp- o - o - o - o - o -Val- o - o -Lys-Glu-Glu-Pro-Gln-Arg-Arg-

                                                                         10                                    20
Trout C    Pro-Lys-Arg-Ser-Lys-Ala- o -Asn- o -Ala- o - o -Asp- o - o - o - o - o -Val-Glu-Ala-Ala- o -Glu-Pro-Lys-Arg-Arg-Ser-Glu-

                                               10                                                            20
CTHMG14    Pro-Lys-Arg- o -Lys-Val-Ser-Ser- o -Ala-Glu-Gly- o - o - o - o - o - o -Ala- o -Ala-Lys-Glu-Glu-Pro-Lys-Arg-Arg-Ser-Ala-

                                                   10                                                          20
CEHMG14    Pro-Lys-Arg- o -Lys-Ala-Pro- o - o -Ala-Glu-Gly-Glu- o - o - o - o - o - o -Ala-Lys-Glu-Glu-Pro-Lys-Arg-Arg-Ser-Ala-

                                                         10                                  20
CTHMG17    Pro-Lys-Arg- o -Lys- o - o - o - o -Ala-Glu-Gly-Asp-Ala-Lys-Gly-Asp-Lys-Ala-Lys-Val-Lys-Asp-Glu-Pro-Gln-Arg-Arg-Ser-Ala

                                                                                       10
Trout H6   Pro-Lys-Arg- o -Lys- o -Ser- o - o -Ala- o - o - o - o - o - o - o - o -Thr-Lys-Gly- o -Asp-Glu-Pro-Ala-Arg-Arg-Ser-Ala-


                                 28
Trout C    ( )-Leu-Ser-Asn-Lys-Pro-Ala-Ile-

                                        30                                    40
CTHMG14    Arg-Leu-Ser-Ala-Lys-Pro-Ala-Pro-Ala-Lys-Val-Glu-Thr-Lys-Pro-Lys-Lys-Ala-Ala-Gly-Lys-Asp-Lys-Ser-Ser- o - o -Asp-Lys-Lys-Val-

                                            30                                    40
CEHMG14    Arg-Leu-Ser-Ala-Lys-Pro-Ala-Pro-Pro-Lys-Pro-Glu-Pro-Lys-Pro-Lys-Lys-Ala-Ala-Pro-Pro-Lys-Glu-Lys-Ala-Ala-Asp-Asp-Lys-Lys-Glu-

                       30                                  40                                   50
CTHMG17    Arg-Leu-Ser-Ala-Lys-Pro-Ala-Pro-Pro-Lys-Pro-Glu-Pro-Lys-Pro-Lys-Lys- o -Ala-Pro-Ala-Lys-Lys-Gly-Glu-Lys-Val-Pro-Lys-Gly-Lys-

                   20                                30                                   40
Trout H6   Arg-Leu-Ser-Ala-Arg-Pro-Val-Pro- o -Lys-Pro-Ala-Ala-Lys-Pro-Lys-Lys-Ala-Ala-Ala-Pro-Lys-Lys-Ala-Val-Lys-Gly-Lys-Lys-Ala-Ala-
```

Figure 4. N-terminal sequences of the low molecular weight HMG proteins. Chicken HMG-17 differs from calf HMG-17 in the N-terminal at two positions, Thr 9 and Ser 48.

Cooper and Spaulding (17) and Walton et al. (18) have studied bovine thyroid tissue and found that HMG-14 is phosphorylated by a cAMP- and cGMP-dependent kinase which is stimulated by thyrotropin. Phosphorylation by this kinase occurs at Ser 6. Other HMG-like proteins are also phosphorylated in this tissue but no phosphorylation of HMG-17 was observed. Similarly, no phosphorylation of HMG-17 was seen in HeLa cells (19) but two proteins running in the HMG-14/HMG-I electrophoretic region were phosphorylated in interphase and metaphase cells. In retrospect, these proteins are probably HMG-I proteins. This phosphorylation occurs by a casein kinase II enzyme. Lund et al. (5) have compared the phosphorylation of HMG-17, HMG-14, and HMG-I in interphase and metaphase cells. All three were phosphorylated in metaphase cells together with the protein HMG-M which is probably a highly phosphorylated form of HMG-I. In interphase only HMG-I was phosphorylated. Ser 24 of HMG-17 is probably the site of phosphorylation in metaphase cells (20). Similarly, in CHO cells a protein similar to HMG-I was observed to be phosphorylated (21). Interestingly, the phosphorylated form migrated faster than the unmodified protein on acid gels. To summarize, it would appear that all the low molecular weight HMG proteins can be phosphorylated at metaphase. In interphase cells HMG-I appears to be the major protein phosphorylated, probably by a casein kinase II enzyme. In addition, a cAMP-dependent kinase can phosphorylate HMG-14 in hormone responsive tissues.

The HMG proteins are modified by acetylated lysines at residues 2, 4, and 10 of HMG-17 and residues 2 and 14 of HMG-14 (22). The importance of ADP-ribosylation of HMG proteins in gene activity is exemplified by the experiments by Tanuma et al. (24). When transcription of the integrated mouse mammary tumor virus is induced by glucocorticoids poly ADP-ribosylation of HMG-14 and HMG-17 is reduced suggesting that removal of this modification of HMG-14/HMG-17 is required for elevated transcription. Further proof for this supposition is provided by the observation that an inhibitor of poly ADP-synthetase, 3-aminobenzamide, can also induce MMTV transcription. There is some evidence that glycosylated HMG proteins are associated with the nuclear matrix (35).

THE HMG PROTEINS AND THE STRUCTURE OF ACTIVE GENES

Early work showed that the HMG-14 and HMG-17 proteins are found associated with monomer nucleosomes when nuclei are digested with micrococcal nuclease, and Ring and Cole (25) provided evidence that these proteins are associated with nucleosomes in the intact nucleus by demonstrating that HMG-14 and HMG-17 can be cross-linked to the histones. Thus, the HMG proteins appear to be bona fide nucleosomal proteins and in vitro binding studies have located two possible attachment sites for the HMG proteins on the core particle (29).

The data presented by Weisbrod, Groudine, and Weintraub (2) provided good evidence that HMG-14 and HMG-17 can bind specifically to the nucleosomes of active genes and on doing so cause a change in the conformation of the nucleosomes which could be detected by DNase I digestion experiments. Thus, when salt-depleted chromatin or nucleosomes from chick red blood cells are complexed with HMG-14 or HMG-17 and then digested with DNase I, the level of globin sequences is reduced by a factor of 3-5 when the DNA is assayed by solution hybridizations. Gazet and Cedar (26), using a different approach, labeled active genes in nuclei by "nick-translation" and then DNase I digested the salt-extracted chromatin with and without reconstituted HMG proteins. The labeled DNA was digested faster than bulk DNA in the HMG reconstituted chromatin, confirming that the HMG proteins are responsible for inducing nuclease sensitivity. An affinity column prepared by immobilizing HMG-17 was used by Weisbrod and Weintraub (27) to selectively isolate active nucleosomes, thus affording a method for investigating what factors are responsible for the specific binding of HMG proteins to active genes. Examination of the bound nucleosomes (40) revealed that the core histones were all present. The levels of histone acetylation were somewhat higher than in the unbound material, but not high enough to account for stoichiometric binding of HMG protein to the bound nucleosomes. The DNA was under-methylated, though this is unlikely to account for specific binding of HMG proteins. The conformation of the active nucleosomes was different from that of the inactive nucleosomes, since the two cysteines in the H3 molecules were more readily oxidized to a dimer in the active nucleosomes. Topoisomerase I was also associated with active nucleosomes.

A third line of evidence to support the notion that HMG-14 and HMG-17 might be associated with active nucleosomes was initially provided by micrococcal nuclease experiments in which it was found that, during brief digestions, nucleosomes soluble in 50-100 mM salt are released which are enriched in active sequences and in HMG proteins (28, 29). However, a more detailed analysis of the nucleosomes in such preparations showed that the HMG proteins were not bound specifically to the active nucleosomes in this population of nucleosomes (30). An example of such an analysis is shown in Figure 5. In this experiment, salt-soluble nucleosomes from chick red blood cells were salt-extracted, trimmed to 145 bp with micrococcal nuclease and then HMG-17 reconstituted onto the nucleosomes. The HMG-bound nucleosomes were separated from unbound nucleosomes by electrophoresis, the DNA electrophoresed into a second dimension gel and blotted onto diazophenylthioether paper. This was analysed by hybridization to β^A-globin and ovalbumin probes. Scanning of the blots showed that there is no preferential association of HMG proteins with active globin sequences. Other experiments using electrophoretic methods to fractionate monomer nucleosomes have similarly failed to demonstrate a specific association of HMG-14 or HMG-17 with active monomer nucleosomes (31, 32). The overriding factor that appears to determine what nucleosomes the HMG proteins bind to in vitro is not the type of DNA sequences but simply the nucleosomal DNA length, since the DNA of HMG bound nucleosomes is generally longer than the unbound nucleosomes (29).

Whilst DNase I digestion of nuclei has clearly shown a DNase I-sensitive structure of active genes, several studies have found that isolated monomer nucleosomes lose most of this sensitivity (29, 33, 34). Figure 6 summarizes some data from DNase I digestion experiments carried out on chick red blood cell (RBC) nuclei (Figure 6A) and nucleosomes (Figure 6B). In these experiments hybridizations were carried out using dot-blot hybridizations with radioactive single-strand probes. It can be seen that DNase I digestion of RBC nuclei from 16 day embryos results in the ratio of α^D-globin to ovalbumin sequences being

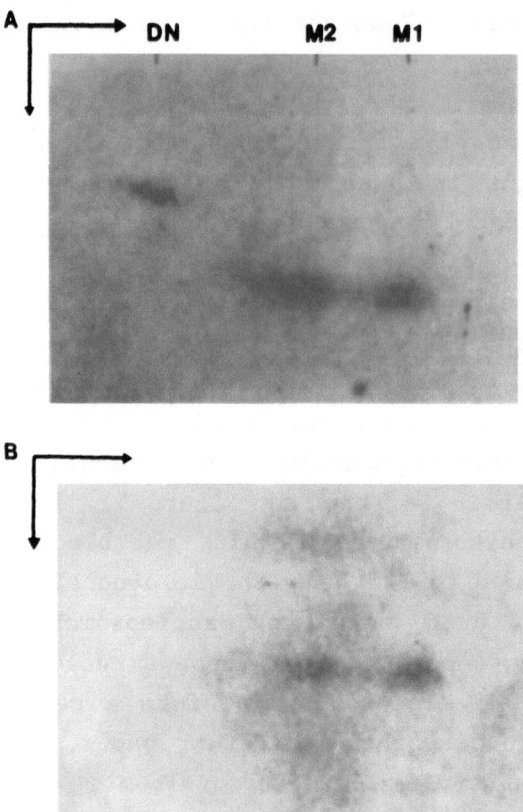

Figure 5. Two-dimensional analysis of nucleosomal core particles reconstituted with HMG-17. Chick erythrocyte
salt-soluble nucleosomes depleted of HMG proteins
were trimmed to 145 base pairs and reconstituted with
HMG-17. The mixture was then resolved on a TBE
electrophoresis gel (1st dimension left to right).
M1 marks the position of uncomplexed core particles,
M2 the position of HMG-core particles. After removal
of protein, the DNA was resolved in a second-
dimension (top to bottom) and then electrophoretically blotted onto DPT-paper. The paper was
analysed by hybridization with (A) the cDNA β globin
plasmid pHB1001 and then (B) with ovalbumin plasmid
(pCROV 2.1).

reduced by a factor of about 12. If the nuclei are first extracted with 0.3 M NaCl at pH 3 (which quantitatively removes
HMG-14, HMG-17, and histone H1), then digested with DNase I,
there is some loss of sensitivity. If RBC nuclei from 6 month
chickens are extracted with 0.34 M NaCl, pH 7, there is no loss
of sensitivity resulting from the removal of HMG proteins
(samples d and e). Salt-soluble nucleosomes, on the other hand
(which have a full complement of HMG-14/HMG-17 proteins), do not

exhibit DNase I-sensitivity. These results suggest that the major structural feature which distinguishes active and inactive sequences in DNase I digestion experiments lies, not in the nucleosomal structure, but in the manner in which active chromatin is assembled into higher order structures in the cell nucleus. What is not clear is whether the HMG proteins contribute at all to the total DNase I-sensitivity of active genes. Senear and Palmiter (34) have argued that the total sensitivity observed is due to additive multiple features. In their experiments, nuclei were extracted with increasing salt concentrations, and then digested with DNase I. At 0.4 M NaCl and above there is partial loss of sensitivity of the ovalbumin gene.

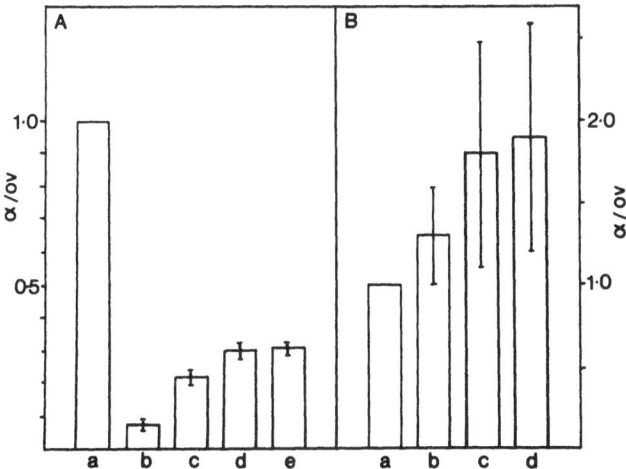

Figure 6. Histogram of dot-blot hybridizations. Ratios of α^D/ovalbumin sequences were obtained from dot-blots probed with single-strand probes. Ratios were normalized so that the α^D/ov ratio for total chicken DNA was 1.0.
A. a: total chicken DNA; b: DNA from DNase I digested 16 day RBC nuclei; c: DNA from DNase I digested 16 day RBC nuclei after 0.3 M NaCl, pH 3, extraction; d: DNA from DNase I digested 6 month RBC nuclei; e: DNA from DNase I digested 6 month RBC nuclei after 0.34 M NaCl, pH 7.4 extraction. Two DNase I digestions experiments were carried out on both 16 day RBC and 6 month RBC and the results are given as a mean of the two.
B. a: total chicken DNA; b: salt-soluble monomer nucleosomal DNA (16 day RBC); c: DNA from DNase I digested salt-soluble monomer nucleosomes (buffer contained Mg^{++}); d: DNA from DNase I digested salt-soluble monomers digested with RSB buffer and Ca^{++}. The samples were analysed at four different loadings and the results are given as a mean of these.

Whether or not this was due to HMG removal was not investigated. The effect of salt extraction on DNase I-sensitivity of monomer nucleosomes and polynucleosomes was also investigated. In monomer nucleosomes and polynucleosomes, the ovalbumin gene was slightly more sensitive to digestion than was the globin gene (1.3-1.5 fold) and this was only minimally affected by salt-extraction. Thus, no clear HMG effect was discernible. Reeves and Chang (35) were also unable to induce DNase I sensitivity of globin genes by the addition of purified HMG proteins.

CONCLUSIONS

From the contradictory data presented in this review, it is apparent that the role of the HMG-14/HMG-17 family of proteins in gene activity is not established and other ideas for their role in chromatin have been entertained. Several studies have considered the possibility that the HMG proteins are associated with satellite sequences. The results have shown that HMG-14 and HMG-17 are either present in depleted quantities (relative to bulk chromatin) or not present at all in satellite chromatin (36, 37), although they are capable of binding to nucleosomes containing these sequences (38). However, recently Strauss and Varshavshy (39) have found that a protein in CVI cells, which is similar to HMG-I, binds preferentially to certain short AT-rich sequences in monkey α-satellite DNA, and they propose that this protein could be responsible for phasing nucleosomes in satellite chromatin. The fact that HMG-I appears only in proliferating cells suggests that it could be required for centromere function during cell division. If this is the case, then it is puzzling why no protein analogous to HMG-I is seen in chick red blood cells transformed by avian erythroblastosis virus (unpublished results). The finding that the HMG proteins are highly phosphorylated at metaphase has led to the suggestion that the proteins could be required for metaphase condensation (5). Thus, HMG-14 and HMG-17 could be required for condensation of single copy sequences and HMG-I for repetitive sequences.

Further studies are required to clarify whether or not the HMG proteins can bind specifically to active chromatin and whether they induce DNase I-sensitivity. The latter will require careful solution hybridization with excess single-strand probes driven to completion. It is possible that the controversy on the binding of HMG proteins to active nucleosomes is due to protein rearrangement occurring during the preparation and fractionation of nucleosomes. This could, in turn, be due to loss of important functional groups on nucleosomal proteins. For example, histone acetylation may be required for specific binding of HMG proteins and acetyl groups can be lost during nucleosome isolation. Further sequencing data will be required to establish the relationships between the HMG proteins of Figure 3, together with cross-linking experiments to identify their binding sites in the nucleus.

ACKNOWLEDGMENTS

The research carried out in the authors' laboratory is funded by the Cancer Research Campaign and the Medical Research Council, U.K.

REFERENCES

1. WEINTRAUB, H. and GROUDINE, M. (1976). Chromosomal subunits in active genes have an altered conformation. Science 193, 848-858.

1a. JOHNS, E.W., ed. (1983). "The HMG Chromosomal Proteins." Academic Press, New York.

2. WEISBROD, S., GROUDINE, M., and WEINTRAUB, H. (1980). Interaction of HMG14 and 17 with actively transcribed genes. Cell 19, 289-299.

3. GOODWIN, G.H., NICOLAS, R.H., and JOHNS, E.W. (1975). An improved large scale fractionation of high mobility group non-histone chromosomal proteins. Biochim. Biophys. Acta 405, 280.

4. GOODWIN, G.H., RABBANI, A., NICOLAS, R.H., and JOHNS, E.W. (1977). The isolation of the high mobility group nonhistone chromosomal protein HMG-14. FEBS Lett. 80, 413-416.

5. LUND, I., HOLTLUND, J., FREDRIKSEN, M., and LALAND, S. (1983). On the presence of two new high mobility group-like proteins in HeLa S3 cells. FEBS Lett. 152, 163-167.

6. GOODWIN, G.H., BROWN, E., WALKER, J.M., and JOHNS, E.W. (1980). The isolation of three new HMG nuclear proteins. Biochim. Biophys. Acta 623, 329-338.

7. WALKER, J.M. (1982). Primary structures. In: "The HMG Chromosomal Proteins," (E.W. Johns, ed.) pp. 69-87. Academic Press, New York.

8. NICOLAS, R.H. and GOODWIN, G.H. (1982). Isolation and analysis. In: "The HMG Chromosomal Proteins" (E.W. Johns, ed.) pp. 41-68. Academic Press, New York.

9. WEN, L., TWETEN, R.K., ISACKSON, P.J., IANDOLO, J.J., and REECK, G.R. (1983). Ionic interacctions between proteins in non-equilibrium pH gradient electrophoresis. Anal. Biochem. 132, 294-304.

10. GOODWIN, G.H., WRIGHT, C.A., and JOHNS, E.W. (1981). The characterization of ISF monomer nucleosomes from hen oviduct and the partial characterization of a third HMG-14/17-like protein in such nucleosomes. Nucleic Acids Res. 9, 2761-2775.

11. RABBANI, A., GOODWIN, G.H., WALKER, J.M., BROWN, E., and JOHNS, E.W. (1980). Trout liver high mobility group non-histone chromosomal proteins. FEBS Lett. 109, 294-398.

12. WATSON, D.C., WONG, N.C.W., and DIXON, G.H. (1979). The complete amino acid sequence of a trout testis non-histone protein H6 localised in a subset of nucleosomes and its similarity to calf thymus non-histone proteins HMG14 and HMG17. Eur. J. Biochem. 95, 193-199.

13. WALKER, J. M., HASTINGS, J. R. B., and JOHNS, E. W. (1977). The primary structure of a non-histone chromosomal protein. Eur. J. Biochem. 76, 461-468.

14. WALKER, J.M., GOODWIN, G.H., and JOHNS, E.W. (1979). The primary structure of the nucleosome-associated chromosomal protein HMG-14. FEBS Lett. 100, 394-399.

15. WALKER, J.M., BROWN, E., GOODWIN, G.H., STEARN, C., and JOHNS, E.W. (1980). Studies on the structures of some HMG-like non-histone chromosomal proteins from trout and chicken tissues comparison with calf thymus proteins HMG-14 and 17. FEBS Lett. 113, 253-257.

16. SAFFER, J.D. and GLAZER, R.I. (1982). The phosphorylation of high mobility group proteins HMG-14 and 17 and their distribution in chromatin. J. Biol. Chem. 257, 4655-4660.

17. COOPER, E. and SPAULDING, S.W. (1983). HMG-14/17-like proteins in calf thyroid. Biochem. J. 215, 643-649.

18. WALTON, G.M., GILL, G.H., COOPER, E., and SPAULDING, S.W. (1984). Thyrotropin-stimulated phosphorylation of high-mobility group protein 14 in vivo at the site catalysed by cyclic nucleoside-dependent protein kinases in vitro. J. Biol. Chem. 259, 601-607.

18a. REECK, G.R. and TELLER, D.C. (1985). High mobility group proteins: purification, properties, and amino acid sequence comparisons. In: "Progress in Nonhistone Protein Research," Vol. II (I. Bekhor, ed.) pp. 1-21. CRC Press, Boca Raton, Florida.

19. WALTON, G.M. and GILL, G.H. (1983). Identity of the in vivo phosphorylation site in high mobility group 14 protein in HeLa cells with the site phosphorylated by casein kinase II in vitro. J. Biol. Chem. 258, 4440-4446.

20. LUND, T., HOLTLUND, J., KRISTENSEN, T., OSTVOLD, A.C., SLETTEN, K., and LALAND, S.G. (1981). HMG-17 in metaphase arrested and interphase HeLa S3 cells. FEBS Lett. 133, 84-89.

21. D'ANNA, J.A., BECKER, R.R., TOBEY, R.A., and GATHEY, L.R. (1983). Composition and synthesis during G and S phase of a high mobility group-E/G component from Chinese hamster ovary cells. Biochim. Biophys. Acta 739, 197-206.

22. ALLFREY, V.G. (1982). Postsynthetic modifications. In: "The HMG Chromosomal Proteins," (E.W. Johns, ed.) pp. 123-148. Academic Press, New York and London.

23. DIXON, G.H. (1978). The HMG proteins of rainbow trout testis nuclei: isolation, structure, and function. In: "The HMG Chromosomal Proteins," (E.W. Johns, ed.) pp. 149-192. Academic Press, New York and London.

24. TANUMA, S., JOHNSON, L.D., and JOHNSON, G.S. (1983). ADP-ribosylation of chromosomal proteins and mouse mammary tumour virus gene expression. J. Biol. Chem. 258, 15371-15375.

25. RING, D. and COLE, R.D. (1979). Chemical cross-linking of H1 histone to the nucleosomal histone. J. Biol. Chem. 254, 11688-11695.

26. GAZIT, B., PANET, A., and CEDAR, H. (1980). Reconstitution of a deoxyribonuclease I-sensitive strucure on active genes. Proc. Natl. Acad. Sci. 77, 1787-1790.

27. WEISBROD, S. and WEINTRAUB, H. (1981). Isolation of activity of transcribed nucleosomes using immobilized HMG-14 and HMG-17 and an analysis of α-globin chromatin. Cell 23, 391-401.

28. LEVY, W.B. and DIXON, G.H. (1978). Partial purification of transcriptionally active nucleosomes from trout testis cells. Nucleic Acids Res. 5, 4155-4167.

29. GOODWIN, G.H. and MATHEW, C.G.P. (1982). Role in gene structure and function. In: "The HMG Chromosomal Proteins," (E.W. Johns, ed.) pp. 193-221. Academic Press, New York.

30. NICOLAS, R.H., WRIGHT, C.A., COCKERILL, P.N., WYKE, J.A., and GOODWIN, G.H. (1983). The nuclease sensitivity of active genes. Nucleic Acids Res. 11, 753-772.

31. BARSOUM, J., LEVINGER, L., and VARSHAVSKY, A. (1982). On the chromatin structure of the amplified, transcriptionally active gene for dihydrofolate reductase in mouse cells. J. Biol. Chem. 257, 5274-5282.

32. SEALE, R.L., ANNUZIATO, A.T., and SMITH, R.D. (1983). High mobility group proteins: abundance, turnover, and relationship to transcriptionally active chromatin. Biochemistry 22, 5008-5015.

33. GAREL, A. and AXEL, R. (1976). Selective digestion of transcriptionally active ovalbumin genes from oviduct nuclei. Proc. Natl. Acad. Sci. 73, 3960-3971.

34. SENEAR, A.W. and PALMITER, R.D. (1981). Multiple structural features are responsible for the nuclease sensitivity of the active ovalbumin gene. J. Biol. Chem. 256, 1191-1198.

35. REEVES, R. and CHANG, D. (1983). Investigations of the possible functions for glycosylation in the high mobility group proteins. J. Biol. Chem. 258, 679-687.

36. MATHEW, C.G.P., GOODWIN, G.H., IGO-KEMENES, T., and JOHNS, E.W. (1981). The protein composition of rat satellite chromatin. FEBS Lett. 125, 25-29.

37. ZHANG, X.Y. and HORZ, W. (1982). Analysis of highly purified satellite DNA containing chromatin from mouse. Nucleic Acids Res. 10, 1481-1494.

38. REUDELHUBER, T. L., BALL, D. J., DAVIS, A. H., and GERRARD, W.T. (1982). Transferring DNA from electrophoretically resolved nucleosomes to DBM-cellulose: properties of nucleosomes along mouse satellite DNA. Nucleic Acids Res. 10, 1311-1325.

39. STRAUSS, F. and VARSHAVSKY, A. (1984). A protein binds to a satellite DNA repeat at three specific sites that would be brought into mutual proximity by DNA folding in the nucleosome. Cell 37, 889 901.

40. WEISBROD, S. (1982). Properties of active nucleosomes as revealed by HMG-14 and 17 chromatography. Nucleic Acids Res. 10, 2017-2042.

POST-TRANSLATIONAL MODIFICATION OF THE LOW MOLECULAR WEIGHT HMG CHROMOSOMAL PROTEINS

S.G. Laland, T. Lund, and J. Holtlund

Department of Biochemistry
University of Oslo
Oslo, Norway

INTRODUCTION

The low molecular weight proteins HMG-14 and 17 (reference 1, and chapter by Goodwin et al., this volume) have been shown to undergo post-translational modifications such as phosphorylation, ADP-ribosylation, glycosylation, and acetylation. We will discuss each of these with particular emphasis on phosphorylation, which is the best studied post-translational modification of these proteins.

IN VITRO PHOSPHORYLATION

Several studies have shown that HMG-14 and 17 can be phosphorylated in vitro. For instance, cyclic GMP dependent protein kinase was found to phosphorylate calf thymus HMG-14 very efficiently but HMG-17 very poorly (2, 3). The catalytic subunit of cyclic AMP dependent protein kinase also phosphorylated HMG-14 more efficiently (4, 5). Ser 6 in HMG-14 was the major modification site for both enzymes. Furthermore, HMG-14 and HMG-17 contain minor phosphorylation sites at Ser 24 and Ser 28, respectively. These sites could be phosphorylated by cGMP and cAMP dependent protein kinase (2). However, chicken erythrocyte HMG-14 was not phosphorylated by cyclic GMP dependent protein kinase from either bovine lung or avian nucleoli. The N-terminal sequence of avian HMG-14 is Pro-Lys-Arg-Lys-Ala-Pro-

Ala-Glu. Hence, Ser 6 is lacking, and this explains why avian HMG-14 is not phosphorylated (3). HMG-14 and HMG-17 from rat C6 glioma cells have been shown to be phosphorylated in both serine and threonine by nuclear protein kinase II from glioma cells (6). Others have reported (7) that nuclear protein kinase II from rat liver preferentially phosphorylates HMG-17 in vitro.

Walton and Gill (8) have studied the phosphorylation of HMG proteins isolated from HeLa cells using cyclic AMP dependent kinase, cyclic GMP dependent protein kinase and casein kinase I and II. The results were compared with those obtained in vivo. Both in vitro and in vivo, HMG-14-like proteins were found to be those phosphorylated. Tryptic maps of the HMG-14-like proteins labeled in vivo with ^{32}P-phosphate gave one labeled acidic peptide. This peptide was found in vitro only after phosphorylation of HMG proteins by casein kinase II (and subsequent trypsinization). This suggests that casein kinase II catalyzed the in vivo phosphorylation of HMG-14-like proteins.

IN VIVO PHOSPHORYLATION

Phosphorylation of HMG-14 and HMG-17 has been most thoroughly studied in proliferating HeLa cells. Based on analysis by SDS gel electrophoresis and subsequent autoradiography, Bhorjee et al. (9) concluded that both HMG-14 and HMG-17 were phosphorylated. Later studies (10, 11) have revealed that SDS gel electrophoresis does not separate all the different low molecular weight HMG proteins and that previous results had to be reevaluated. Three different groups (8, 10, 12) have recently analyzed ^{32}P-phosphate labeled HMG proteins from HeLa cells by acetic acid-urea gel electrophoresis and obtained similar results. There were several protein bands with mobility in the region of calf thymus HMG-14, two of which were labeled. In none of these studies was HMG-17 found to be labeled. The phosphorylated proteins were designated HMG-14$_a$ and HMG-14$_b$ by Bhorjee et al. (12), hHMG-14$_1$ and hHMG-14$_2$ by Walton and Gill (8), and HMG-I and HMG-Y by Lund et al. (10). The two former groups describe these two proteins as different forms of HMG-14 without any further characterization. We have purified HMG-I and HMG-Y and found that they have amino acid compositions typical for HMG proteins but different from HMG-14 from calf thymus (10 and unpublished results). The low molecular weight HMG proteins in HeLa cells seem to be a more complex class of proteins than realized, although the complexity of HMG-14 proteins has been reported (13-15).

Cell cycle studies have shown that there is an increase in specific phosphorylation of hHMG-14_1 and hHMG-14_2 in metaphase arrested cells compared to interphase cells (8). Similar results are reported for G2 cells by Bhorjee et al. (12). Results from this laboratory differ, since we have shown that the phosphorylated protein HMG-I in interphase cells is absent in metaphase arrested cells. Instead we find a phosphorylated low molecular weight protein designated HMG-M with a lower mobility in acetic acid-urea gel electrophoresis than HMG-I. HMG-M may be super-phosphorylated HMG-I. This is supported by the finding that HMG-M and HMG-I are not significantly different in amino acid composition (10). It is possible that HMG-M may play some role in chromatin condensation during mitosis. An increase in the phosphorylation of HMG-14-like proteins in metaphase arrested HeLa cells was reported by Paulson and Taylor (16). Furthermore, they found that isolated metaphase chromosomes contained an endogenous protein kinase which catalyzed the phosphorylation of serine and threonine residues in the HMG-14-like protein. Paulson and Taylor have suggested that the increased phosphorylation in metaphase could play a role in shutting off transcription during mitosis.

Saffer and Glazer (17, 18) have examined the phosphorylation of HMG-14 and HMG-17 in proliferating Ehrlich ascites, L 1210 and P 388 leukemia cells, human colon carcinoma cells (HT 29) and chinese hamster ovary cells. They claim that these proteins were phosphorylated in all cell lines. The identification of the phosphorylated proteins was based on acetic acid-urea gel electrophoresis, but the lack of standard HMG proteins casts doubt on their conclusions. Recent work from our laboratory (19) has shown that there are three protein bands on acetic acid-urea gels in the region of calf thymus HMG-14. Two of these were phosphorylated and corresponded to HMG-I and HMG-Y from HeLa cells. Amino acid analysis and peptide maps of the three proteins showed that the phosphorylated proteins differed significantly from the unphosphorylated protein, which resembled HMG-14 from calf thymus. In contrast to the results of Saffer and Glazer, we did not find HMG-17 to be phosphorylated.

Phosphorylation of HMG-14 and HMG-17 in chinese hamster ovary cells have also been studied by Arfmann et al. (20, 21), who report that both proteins are phosphorylated in interphase cells and that in metaphase arrested cells the phosphorylation of HMG-14 was increased compared to HMG-17. These results will also have to be re-evaluated since SDS gel electrophoresis was used; as mentioned above, this method does not separate the various members of the low molecular weight HMG proteins very well. D'Anna et al. (22) have isolated and analyzed a phosphorylated low molecular weight HMG protein designated CHO HMG-E/G from chinese hamster cells. Its amino acid composition differed significantly from bovine HMG-14 and HMG-17 but was very similar to HMG-Y from HeLa and Ehrlich ascites cells and may therefore be the same protein.

HMG-14 and HMG-17 in rat C_6 glioma cells are reported to be phosphorylated in vivo on both serine and threonine residues (6). Several unidentified phosphorylated HMG proteins have been shown to be present in lymphosarcoma P1798 (in mice). Following injection of cortisol, which results in tumor regression, a marked suppression of phosphorylation resulted (23). Nerve growth factor which induces extension of neurites and cessation of cell division in a clone PC 12 (pheochromocytoma), was found to simultaneously induce increased phosphorylation of HMG-17 (24).

Recent work using bovine thyroid slices and thyrotropin stimulation have led to interesting results. Two-dimensional polyacrylamide gel electrophoresis has shown that, in addition to proteins co-migrating with standard HMG-14 and 17, there were several other proteins with mobility similar to low molecular weight HMG proteins. Two of these and HMG-14 were phosphorylated when thyroid slices were incubated with ^{32}P-phosphate. Furthermore, addition of thyrotropin led to a rapid two-fold increase in the labeling of HMG-14 (11). The site of phosphorylation was Ser 6, which is the major site of in vitro phosphorylation by cyclic nucleotide dependent protein kinase

(25). Addition of analogs of cAMP, but not of cGMP, enhanced phosphorylation. The results lend strong support to the idea that thyrotropin-stimulated phosphorylation of HMG-14 is catalyzed by the cAMP dependent protein kinase.

The studies reported on the phosphorylation of the low molecular weight HMG proteins have shown that they can be phosphorylated in vivo and in vitro. The in vivo incorporated ^{32}P-phosphate is present in these proteins as phospho-serine or in some cases as phospho-threonine, but not as ADP-ribose residues since phosphatase have been shown to remove all phosphate (10). At present it seems that cell cycle specific phosphorylation is catalyzed by casein kinase II (8) whereas the hormone-regulated phosphorylation is catalyzed by cAMP dependent protein kinase (25). Furthermore, it is interesting that growth factors and steroid hormones may affect the phosphorylation of the low molecular weight HMG proteins.

ADP-RIBOSYLATION

ADP-ribosylation of low molecular weight HMG proteins has been reported in HeLa nuclei and in nuclei from Friend erythroleukemia cells (26, 27). Protein H6 (which is a trout testis homolog of HMG-17) was found to be ADP-ribosylated with a mean chain length of 4.5 units (28). HMG-14 from chicken erythrocytes could be ADP-ribosylated by poly(ADP-ribose) polymerase (29).

HMG-14 and HMG-17 in the 34I (C3H mouse mammary carcinoma) cell line were found to incorporate ^{3}H-adenosine as ADP-ribose with a chain length of 1.5 to 1.8. Only 0.03% of the HMG proteins were modified (30). The incorporation was reduced in the presence of 3-aminobenzamide, an inhibitor of poly(ADP-ribose) polymerase. Furthermore, treatment of the cells with glucocorticoids which induce formation of mouse mammary tumor RNA (with information for gag, pol, and env proteins), led to a simultaneous removal of ADP-ribose from HMG-14 and HMG-17. Similar results were also obtained in the presence of 3-aminobenzamide. One explanation may be that the ADP-ribose residues on HMG-14 may serve as negative regulation of mouse mammary tumor virus expression (31).

GLYCOSYLATION

HMG-14 and HMG-17 from calf thymus and Friend erythroleukemia cells have been found to be modified by glycosylation and shown to contain fructose, galactose, mannose, and N-acetyl glucosamine (27). What percentage of the protein was glycosylated and whether there is heterogeneity is the glycosylation pattern between different molecules are not known. It has also been found (32) that HMG-14 and HMG-17 from Friend cells preferentially bind to the nuclear matrix proteins and that removal of the carbohydrate greatly reduced the binding.

ACETYLATION

Acetylation of HMG-14 and HMG-17 in duck erythrocytes has been shown to occur on lysine residues at positions 2 and 4 in HMG-14 and positions 2, 4, and 10 in HMG-17 (33). In order to isolate the multiacetylated forms, butyrate had to be added to the cells. Recently, Sterner and Allfrey (34), using duck erythrocytes, have shown that HMG-14 and HMG-17 can carry modifications in the form of acetylation and phosphorylation at the same time.

CONCLUSIONS

The present review shows that the low molecular weight proteins can undergo a number of post-translational modifications, as has been found for the histones. Being chromosomal proteins, modifications may introduce changes in chromatin structure. This might result in chromatin condensation or greater DNA accessibility and subsequent changes in transcription. For instance, some workers (18, 35) have suggested that phosphorylated HMG-14 and 17 are preferentially located in DNase I sensitive genes.

Another interesting aspect of the modulation of the biological function of these proteins is the finding by Sterner and Allfrey (34) that multitype postsynthetic modification may be present simultaneously in these proteins. The possible presence of several different types of modifications increases in the potential number of different biological functions for these proteins.

REFERENCES

1. NICOLAS, R.H. and GOODWIN, G.H. (1982). Isolation and Analysis. In: "The Chromosomal Proteins," (E.W. Johns, ed.) pp. 41-68. Academic Press, New York.

2. LINNALA-KANKKUNEN, A. and MAENPAA, P.H. (1981). Phosphorylation of high mobility group protein HMG 14 by a cyclic GMP-dependent protein kinase from avian liver nucleoli. Biochim. Biophys. Acta 654, 287-291.

3. PALVIMO, P., LINNALA-KANKKUNEN, A., and MAENPAA, P.H. (1983). Differential phosphorylation of high mobility group protein HMG 14 from calf thymus and avian erythrocytes by a cyclic GMP-dependent protein kinase. Biochem. Biophys. Res. Commun. 110, 378-382.

4. WALTON, M.W., SPIESS, J., and GILL, G.N. (1982). Phosphorylation of high mobility group 14 protein by cyclic nucleotide-dependent protein kinase. J. Biol. Chem. 257, 4661-4668.

5. TAYLOR, S.S. (1982). The in vitro phosphorylation of chromatin by the catalytic subunit of cAMP-dependent protein kinase. J. Biol. Chem. 257, 6056-6063.

6. HARRISON, J.J. and JUNGMANN, R.A. (1982). Phosphorylation of high mobility group proteins 14 and 17 by nuclear protein kinase NII in rat C_6 glioma cells. Biochem. Biophys. Res. Commun. 108, 1204-1209.

7. INOUE, A., TEI, Y., HASUMA, T., YUKIOKA, M., and MORISAWA, S. (1980). Phosphorylation of HMG 17 by protein kinase NII from rat liver nuclei. FEBS Lett. 117, 68-72.

8. WALTON, G.M. and GILL, G.N. (1983). Identity of the in vivo phosphorylation site in high mobility group 14 protein in HeLa cells with the site phosphorylated by casein kinase II in vitro. J. Biol. Chem. 258, 4440-4446.

9. BHORJEE, J.S. (1981). Differential phosphorylation of nuclear nonhistone high mobility group proteins HMG 14 and HMG 17 during cell cycle. Proc. Natl. Acad. Sci. USA 78, 6944-6948.

10. LUND, T., HOLTLUND, J., FREDRIKSEN, M., and LALAND, S.G. (1983). On the presence of two new high mobility group proteins in HeLa S3 cells. FEBS Lett. 152, 163-167.

11. COOPER, E. and SPAULDING, S.W. (1983). HMG (high mobility group)-14/17-like proteins in calf thyroid. Thyrotropin-

dependent phosphorylation and comparison with calf thymus proteins. Biochem. J. 215, 643-649.

12. BHORJEE, J.S., MELLON, I., amd KIFLE, L. (1983). Is high mobility group protein 17 phosphorylated in vivo? Re-examination of the HeLa cell cycle data. Biochem. Biophys. Res. Commun. 111, 1001-1007.

13. ISACKSON, P.J., DEBOLD, W.A., and REECK, G.R. (1980). Isolation and separation of chicken erythrocyte high mobility group non-histone chromatin proteins by chromatography on phosphocellulose. FEBS Lett. 119, 337-342.

14. WEN, L., TWETEN, R.K., ISACKSON, P.J., IANDOLO, J.J., and REECK, G.R. (1983). Ionic interactions between proteins in nonequilibrium pH gradient electrophoresis: histones affect the migration of high mobility group nonhistone chromatin proteins. Anal. Biochem. 132, 294-304.

15. WEN, L. and REECK, G.R. (1984). Purification of high mobility group nonhistone chromosomal proteins by liquid chromatography on a column containing immobilized H5. J. Chromatog. 314, 436-444.

16. PAULSON, J.R. and TAYLOR, S.S. (1982). Phosphorylation of histones 1 and 3 and non-histone high mobility group 14 by an endogenous kinase in HeLa metaphase chromosomes. J. Biol. Chem. 257, 6064-6072.

17. SAFFER, D.J. and GLAZER, R.I. (1980). The phosphorylation of high mobility group proteins 14 and 17 from Ehrlich ascites and L 1210 in vitro. Biochem. Biophys. Res. Commun. 93, 1280-1285.

18. SAFFER, J.D. and GLAZER, R.I. (1982). The phosphorylation of high mobility group proteins 14 and 17 and their distribution in chromatin. J. Biol. Chem. 257, 4655-4660.

19. LUND, T., HOLTLUND, J., LALAND, S.G. (1985). On the phosphorylation of low molecular mass HMG (high mobility group) proteins in Ehrlich ascites cells. FEBS Lett. 180, 275-279.

20. ARFMANN, H.A. and BAYDOUN, H. (1981). Preferential phosphorylation of high mobility group protein 17 in vitro by a nuclear protein kinase. Z. Naturforsch. 36, 319-322.

21. ARFMANN, H.A., HAASE, E., and SCHROTER, H. (1981). High mobility group proteins from CHO cells and their modification during cell cycle. Biochem. Biophys. Res. Commun. 101, 137-143.

22. D'ANNA, J.A., BECKER, R.R., TOBEY, R.A., and GURLEY, L.G. (1983). Composition and synthesis during G1 and S phase of a high mobility group-E/G component from Chinese hamster ovary cell. Biochem. Biophys. Acta. 793, 197-206.

23. MILHOLLAND, R.J., IP, M.M., and ROSEN, R. (1979). The effect of hydrocortisone treatment on the in vivo phosphorylation of a subgroup of non-histone nuclear proteins in the mouse lymphosarcoma P 1798. Biochem. Biophys. Res. Commun. 88, 993-997.

24. HALEGOUA, S. and PATRICK, J. (1980). Nerve growth factor mediates phosphorylation of specific proteins. Cell 22, 571-581.

25. WALTON, G.M., GILL, E.C., and SPAULDING, S.W. (1984). Thyrotropin-stimulated phosphorylation of high mobility group protein 14 in vivo at the site catalyzed by cyclic nucleotide protein kinases in vitro. J. Biol. Chem. 259, 601-607.

26. GIRI, C.P., WEST, M.H., and SMULSON, M. (1978). Nuclear protein modification and chromatin structure. 1. Differential poly(adenosin diphosphate) ribosylation of chromosomal proteins in nuclei versus nucleosomes. Biochemistry 17, 3495-3500.

27. REEVES, R., CHANG, D., and CHUNG, S.C. (1981). Carbohydrate modification of high mobility group proteins. Proc. Natl. Acad. Sci. USA 78, 6704-6708.

28. WONG, N.C.W., POIRIER, G.G., and DIXON, G.H. (1977). Adenosine di-phosphoribosylation of certain basic chromosomal proteins in isolated trout testis nuclei. Eur. J. Biochem. 77, 11-21.

29. POIRIER, G.G., NIEDERGANG, C., CHAMPAGNE, M., MAZEN, A., and MANDEL, P. (1982). Adenosine diphosphate ribosylation of chicken-erythrocyte histones H1, H5, and high mobility group proteins by purified calf thymus poly(adenosinediphosphate-ribose) polymerase. Eur. J. Biochem. 127, 437-442.

30. TANUMA, S. and JOHNSON, G.S. (1983). ADP-ribosylation of nonhistone high mobility group proteins in intact cells. J. Biol. Chem. 258, 4067-4070.

31. TANUMA, S., JOHNSON, L.D., and JOHNSON, G.S. (1983). ADP-ribosylation of chromosomal proteins and mouse mammary tumor virus gene expression. J. Biol. Chem. 258, 15371-15375.

32. REEVES, R. and CHANG, D. (1983). Investigation of the possible functions for glycosylation in the high mobility group proteins. J. Biol. Chem. 258, 679-687.

33. STERNER, R., VIDALI, G., and ALLFREY, V.G. (1981). Studies of acetylation and deacetylation in high mobility group proteins. Identification of the sites of acetylation in high mobility group proteins HMG 14 and 17. J. Biol. Chem. 256, 8892-8895.

34. STERNER, R. and ALLFREY, V.G. (1983). Selective isolation of polypeptide chains bearing multiple types of postsynthetic modification. J. Biol. Chem. 258, 12135-12138.

35. LEVY-WILSON, B. (1981). Enhanced phosphorylation of high-mobility group proteins in nuclease-sensitive mononucleosomes from butyrate-treated HeLa cells. Proc. Natl. Acad. Sci. USA 78, 2189-2193.

STRUCTURE AND FUNCTION IN THE HMG-1 FAMILY OF CHROMOSOMAL
PROTEINS

Pere Puigdomènech and Matilde Jose

Institut de Biologia de Barcelona del CSIC
Jordi Girona Salgado, 18
08034 Barcelona,   Spain

INTRODUCTION

Analysis of HMG chromosomal proteins by SDS-polyacrylamide
gel electrophoresis reveals a group of polypeptides in the
region around 26,000 daltons. These are the proteins of the
high mobility group that have the lowest electrophoretic mobility in both SDS and acid/urea gels and they have therefore been
called HMG-1 or 2 in most tissues. For historical reasons,
homologous proteins from trout testis and chicken erythrocytes
are called HMG-T and HMG-E, respectively. In the rest of the
present article HMG-1, 2, T, or E are considered to comprise the
HMG-1 family. Apart from their apparent molecular weights, all
these proteins share many common structural features, as can be
expected from their similar amino acid compositions and sequences. In fact, the information that we have about the
proteins of the HMG-1 family is mostly structural. Very little
functional information is available and most of it is inferred
from in vitro studies. In this respect, the situation of our
knowledge on the HMG-1 family of proteins is very different than
that for HMG-14 and 17. In the case of the latter, low molecular weight HMG proteins, there is considerable literature on
their possible function, although firm conclusions are not yet
available (see chapter by Goodwin et al., this volume). The low
molecular weight HMG proteins do not display any evidence of
secondary or tertiary folding. For the HMG-1 family the most
interesting results have, in contrast, been in structural

studies of the proteins and studies on their interactions with other macromolecules.

HMG-1 INTERACTIONS AND FUNCTIONS

Several properties of HMG-1 family proteins were established as purified fractions became available. The calf thymus HMG-1 fraction is a single band in SDS gel electrophoresis, whereas heterogeneity appears for the HMG-2 fraction when analysed by either electrofocusing or SDS gel electrophoresis (see reference 1 for a review). Thus, subfractions have been defined for HMG-2. Some of the heterogeneity observed may be due to post-translational modifications. Acetylation (2, 3), methylation (4), phosphorylation (5), ADP-ribosylation (6), and glycosylation (7) of the proteins of this family have been observed either in vivo or in vitro.

HMG-1 proteins are able to interact with DNA (8), and it has been found that in some conditions they have a preferential binding to single-stranded DNA (9-11). The formation of complexes of HMG-1 and HMG-2 with DNA has been studied and the parameters of the interaction have been measured as well as their dependence on ionic strength and chemical composition of the nucleic acid (12, 13). It appears that the interaction is non-cooperative and that both the affinity constant and site size are dependent on salt concentration. By NMR spectroscopy (14) it was observed that a zone of the molecule can interact with DNA forming a complex that, when studied by electron microscopy after glutaraldehyde fixation, has a globular appearance (15). Another characteristic of HMG-1 and 2 is their ability to change the topology of DNA, an effect that can be measured by observing the variation of the superhelicity of closed-circular DNA in their presence (16-18). All these results point towards a behavior of these proteins similar to certain prokaryotic DNA-unwinding proteins, such as the gene 32 protein (9, 19). However, this similarity has been put into question (12) considering the non-cooperative binding of HMG-1 to DNA and contradictory results on the effects of HMG-1 on thermal denaturation of DNA. It has been proposed that the fact that different authors have found that HMG-1 has either a protective or a disrupting effect against DNA melting could be due to different methods for extraction of the protein (12). A decrease in the ability of these proteins to form α-helix and tertiary folding when extracted with acid solutions as compared with salt-extracted methods has been observed (20), however, an overall rearrangement of the structure does not seem to occur.

Results with HMG-14 and 17 showed that these proteins interact with nucleosomes (21, 22) mainly in the last 20 base pairs of the core particle DNA (23). It has also been shown that they have a preferential binding to mononucleosomes that have long tails (22, 24, 25). The fact that HMG-1 and 2 are released after short nuclease digestion (26), that in these conditions they seem to form a complex with short polynucleotides (27), and that they bind to long nucleosomes (28) has led to a proposal that they bind to linker DNA. These indications and the fact that HMG-1 and its homologs have a molecular weight similar to that of histone H1 have been taken as arguments in favor of similar function for H1 and the HMG-1 family. It has been suggested that HMG-1 might could substitute for histone H1 (29). The lack of similarity in the sequences of these proteins (30), however, does not seem to support such suppositions.

Little is known about the effect of HMG-1 proteins on higher order chromatin structure. Experiments were carried out taking advantage of the fact that it is possible to prepare chromatin fragments that are stable in solution even if their DNA is internally cleaved (see reference 31 and the chapter by Muyldermans et al., this volume). Such preparations have been used to study different variables that may affect higher order structure of chromatin (32). When using immunological detection methods, such as immuno-dot or ELISA, the distribution of HMG-1 and 2 is measured after sucrose gradient centrifugation of such stable aggregates, a different distribution is found as compared with the bulk of the chromatin (Figure 1). This indicates either that the HMG-1/2-containing chromatin is less stable in higher order structure than the majority of chromatin or that HMG-1 and HMG-2 tend to concentrate in smaller oligonucleosomes. In any case, no HMG protein is found in large (about 15-nucleosome) chromatin fragments stable at physiological ionic strength. The association of HMG proteins with chromatin has been studied and it has been proposed that, while a proportion of these proteins is easily solubilized at physiological ionic strength, a population remains tightly bound to chromatin (33-36). In light of such observations, our results presented above would indicate that the presence of HMG-1 or 2 produces an increased instability in higher order structure of chromatin. Whether these proteins are distributed at random along the chromatin fiber, periodically giving rise to discontinuities that may explain a "superbead" structure (37), or whether they concentrate in specific zones of the chromatin is an unresolved

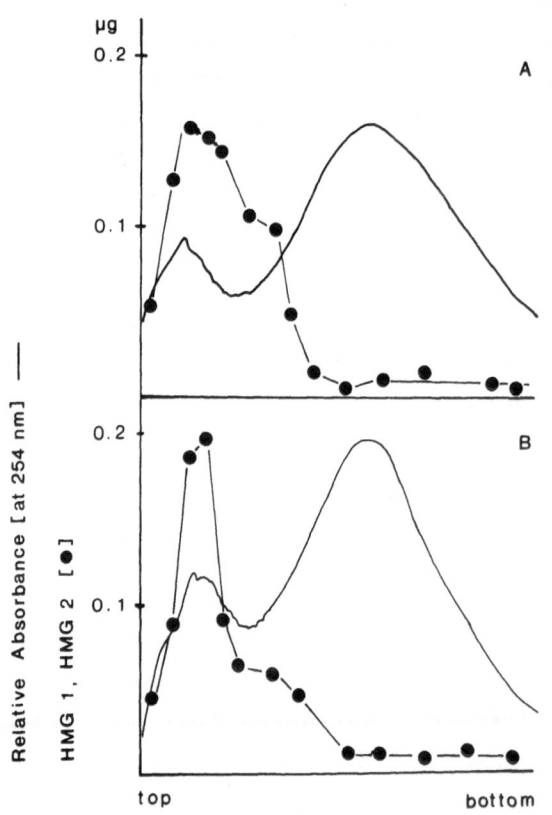

Figure 1. Distribution of HMG-1 and 2 in fractions of chromatin prepared by micrococcal nuclease digestion of rat liver nuclei at 140 mM NaCl, 0.1 mM PMSF, 0.2 mM EDTA, 10 mM Tris-HCl, pH 7.4, and 4°C and analyzed by sucrose gradient (5-20%) centrifugation. The measurement of HMG-1 and 2 in the different chromatin fractions has been done by quantitating the reaction of antisera against HMG-1 (that has strong cross-reaction with HMG-2) in dots and incubation with radiolabelled protein A. The two gradients correspond to two times of nuclease digestion, A (45 minutes) and B (60 minutes) with 200 units of enzyme per milligram of DNA.

question. Transcriptionally active chromatin shows an enhanced accessibility to DNases indicating a different structure as compared with the bulk of chromatin. In this sense HMG-1 could take part in the maintenance of the structural requirements for transcription. In fact, it was found that these proteins may be required for transcription in mouse myeloma cell nuclei (38).

HMG-1 and 2 have been shown to interact with histones in vitro. This interaction has been found with histone H1 in solution (39, 40) and in blots (41). Cross-linking experiments, in solution, of mixtures of HMG-1 or 2 with oligomers of core histones indicate that these HMG proteins are also able to form stoichiometric complexes with histones H2A-H2B and H3-H4 (42). A result that may be related to this kind of interaction is the observation that these proteins may act as an assembly factor of nucleosomes at physiological ionic strength (43). In this respect, members of the HMG-1 family may behave similarly to nucleoplasmin (44, 45) or poly-glutamic acid (46) in favoring the formation of the histone octamer in solution, thus allowing histones to be correctly placed on DNA to form nucleosomes. Their nucleosome assembly activity has been taken as an indication of the possible involvement of HMG-1 and 2 in replication. Nevertheless, while a correlation of the presence of HMG-2 with cell cycle length has been described (47) no such effect has been found for HMG-1. Immunological and oocyte microinjection studies apparently show the accumulation of HMG-1 and HMG-2 in the nucleus (48, 49) but a periodical accumulation of these proteins in the nucleus and cytoplasm (50-52) might explain all these results. In any case, the involvement of HMG-1 in replication is still only a hypothesis.

STRUCTURAL STUDIES

The amino acid composition of HMG-1 shows a high proportion of charged residues, both basic and acidic. Calf thymus HMG-1 for instance, has 15% lysine, 5% arginine, 18% glutamic acid, and 12% aspartic acid (53). The largely complete sequence of Walker et al. (54) showed that these residues are not evenly distributed in the sequence. Rather an asymmetrical distribution can be observed. Two parts can be clearly distinguished.

The N-terminal two-thirds of the molecule have a rather normal amino acid composition with a predominantly basic character, while the last third of the polypeptide chain has a marked acidic character and contains a long stretch of glutamic acid and aspartic acid residues. A cDNA corresponding to bovine HMG-1 has been cloned (55) and the protein sequence (54) has been corrected in a number of positions. In particular, the protein has been found to terminate with a continuous stretch of 30 acidic residues (21 Glu and 9 Asp).

The sequences of HMG-1 and 2 show a high degree of similarity (54). In fact, the main differences are observed in the residues preceding the acidic C-terminal part of the molecule. In general, the amino acids in this zone have a more structure-disrupting character in HMG-2 than in HMG-1 and that might affect interactions of the different domains (see below).

Initial studies using nuclear magnetic resonance spectroscopy showed that the spectrum of HMG-1 can be interpreted as that of a protein with domains having different stabilities in their tertiary structures (14). An analysis of the protein sequence further indicated that, in the N-terminal part of the molecule, two domains may be distinguished which have a high degree of sequence similarity. An internal gene duplication was proposed as the origin of this feature (56). Therefore, the structure of a member of the HMG-1 family appears to be formed from three main domains (56): Domains A and B are the N- terminal and central part of the molecule and domain C in the C- terminal third. Domains A and B have amino acid compositions typical of globular proteins and they are apparently homologous units. Domain C has a high proportion of acidic residues and is unrelated to the other domains.

The action of proteolytic enzymes has been widely used to study the domain structure of histones (e.g., see references 57 and 58). A similar approach has been taken with members of the HMG-1 family. HMG-E (56) and HMG-1 (59) were digested with trypsin and it was shown that limit peptides were produced. These peptides were isolated by chromatography. By amino acid analysis and partial sequencing, they were found to correspond roughly to the two N-terminal domains A and B of the molecule or to domains A+B, equivalent to HMG-3 (60). A possible extra domain of 11 residues at the N-terminus was hydrolyzed by the enzyme in HMG-1. The two large limit peptides were folded, thus producing evidence of secondary and tertiary structures.

We have found that other peptides are produced using protease of Staphylococcus aureus V8 (41). The main cutting point of this enzyme at short times of digestion seems to be located at the boundary between domains A and B. This is also a target for tryptic attack. Thus, this point (around residue 75) is susceptible to attack by two proteases of very different specificities (trypsin and the S. aureus protease). It appears as a main hinge between domains A and B in the protein. At longer times of digestion V8 protease also cuts at the boundary between domains B and C. We also found that whereas domains A and B bind DNA in blots, the peptide corresponding to domains B+C does not. This peptide, which corresponds to the last two thirds of the molecule, appears to be the point of interaction of HMG-1 with histone H1 and probably with H2A-H2B (42). We have also observed that domain B is the responsible for the change in the superhelicity of closed circular DNA. This effect is lost, however, when domain B is attached to domain C (18). This behavior led us to examine the interactions between the three domains of HMG-1. It appears, from studies with different proteases and cleaving reagents (41, 56, 58), that domain A is the one having the highest degree of structure, that fragment A+B seems to retain all the tertiary and possibly secondary structure of the molecule, and that when domain B is attached to

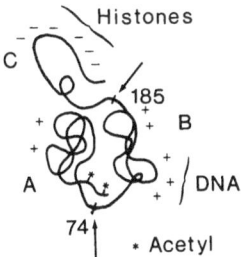

Figure 2.   Schematic representation of the structure of HMG-1. The main cutting points for Staphlyococcus aureus V8 protease are shown (74 and 185) as well as the three structural domains the zones of interaction with DNA and histones and the main acetylated residues.

domain C the secondary and tertiary structure of the former is abolished.

CONCLUSIONS

The results obtained on the structure of the proteins of the HMG-1 family allow us to propose a schematic model such as the one presented in Figure 2. The protein chain is divided into three main domains corresponding roughly to the three thirds of the molecule, with a possible extra short domain in the N-terminal eleven residues where the main acetylation points for the protein have been detected. Domain A appears as the most structured unit. Its absence allows the folding of domain C onto domain B, the DNA-interaction domain, thus abolishing domain B's ability to change DNA topology. The existence of interactions between domains that abolish possible functions of the proteins, such as the action on DNA superhelicity, suggests that such interdomain interactions may be a mechanism to modulate HMG-1 function.

Proposals for HMG-1 biological function point mainly towards an essentially structural function, either as an assembly factor in replication or as a protein that produces an adequate superstructure for transcription. The absence of HMG proteins from fragments of chromatin having a stable higher-order structure in solution (see above) may also indicate a role for these proteins in the folding of chromatin superstructure. The importance of the topological state of DNA for transcription, for instance, has been stressed in recent results (61). The loop structure of chromatin (62) may allow comparison of the results obtained with closed circular DNA with fragments of chromatin fiber attached to the nuclear scaffold. HMG-1 proteins may be important in maintaining such a structure.

REFERENCES

1. NICOLAS, R.N. and GOODWIN, G.H. (1982). Isolation and analysis. In: "The HMG Chromosomal Proteins," (E.W. Johns, ed.) pp. 41-68. Academic Press, New York.

2. STERNER, R., VIDALI, G., HEINRICKSON, R.L., and ALLFREY, V.G. (1978). Post-synthetic modification of high mobility group proteins. Evidence that HMG proteins are acetylated. J. Biol. Chem. 253, 7601-7604.

3. STERNER, R., VIDALI, G., and ALLFREY, V.G. (1979). Studies of acetylation and deacetylation in high mobility group proteins. Identification of the sites of acetylation in HMG-1. J. Biol. Chem. 254, 11577-11583.

4. BOFFA, L.C., STERNER, R., VIDALI, G., and ALLFREY, V.G. (1979). Post-synthetic modifications of nuclear proteins. High mobility group proteins are methylated. Biochem. Biophys. Res. Commun. 89, 1322-1327.

5. SUN, I.V., JONHSON, E.M., and ALLFREY, V.G. (1980). Affinity purification of newly phosphorylated protein molecules. J. Biol. Chem. 255, 742-747.

6. POIRIER, G.G., NIEDERGANG, C., CHAMPAGNE, M., MAZEN, A., and MANDEL, P. (1982). Adenosine diphosphate ribosylation of chicken erythrocyte histones H1, H5, and high mobility group proteins by purified calf thymus poly(adenosine-diphosphateribose) polymerase. Eur. J. Biochem. 127, 437-442.

7. REEVES, R., CHANG, D., and CHUNG, S.C. (1981). Carbohydrate modifications of the high mobility group proteins. Proc. Natl. Acad. Sci. USA 78, 6704-6708.

8. SHOOTER, K.V., GOODWIN, G.H., and JOHNS, E.W. (1974). Interactions of a purified non-histone chromosomal protein with DNA and histones. Eur. J. Biochem. 47, 263-270.

9. BIDNEY, D.L. and REECK, G.R. (1978). Purification from cultured hepatoma cells of two nonhistone chromatin proteins with preferential affinity for single-stranded DNA: apparent analogy with calf thymus HMG proteins. Biochem. Biophys. Res. Commun. 85, 1211-1218.

10. ISACKSON, P.J., FISHBACK, J.L., BIDNEY, D.L., and REECK, G.R. (1979). Preferential affinity of high molecular weight high mobility group non-histone chromatin proteins for single-stranded DNA. J. Biol. Chem. 254, 5569-5572.

11. DUGUET, M. and DE RECONDO, A.M. (1978). A DNA unwinding protein isolated from regenerating rat liver. Physical and functional properties. J. Biol. Chem. 253, 1660-1666.

12. ISACKSON, P.J., COX, D.J., MANNING, D., and REECK, G.R. (1985). Studies on the interactions of HMG-1 and its homologs with DNA. In: "Progress in Non-histone Protein Research," (I. Bekhor, ed.) pp. 23-39. CRC Press, Boca Raton, Florida.

13. SCHEPELEV, V.A., KOSAGANOV, Y.N., and LAZURKIN, Y.S. (1984). Interaction of the HMG1 protein with nucleic acids. FEBS Lett. 172, 172-176.

14. CARY, P.D., CRANE-ROBINSON, C., BRADBURY, E.M., JAVAHERIAN, K., GOODWIN, G.H., and JOHNS, E.W. (1976). Conformational studies of two non-histone chromosomal proteins and their interactions with DNA. Eur. J. Biochem. 62, 583-590.

15. MATHIS, D.J., KINDELIS, A., and SPADAFORA, C. (1980). HMG proteins (1+2) form beaded structures when complexed with closed circular DNA. Nucleic Acids Res. 8, 2577-2590.

16. JAVAHERIAN, K., LIU, L.F., and WANG, J.C. (1978). Non-histone proteins HMG1 and HMG2 change the DNA helical structure. Science 199, 1345-1346.

17. DUGUET, M., BONNE, C., and DE ROCONDO, A.M. (1981). Single-stranded DNA binding protein from rat liver changes the helical structure of DNA. Biochemistry 20, 3598-3603.

18. JAVAHERIAN, K. and SADEGHI, M. (1979). Nonhistone proteins HMG1 and HMG2 unwind DNA double helix. Nucleic Acids Res. 6, 3569-3580.

19. CARBALLO, M., PUIGDOMENECH, P., TANCREDI, T., and PALAU, J. (1984). Interaction between domains in chromosomal protein HMG-1. EMBO J. 3, 1255-1261.

20. COCKERILL, P.N., GOODWIN, G.H., CARY, P.D., TURNER, C., and JOHNS, E.W. (1983). Comparison of the structures of the chromosomal high mobility group proteins HMG1 and HMG2 prepared under conditions of neutral and acidic pH. Biochim. Biophys. Acta 745, 70-81.

21. SANDEEN, G., WOOD, W.I., and FELSENFELD, G. (1980). The interaction of high mobility proteins HMG14 and 17 with nucleosomes. Nucleic Acids Res. 8, 3757-3778.

22. ALBRIGHT, S.C., WISEMAN, J.M., LANGE, R.A., and GARRARD, W.T. (1980). Subunit structure of different electrophoretic forms of nucleosomes. J. Biol. Chem. 255, 3673-3684.

23. MARDIAN, J.K.W., PATON, A.E., BUNICK, G.J., and OLINS, D.E. (1980). Nucleosome cores have two specific binding sites for nonhistone chromosomal proteins HMG-14 and HMG-17. Science 209, 1534-1536.

24. STEIN, A. and TOWNSEND, T. (1983). HMG14/17 binding affinities and DNase I sensitivities of nucleoprotein particles. Nucleic Acids Res. 11, 6804-6819.

25. SWERDLOW, P.S. and VARSHAVSKY, A. (1983). Affinity of HMG17 for a mononucleosome is not influenced by the presence of ubiquitin-H2A semihistone but strongly depends on DNA fragment size. Nucleic Acids Res. 11, 387-401.

26. LEVY-WILSON, B., WONG, N.C.W., and DIXON, G.H. (1977). Selective association of the trout-specific H6 protein with chromatin regions susceptible to DNase I and DNase II: possible location of HMG-T in the spacer region between core nucleosomes. Proc. Natl. Acad. Sci. USA 74, 2810-2814.

27. LEVY-WILSON, B., CONNOR, W., and DIXON, G.H. (1979). A subset of trout testis nucleosomes enriched in transcribed DNA sequences contains high mobility group proteins as major structural components. J. Biol. Chem. 254, 609-620.

28. SCHROTER, H. and BODE, J. (1982). The binding sites for large and small high mobility group (HMG) proteins. Studies on HMG-nucleosome interactions in vitro. Eur. J. Biochem. 127, 429-436.

29. JACKSON, J.B., POLLOCK, J.M., and RILL, R.L. (1979). Chromatin fractionation procedures that yield nucleosomes containing near-stoichiometric amounts of high mobility group nonhistone chromosomal proteins. Biochemistry 18, 3739-3748.

30. REECK, G.R. and TELLER, D.C. (1985). High mobility group proteins: purification, properties and amino acid sequence comparisons. In: "Progress in Nonhistone Protein Research," (I. Bekhor, ed.) pp. 1-21, CRC Press, Boca Raton, Florida.

31. RUIZ-CARRILLO, A., PUIGDOMENECH, P., EDER, G., and LURZ, R. (1980). Stability and reversibility of the higher ordered structure of interphase chromatin. Continuity of deoxyribonucleic acid is not required for maintenance of folded structure. Biochemistry 19, 2544-2554.

32. PUIGDOMENECH, P. and RUIZ-CARRILLO, A. (1982). Effect of histone composition on the stability of chromatin structure. Biochim. Biophys. Acta 696, 267-274.

33. MATTHEW, C.G.P., GOODWIN, G.H., and JOHNS, E.W. (1979). Studies on the association of the high mobility group nonhistone chromatin proteins with isolated nucleosomes. Nucleic Acids Res. 6, 167-179.

34. KUEHL, L., BARTON, D.J., and DIXON, G.H. (1980). Binding of the high mobility group protein H6 to trout testis chromatin. J. Biol. Chem. 255, 10671-10675.

35. KUEHL, L., LYNESS, T., DIXON, G.H., and LEVY-WILSON, B. (1980). Distribution of high mobility group proteins among domains of trout testis chromatin differing in their susceptibility to micrococcal nuclease. J. Biol. Chem. 255, 1090-1095.

36. ISACKSON, P.J., CLOW, L.G., and REECK, G.R. (1981). Comparison of the salt dissociations of high molecular weight HMG non-histone chromatin proteins from double-stranded DNA and from chromatin. FEBS Lett. 125, 30-34.

37. ZENTGRAF, H. and FRANKE, W.W. (1984). Differences of supranucleosomal organization in different kinds of chromatin: cell type ecific globular subunits containing different numbers of nucleosomes. J. Cell Biol. 99, 272-286.

38. STOUTE, J.A. and MARZLUFF, W.F. (1982). HMG proteins 1 and 2 are required for transcription of chromatin by endogenous RNA polymerase. Biochem. Biophys. Res. Comm. 107, 1279-1284.

39. SMERDON, M.J. and ISENBERG, I. (1976). Interactions between subfractions of calf thymus H1 and nonhistone chromosomal proteins HMG1 and HMG2. Biochemistry 15, 4242-4247.

40. CARY, P.D., SHOOTER, K.V., GOODWIN, G.H., JOHNS, E.W., OLAYEMI, J.Y., HARTMAN, P.G., and BRADBURY, E.M. (1979). Does high mobility group non- histone protein HMG1 interact specifically with histone H1 subfractions? Biochem. J. 183, 657-662.

41. CARBALLO, M., PUIGDOMENECH, P., and PALAU, J. (1983). DNA and histone H1 interact with different domains of HMG1 and 2 proteins. EMBO J. 2, 1759-1764.

42. BERNUES, J., QUEROL, E., MARTINEZ, P., BARRIS, A., ESPEL, E., and LLOBERAS, J. (1983). Detection by chemical cross-linking of interaction between high mobility group protein 1 and histone oligomers in free solution. J. Biol. Chem. 258, 11020-11024.

43. BONNE-ANDREA, C., HARPER, F., SOBCZAK, J., and DE RECONDO, A. (1984). Rat liver HMG1: a physiological nucleosome assembly factor. EMBO J. 3, 1193-1199.

44. LASKEY, R.A., HONDA, B.M., MILLS, A.D., and FINCH, J.T. (1978). Nucleosomes are assembled by an acidic protein which binds histones and transfers them to DNA. Nature 275, 416-420.

45. EARNSHAW, W.C., HONDA, B.M., LASKEY, R.A., and THOMAS, J.O. (1980). Assembly of nucleosomes: the reaction involving X. laevis nucleoplasmin. Cell 21, 373-383.

46. STEIN, A. and KUNZLER, P. (1983). Histone H5 can correctly align randomly arranged nucleosomes in a defined in vitro system. Nature 302, 549-550.

47. SEYEDIN, S.M. and KISTLER, W.S. (1979). Levels of chromosomal protein high mobility group 2 parallel the proliferative activity of testis, skeletal muscle, and other organs. J. Biol. Chem. 254, 11266-11271.

48. SMITH, B.J., ROBERTSON, D., BIRBECK, G.H., GOODWIN, G.H., and JOHNS, E.W. (1978). Immunochemical studies of high mobility group non-histone chromatin proteins HMG1 and HMG2. Exp. Cell Res. 115, 420-423.

49. WU, L., RECHSTEINER, M., and KUEHL, L. (1981). Comparative studies on microinjected high mobility group chromsomal proteins HMG-1 and HMG-2. J. Cell Biol. 91, 488-496.

50. BUSTIN, M. and NEIHART, N.K. (1979). Antibodies against chromosomal HMG proteins stain the cytoplasm of mammalian cells. Cell 16, 181-186.

51. ISACKSON, P.J., BIDNEY, D.L., REECK, G.R., NEIHART, N.K., and BUSTIN, M. (1980). High mobility group chromosomal proteins isolated from nuclei and cytosol of clutured hepatoma cells are similar. Biochemistry 19, 4466-4471.

52. EINCK, L., SOARES, N., and BUSTIN, M. (1984). Localization of HMG high mobility group chromosomal proteins in the nucleus and cytoplasm by microinjection of functional antibody gragments into living fibroblasts. Exp. Cell Res. 152, 287-301.

53. WALKER, J.M. (1982). Primary Structures. In: "The HMG Chromosomal Proteins," (E.W. Johns, ed.) pp. 69-87. Academic Press, New York.

54. WALKER, J.M., GOODERHAM, K., HASTINGS, J.R.B., MAYES, E., and JOHNS, E.W. (1980). The primary structures of non-histone chromosomal proteins HMG1 and 2. FEBS Lett. 122, 264-270.

55. PENTECOST, B. and DIXON, G.H. (1984). Isolation and partial sequence of bovine cDNA clones for the high-mobility-group protein (HMG-1). Bioscience Rep. 4, 49-57.

56. REECK, G.R., ISACKSON, P.J., and TELLER, D.C. (1982). Domain structure in high molecular weight high mobility group nonhistone chromatin proteins. Nature 300, 76-78.

57. CRANE-ROBINSON, C., BOHM, L., PUIGDOMENECH, P., CARY, P.D., HARTMAN, P.G., and BRADBURY, E.M. (1980). Structural domains in histones. In: "DNA-recombination, interactions, and repair," (S. Zadrazil and J. Sponar, eds.) pp. 293-300. Pergamon Press, New York.

58. BOHM, L. and CRANE-ROBINSON, C. (1984). Proteases as structural probes for chromatin. The domain structure of histones. _Bioscience Rep_. _4_, 365-368.

59. CARY, P.D., TURNER, C.H., MAYES, E., and CRANE-ROBINSON, C. (1983). Conformation and domain structures of the non-histone chromosomal proteins HMG-1 and 2. Isolation of two folded fragments from HMG-1 and 2. _Eur. J. Biochem_. _131_, 367-374.

60. ISACKSON, P.J., BEAUDOIN, J., HERMODSON, M.A., and REECK, G.R. (1983). Production of HMG-3 by limited trypsin digestion of purified high-mobility-group nonhistone chromatin proteins. _Biochim. Biophys. Acta_ _748_, 436-443.

61. RYOGI, M. and WORCEL, A. (1984). Chromatin assembly in Xenopus oocytes: in vivo studies. _Cell_ _37_, 21-32.

62. PAULSON, J.R. and LAEMMLI, U.D. (1977). The structure of histone-depleted metaphase chromosomes. _Cell_ _12_, 817-828.

STACKING INTERACTIONS: THE KEY MECHANISM FOR BINDING OF PROTEINS TO SINGLE-STRANDED REGIONS OF NATIVE AND DAMAGED NUCLEIC ACIDS?

Jean-Jacques Toulmé

Laboratoire de Biophysique, INSERM U.201
Muséum National d'Histoire Naturelle
61, Rue Buffon
F-75005 Paris, France

INTRODUCTION

The enzymatic removal of chemical damage in DNA, the regulation of gene expression by a repressor molecule, and the binding of RNA polymerase to a promoter are some examples of processes which involve specific interactions between proteins and nucleic acids. Actually, most of the steps of replication, transcription, and translation of the genetic information, as well as DNA repair, RNA maturation, or building of nucleosomes or ribosomes require the specific recognition of a nucleic acid structure or base sequence by proteins. Besides gross structural complementarity between the two interacting macromolecules, interactions between individual amino acids and nucleotides can provide the required specificity. We may consider the active center of a nucleic acid-binding protein as a three dimensional distribution of functional groups able to form ionic bonds, hydrogen bonds, and hydrophobic bonds (including stacking). Specificity will be attained when this distribution corresponds to a complementary one on a polynucleotide chain. This might also involve conformational changes of the macromolecule(s).

One may distinguish two different classes of nucleic acid-binding proteins. A protein in the first class recognizes a particular sequence of nucleic acid bases. The bases may be contiguous to each other in the sequence or they may move into juxtaposition after folding of the nucleic acid molecule. Restriction endonucleases, the lac repressor of E. coli, and mRNA splicing enzymes (for instance) belong to this category. A protein of the second class specifically interacts with particular physical forms of nucleic acids. For example, the enzymes involved in DNA repair appear to recognize a locally distorted region in the DNA double helix. Proteins which bind specifically to single-stranded nucleic acids (SSB proteins) also fall into this category. Some proteins clearly exhibit properties of both classes. For example RNA polymerases are able to recognize promoter sequences and to also discriminate between single-stranded and double-stranded structures of terminators and attenuators.

The mechanisms employed by each of the above two classes of proteins to ensure their binding specificity are certainly not the same. The recognition of a particular base sequence probably requires structural complementarity and hydrogen bonding interactions. On the other hand, when a protein recognizes a single-stranded nucleic acid, stacking interactions between aromatic amino acids and nucleic acid bases are certainly important. Analyzing the exact role played by each type of interaction in macromolecular complexes has been rather difficult and therefore much work has been devoted to model systems.

This chapter will focus on proteins that recognize single-stranded nucleic acids. The involvement of stacking interactions in these complexes will be discussed. Results will be presented from studies both on models (synthetic oligopeptides) and natural proteins.

POTENTIAL MOLECULAR MECHANISMS FOR THE RECOGNITION OF SINGLE-STRANDED NUCLEIC ACIDS BY PROTEINS

The types of interactions between functional groups of amino acids and nucleic acid bases have been recently reviewed (1). Most if not all nucleic acid-protein complexes involve electrostatic interactions. Generally this contribution accounts for the major part of the association energy (1). Such interactions result in a stabilization of the nucleic acid structure (2, 3). In addition, small conformational changes of the nucleic acids can be detected by UV absorption and circular dichroism (4). In a first approximation, electrostatic interactions will weakly contribute to the specificity of association. Nevertheless, it has been shown that poly-lysine binds preferentially to A-T rich DNAs, whereas poly-arginine has a higher affinity for G-C rich DNAs (5, 6). Hydrogen bonding may involve the nucleic acid bases, the carbohydrate, or the phosphate group. On the other hand, the peptide bond itself and many amino acids possess groups which can engage in hydrogen bonds. Models have been proposed for molding of antiparallel β sheets into the grooves of A-RNA or B-DNA (8, 9, and for a review see reference 10). Several possibilities have been proposed for specific recognition of base pairs by amino acid side chains through hydrogen bonding (10, 11). Crystallographic studies have confirmed that hydrogen bonding may take place in nucleic acid-protein complexes (12).

Numerous studies have been devoted to aromatic amino acids (tryptophan, tyrosine, and phenylalanine) and to complexes they are able to form with nucleic acid bases. Several dyes, antibiotics, and other polycyclic aromatic compounds are known to form intercalated complexes with DNA. In contrast, the intercalation of tryptophan, phenylalanine, or tyrosine between nucleic acid bases is not likely to occur due to the sizes of these aromatic molecules; but, the aromatic amino acids may partially stack with bases on one strand. First evidence for stacking interactions was provided by luminescence (13, 14). Stacked complexes between indole derivatives and purines have also been demonstrated in concentrated aqueous solutions (15, 16). In oligopeptide-nucleic acid complexes, stacking interactions were suggested by Brown (17), who investigated the melting of DNA in the presence of a series of basic dipeptide methyl esters. He found that the peptide Arg-Trp-O-methyl led to a higher stabilization of DNA than Arg-Arg-O-methyl.

Various spectroscopic and hydrodynamic methods have been used to follow the behavior of the aromatic residues of peptides upon association with polynucleotides (10). NMR and fluorescence are the most powerful techniques for this purpose and therefore have been the most widely used. Stacking interactions result in an upfield shift of the proton resonances of aromatic rings due to ring current effects. Binding of an oligopeptide containing an aromatic amino acid to a single-stranded poly-nucleotide generally leads to such a shift of both aromatic and $(CH_2)\beta$ proton resonances of the amino acid. This is accompanied by a broadening of the peaks due to the restricted mobility of the ligand in the complexes(18-20). With sonicated double-stranded DNA, oligopeptides containing either a tryptophyl or a phenylalanyl residue exhibit a similar behavior as compared with single-stranded polynucleotides even though the upfield shifts are much smaller. In contrast, it was reported that no upfield shift could be detected in the case of tyrosyl residues with DNA (20, 21). This suggests that tyrosine does not stack with nucleic acid bases in double-stranded nucleic acids. However, it must be pointed out, as discussed in reference 10, that the amplitude of the upfield shifts depends on: 1) the relative location of the stacked rings; 2) the amplitude of the ring current effect, which is higher for purines than for pyrimidines; and 3) the relative proportion of stacked and unstacked molecules.

Binding of an oligopeptide containing an aromatic amino acid to a nucleic acid generally results in a partial quenching of the fluorescence emission of the peptide. However, Phe-containing peptides have been studied little by fluorescence spectroscopy, due to a significant overlap between the absorption of nucleic acids and the fluorescence emission of phenyl-alanine, and to the very low molar extinction coefficient of Phe (20). The fluorescence of tyrosine can be quenched either by a direct interaction with nucleic acid bases (hydrogen bonding or stacking) or by a singlet-singlet energy transfer to nearby nucleic acid bases. The latter mechanism does not require a contact between the two partners: the critical Forster distance ranges up to about 20 Å (22). In many cases, the fluorescence is quenched but there is no indication of stacking interactions (e.g., when tyrosyl-containing peptides are bound to native DNA (20)).

As stated above, stacking interactions of tryptophan derivatives with nucleic acid bases results in the quenching of indole fluorescence. It is crucial to determine whether other kinds of interactions can also lead to such a quenching. It has been shown that anions of phosphate diesters such as $(CH_3O)_2PO_2^-$ are poor quenchers of tryptophan fluorescence (in contrast to dianions of phosphate monoesters) (23). This suggests that quenching of tryptophan by phosphate groups of polynucleotides is not expected unless the indole ring is close to a terminal phosphate group. Similarly, hydrogen bonding by N-H of indole group is not very likely to result in fluorescence quenching <u>via</u> a proton transfer (1). In contrast to tyrosine, singlet-singlet energy transfer from tryptophan to nucleic acid bases requires a very short distance ($\sim$ 5 Å) to occur (22). At that close range, other mechanisms (electron transfer) are also possible for fluorescence quenching. Thus, in the case of tryptophan-nucleic acid complexes, stacking interactions are presumably the main mechanism responsible for fluorescence quenching in peptide-nucleic acid complexes.

The binding to single-stranded and double-stranded poly-nucleotides of oligopeptides that contain both basic residues and tryptophan has been investigated in several laboratories (10). A thoroughly studied peptide of this sort is lysyl-tryptophyl-lysine (KWK). Binding results in an upfield shift of the tryptophan proton resonances (19). Modifications in the polynucleotide circular dichroism spectrum also indicate a con-formational change of the macromolecule (24). Moreover, this is accompanied by a decrease of the tryptophan fluorescence inten-sity, corrected for screening effect (10), although the average fluorescence lifetime of the peptide is not different from that of the free molecule (25). These results led to a two step model for binding of KWK to nucleic acids. A first complex, C_1, involves only electrostatic interactions between amino groups of the peptide and phosphate groups of the polynucleotide. This initial complex can then convert into a second complex, C_2, in which besides the electrostatic interactions, the indole ring is stacked with the nucleic acid bases (26). The unitless equilib-rium constant, K_2, between unstacked and stacked complexes describes the "stacking tendency" of the system. A value of K_2 that is less than 1 means that the formation of the outside complex is favored and, conversely, a value larger than 1 is characteristic of a system in which stacked complexes form preferentially. Investigations of various KWK-polynucleotide complexes have shown that K_2 ranges from 0.3-0.6 for double-stranded polynucleotides and up to 2-5 for single-stranded ones (26, 27). Therefore, stacked complexes are strongly favored in single-stranded structures. On the other hand, the equilibrium constant K_1 for formation of the outside complex appears not to depend on the polynucleotide structure (26). Therefore, the overall constant $K = K_1 (1 + K_2)$ will be higher for single-stranded than for double-stranded polynucleotides. In other words, stacking interactions are responsible for the preferential binding of the peptide KWK to single-stranded nucleic acids.

Several other results agree in indicating stacking of indole with nucleic acid bases in KWK-polynucleotide complexes. Electric dichroism measurements have shown that the tryptophyl ring is nearly parallel to adenylic residues in KWK-poly(A) complexes (28). Moreover, from triplet-triplet energy transfer in such complexes, it was concluded that a peptide molecule traps the energy from about 60 residues (29). Such a process requires a good orbital overlap as is encountered in stacked complexes. Finally, it has been reported that the luminescence properties of KWK at low temperature are strongly perturbed upon binding to poly(5-mercuriuracil) (30-32). In particular, the phosphorescence lifetime is shortened from about 6 s to 5 ms. This is characteristic of a so-called heavy atom effect, which implies a van der Waals contact between the indole ring and 5-mercuriuracil.

Using synthetic random copolymers (of A and U, for instance), it has been shown that stacking of tryptophan can depend on the base sequence: stacked complexes are favored in UU as compared to AA sequences (10, 33). This base sequence specificity actually reflects the internal base-base stacking efficiency of the polymers. Moreover, stacking of the indole ring is expected to be more important with purines than with pyrimidines. Therefore, stacking interactions can lead to sequence specificity. It has also been reported that diastereomeric peptides exhibit different behaviors with respect to DNA binding: the aromatic ring of L-Lys-L-Phe is partially inserted between the DNA base pairs, whereas, the ring of L-Lys-D-Phe points outward (34, 35). Finally, it should be added that stacking of the indole ring with nucleic acid bases still occurs in complexes between tryptophan-containing tetrapeptides and nucleosome core particles. Thus, DNA interactions with histones do not prevent stacking (36).

SINGLE-STRANDED DNA BINDING PROTEINS

Proteins which bind preferentially to single-stranded nucleic acids have been isolated from various organisms ranging from phages to higher eukaryotes (37-41; also see the chapter by Puigdomènech and Jose, this volume). These single-strand binding (SSB) proteins are essential to several physiological functions including replication, recombination and repair of DNA (SSB proteins are also termed helix-destabilizing (HD) proteins due to the fact that they lower the melting temperature of DNA by selectively binding and trapping single-stranded regions, or at least shift the conformational equilibrium from helix to coil). Although these proteins are devoid of any enzymatic activity, they control several steps of the biological processes in complicated ways (see 42-44 for reviews). I will focus here on two representative members of this class: the gene 5 product from bacteriophages fd or M13 and the gene 32 product from bacteriophage T4. For both proteins, binding parameters and details regarding the molecular structures are available.

Characterization of the binding requires the determination of both thermodynamic and kinetic parameters. General approaches have been developed to determine the number of residues, n, covered by a protein (the size of the binding site), K (the association constant of the protein to an isolated binding site), and ω (the ratio of the association constants for isolated and contiguous binding sites) (43). If ω is greater than one, the binding is cooperative, which is generally observed for SSB proteins on single-stranded nucleic acids. Additional information on the number of charge-charge interactions involved in the nucleic acid-protein complexes can be obtained from an analysis of the variation of the intrinsic association constant K (or of the apparent association constant $K\omega$, for cooperative binding) with salt concentration (7). Finally, the complete description of the system requires the determination of association and dissociation rate constants for various pathways.

The binding of proteins to nucleic acids can be monitored by various techniques. Signals related to nucleic acids (e.g., UV absorption, circular dichroism, fluorescence emission of modified polynucleotides) can be used if binding is accompanied by a structural change of the polynucleotide chain. Proteins containing tryptophan and/or tyrosine can be followed by emission spectroscopy using the intrinsic luminescence of these chromophores. Nuclear magnetic resonance studies and chemical modification of the protein or of the complex can bring additional information about the nature of the interacting domains.

Gene 5 Protein from Bacteriophage fd

The gene 5 product (gp5) from filamentous phages fd (M13) is a protein of 9700 daltons. It plays a central role in the production of single-stranded DNA. An infected E. coli cell contains about 10^5 molecules of gp5 to protect the nucleic acid against nucleases. Moreover, gp5 collapses the viral DNA circle into two antiparallel strands, which is the conformation required for encapsidation by the native virus (42, 43).

Nucleic acid binding parameters. Gp5 exists in solution mainly in a dimer-monomer equilibrium. The dimer is the active binding form. The spectroscopic properties of the nucleic acid are changed upon gp5 association: both an hyperchromism in the UV absorption band of the bases and a dramatic modification of the circular dichroism spectrum are observed (46). This suggests that single strands adopt an extended conformation in the complex resulting from unstacking of the nucleic acid bases. These effects are saturated at about four nucleotide residues per protein monomer. Gp5 protein destabilizes duplex DNA molecules. This effect is relatively nonspecific with respect to DNA base composition (38).

It is well known that gp5 binding to single-stranded nucleic acids is cooperative (38). The apparent equilibrium constant Kω was determined using fluorescence quenching of the protein upon nucleic acid binding. For poly(dA), Kω varies from

10^7 M^{-1} (in 0.06 M Na^+) to 10^5 M^{-1} (in 0.22 M Na^+) (47). Part of the affinity is due to the cooperativity factor, but the major salt dependence resides essentially in K. ω increases slightly from 50 to 300 over the same range of Na^+ concentration. Therefore, electrostatic interactions play the major role in gp5-poly(dA) complexes, whereas protein-protein interactions responsible for co-operativity are primarily nonionic (47). Moreover, it has been shown that the gp5 affinity for single strands depends on the base composition of the polymers. Poly(dT) is the preferred substrate for the protein (48).

In contrast to the results obtained with larger polynucleotides, a stoichiometry of 3 nucleotides per monomer is obtained for binding to the short oligonucleotides $(dA)_8$ to $(dA)_{16}$. Also, the ionic strength dependence of K is different and the apparent association constant is lower than that for polynucleotides. The suggestion that complexes formed by gp5 with oligo(dA) differ from that with poly(dA) was confirmed by results obtained with oligonucleotides of intermediate lengths (25 to 30 residues). Alma et al. (49) found that two types of complexes can exist with such substrates, depending on the relative concentrations of gp5 and oligo(dA)$_{25-30}$. The authors proposed that these effects could be related to additional protein-protein interactions causing an helical structure of the complexes (49). Kinetic experiments have also been performed to determine rate constants of association to isolated sites and dissociation from clusters (48, 50).

Structure of the gene 5 protein and its nucleic acid binding site. X-ray analysis of crystalline gp5 reveals a T-shaped molecule composed primarily of β structures organized as one three-stranded sheet and one two-stranded ribbon (12). Along the three-stranded β sheet is a putative DNA binding channel about 10 Å wide and 40 Å long. Therefore, the active DNA binding species (the gp5 dimer) provides two antiparallel binding grooves with opposite polarities separated by 30 Å. Each channel is long enough to accommodate about five nucleotide residues. Located in this region are several aromatic amino acids (Tyr and Phe) and basic residues (Lys and Arg) (51).

271

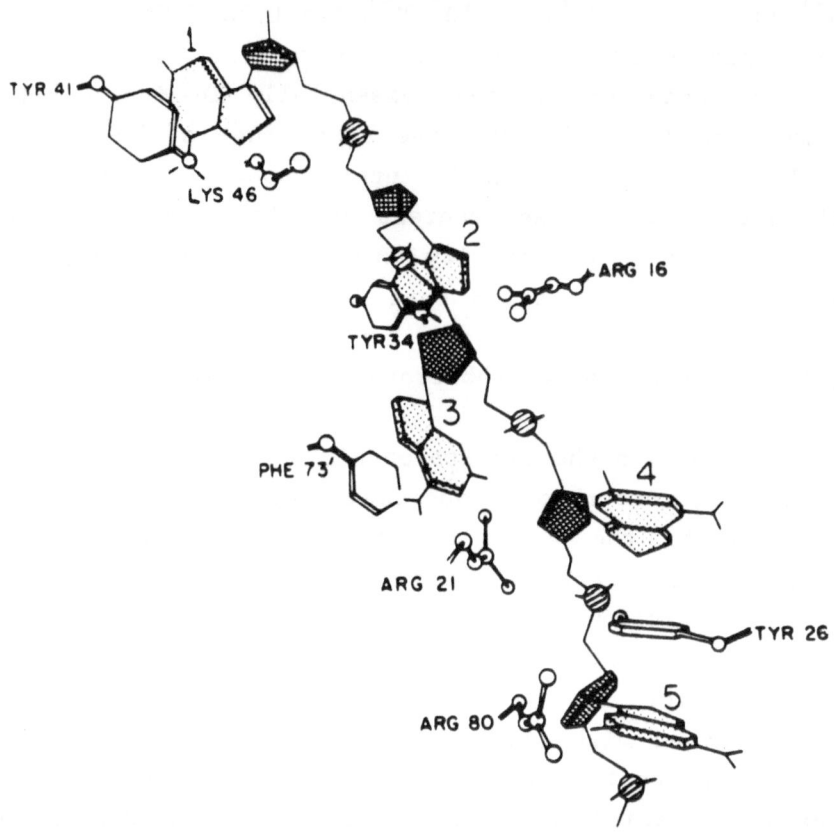

Figure 1.   Schematic drawing of one DNA binding site of the gp5
dimer.   The DNA strand is composed of a pentanucleo-
tide $(dAp)_5$.   Electrostatic   interactions   occur
between phosphate groups and basic amino acids (Arg
16, Arg 21, Arg 80, and Lys 46).   Aromatic amino
acids are stacked with nucleic acids bases (Tyr 41,
Tyr 34 and Phe 73') or sandwiched between two adenines (Tyr 26).   The Phe residue (indicated with a
prime) belongs to the symmetry-related gp5 monomer.
Figure is taken from reference 52 by permission of
the American Chemical Society and Dr. A. McPherson.

Both the salt dependence of the gp5 binding to polynucleotides (see above) and acetylation of lysine (46) suggest interactions between positively charged amino acids and phosphate
backbone.   Two residues (Lys 46 and Arg 80) are immediately
adjacent to the DNA in the structure deduced from X-ray data.
Furthermore, a minor structural rearrangement proposed by Brayer
and McPherson (52) brings two additional basic residues (Arg 16
and Arg 21) into the DNA binding groove.   This agrees with the
value of 4 ± 0.5 charges involved in complex formation (48).

Several methods have also suggested the involvement of aromatic residues in binding of gp5 to oligonucleotides or DNA. In the free protein, three of five tyrosines can be reacted with tetranitromethane. DNA binding is lost upon nitration of these residues. Conversely, the presence of DNA protects these residues from chemical modification (46). Marked changes in the tyrosyl CD band and in the NMR spectra are observed upon complexation of gp5 (53, 54). From photochemically induced dynamic nuclear polarization effects it was concluded that the hydroxyl groups of three tyrosines are not accessible for reaction with the dye even when gp5 is bound to as small a fragment as a tetranucleotide (55). Chemical shifts in the proton resonances of tyrosine and phenylalanine in gp5-oligonucleotide complexes have been interpreted as due to stacking of three tyrosyl and one phenylalanyl residues with nucleic acid bases. On the basis of the crystal structure, the shifted proton resonances have been assigned to Tyr 26, Tyr 34, Tyr 41, and Phe 73 (56). A model which accounts for both NMR and X-ray data has been recently proposed (52) (Figure 1).

Complexes of gp5 and various oligodeoxynucleotides have been crystallized. Analysis of all the crystals shows an arrangement of 12 gp5 monomers per unit cell with a hexagonal symmetry (12). From these results, a double-helical model was postulated for the complex, in which the DNA is wrapped around the gp5 core. However, neutron small-angle scattering studies of fd DNA-gp5 complexes in solution would rather suggest a central position for the DNA (57).

Gene 32 Protein from Bacteriophage T4

The protein encoded by gene 32 from phage T4 (gp32) was the first single-strand binding protein to be purified (37). It is involved in several steps of the viral lifecycle: amber or ts mutants exhibit an abnormal or stopped lytic cycle (58). Gp32 is part of the T4 multi-enzymatic complex of replication (59). It specifically stimulates T4 DNA polymerase (60) and improves the accuracy of in vitro replication (61). A functional gene 32 product is required for phage DNA recombination (62), and it also protects single-stranded regions of DNA against nucleases (63). The overproduction of gp32 following UV-irradiation of

T4-E. coli infected cells suggests that it plays a role in excision repair of photodamage (64). It has been demonstrated that it autogenously regulates its synthesis by specific binding to its own mRNA (65-67).

Different domains exist in this protein of about 34,000 daltons (68). Limited tryptic digestion gives rise to three different overlapping products, termed gp32*I, gp32*II, and gp32*III. The gp32*I fragment has lost about 60 amino acids (the A peptide) from the carboxy terminus, the gp32*II product is obtained by removing about 20 residues (the B peptide) from the amino terminus, and the gp32*III fragment results from the removal of both the A and B peptides (69). These three proteolytic products each have different binding properties (see below). The amino and carboxy terminal domains are also apparently involved in the formation of gp32-nucleic acid complexes. This is suggested by the fact that the rates of limited proteolysis depend upon whether the protein is free or bound to a polynucleotide (69, 70).

Binding parameters. From several points of view, gp32 exhibits a behavior similar to that of the M13 SSB protein (gp5). The binding to single-stranded polynucleotides is highly cooperative ($\omega = 10^3$-10^4). The intrinsic association constant K is ionic strength-dependent whereas the co-operativity factor is not. The major part of salt dependence is due to anion displacement effects (71). As expected from the proteins' respective molecular weights, the size of the binding site is larger for gp32 (n = 7 $\pm$ 1 nucleotides) than for the gp5 monomer (72, 73). As in the case of gp5, the polynucleotide chain is fully extended upon binding to gp32 (74). Unstacking of the nucleic acid bases in the complexes results in hyperchromicity (73). When poly(1,N^6-ethenoadenylic acid) is used as a substrate, there is an increase of its fluorescence (71, 75). As already observed with M13 gene 5 product, poly(dT) is the polynucleotide for which gp32 exhibits the highest affinity: K ranges from 10^6 M^{-1} for poly(rC) to 10^{11} M^{-1} for poly(dT) in 0.2 M NaCl, but ω does not depend on the polynucleotide composition (76). Moreover, the apparent association constant is higher for polydeoxyribonucleotides than for the homologous polyribonucleotides. These results extrapolated to T4 DNA and mRNAs indicate that autogenous regulation of gene 32 might occur through a DNA-RNA competition (77). However, there is evidence that the mechanism of translational regulation also involves some sequence specificity: gp5 does not regulate gene 32 expression in spite of the similarity between the binding properties of these two SSB proteins (78).

Although the affinity of gp32 for native DNA is quite weak as compared to single-stranded DNA, no destabilization of duplex DNA has been observed, except for synthetic poly(dAT)/poly(dAT) (73). In contrast, the gp32*I fragment is able to melt double-stranded DNA (69, 79). This suggests that the melting of double-strands by intact gp32 is kinetically blocked and that the carboxy terminal domain is responsible for this block. The thermodynamic parameters n, K, and ω for the association of the gp32*I fragment with polynucleotides are essentially the same as that for the native gp32 (80).

The most striking difference between the gp32*III fragment and the native protein is that the fragment does not bind co-operatively to single-stranded nucleic acids (i.e., ω = 1). Nevertheless, similar modifications in the circular dichroism and UV absorption spectra of polynucleotides are observed upon binding either gp32*III or native gp32 (80). This suggests that gp32 co-operativity is predominantly due to protein-protein interactions rather than to induced structural changes in the nucleic acids. Amino acid residues in the N-terminal part of the protein are engaged in these interactions.

Binding of gp32 to short oligonucleotides differs from that observed with longer polynucleotides. The association does not show a salt dependence and the association constant K does not vary with the length of the oligonucleotide. This last observation indicates that the protein is not free to bind to the lattice in multiple ways (71). Moreover, K does not vary between different nucleic acid bases. These results indicate a different mode of binding of gp32 to oligo- and polynucleotides. A model has been proposed by von Hippel and co-workers which can account for the results obtained both with poly- and oligonucleotides (43, 71).

Association and dissociation kinetics of gp32 with various polynucleotides have been investigated. From fluorimetric stopped-flow experiments, it has been concluded that co-operative binding of gp32 consists of a nucleation step (binding of isolated protein molecules) followed by an elongation step (association of additional molecules adjacent to already-bound gp32 (81)). The rate-limiting step depends on the salt conditions. Moreover, gp32 is able to slide along the polynucleotide strand. It was recently shown that the polynucleotide specificity and the cooperativity reside in the dissociation constants (82).

Interactions involved in gp32-nucleic acid complexes. In contrast to gp5, refined crystallographic data are not available for gp32. Gp32*I fragment has been crystallized but the crystals diffract poorly (83, 84).

Evidence for both electrostatic and stacking interactions in gp32-polynucleotide complexes has been reported. Equilibrium binding studies have indicated that 2 to 3 phosphate residues interact with each gp32 monomer in these complexes. With short oligonucleotides, on the other hand, there is at most 1 electro-static interaction (71). Dinucleotides have been used as reporter molecules to map the gp32 binding site. Upon binding d(ApAp), a larger fluorescence quenching was observed than with d(pApA). This indicates that the binding of oligonucleotides to gp32 is polarized. A tryptophan whose fluorescence was quenched by the 3' phosphate group of d(ApAp) is located in the binding site (85). A similar conclusion was also drawn for gp32-poly-nucleotide complexes (75). These results and that obtained from chemical modification of gp32 by tetranitromethane (86) were the first indications that aromatic residues were important compo-nents in the binding site of the T4 SSB protein.

The determination of the gp32 primary sequence revealed an unusual distribution of tyrosyl and tryptophyl residues in the part of the protein that contains the nucleic acid binding domain (68). The region between positions 72 and 116 contains 6 of the 8 tyrosines and 2 of the 5 tryptophans. The possibility that some of these residues might interact with nucleic acid constituents has been examined using various techniques. Selective photochemical and radiochemical modifications of tryptophan in gp32 or gp32-polynucleotide complexes have been carried out (87, 88). UV irradiation of gp32 in the presence of trichloroethanol leads to the modification of 3 tryptophyl residues. Although no conformational change can be detected using circular dichroism, the photochemically modified protein is no longer able to induce hyperchromism of poly(rA) and does not promote the retention of heat-denatured DNA on nitrocellulose filters, as does the native protein. This suggests that some of the modified residues could be involved in crucial interactions with the nucleic acid. A contact between tryptophan and uracil has been demonstrated using poly(5-mercuriuridylic acid). Binding of this polynucleotide to gp32 (whose accessible cysteinyl residues have been previously blocked by N-ethylmaleimide to prevent the formation of thiol-mercury bonds) results in the appearance of a short-lived component ($\tau \sim 5$ ms) in the phosphorescence emission of the complex (89). This heavy-atom effect is similar to that observed in stacked complexes of KWK and poly(5-mercuriuridylic acid) (see above and reference 30). This suggests that at least one tryptophan interacts. It is probably stacked with nucleic acid bases. Such a heavy atom effect has been also reported for SSB protein from E. coli (98). The photosensitized cleavage of pyrimidine dimers in UV irradiated DNA-gp32 complexes also supports the existence of stacking interactions between tryptophanyl residues and nucleic acid bases (72). Recently, covalent adducts between gp32 and DNA containing 5-bromouracil residues have been formed upon UV irradiation (Toulme, Loreau, and Le Doan, unpublished results). The photoproduct exhibits a fluorescence emission similar to that of tryptophanbromouracil photoadduct (90), again confirming the intimate association of the tryptophan with the DNA.

The involvement of tyrosine residues also has been demonstrated in several laboratories. In the free protein, 4 to 5 tyrosine residues can be reacted with tetranitromethane, whereas none can be nitrated in complex with single-stranded DNA. Moreover, nitration of the gp32 tyrosyl residues abolishes the DNA binding activity of the protein (86). The results from pulse-radiolysis experiments obtained either on both the free molecule and on gp32-nucleic acid complexes confirm the presence of tyrosine(s) in the gp32 binding site (91).

In contrast to gp5 from filamentous phages, gp32 in concentrated solutions forms long oligomers even in the absence of polynucleotides. The resulting broadening of the proton resonances prevents NMR studies of native gp32 and gp32-oligonucleotide complexes. This difficulty has been overcome by using the gp32*III fragment which exhibits binding properties for isolated sites very similar to that of intact protein but does not self-associate (see above). Proton NMR studies of gp32*III and its complexes with oligoadenylic acids have revealed changes in the chemical shifts of several of the aromatic residues (92). Comparison with NMR spectra of gp5-oligonucleotides complexes led the authors to conclude that several tyrosyl and phenylalanyl residues and probably one tryptophan were stacked with nucleic acid bases in gp32-oligo(dA) complexes.

CONCLUSIONS

Studies of oligopeptide binding to polynucleotides clearly demonstrate that stacking interactions between aromatic amino acids and nucleic acid bases are strongly favored in single-stranded nucleic acids. This allows a tripeptide such as KWK to have a higher affinity for single-stranded than for double-stranded polynucleotides. Moreover, these interactions allow the peptide to recognize locally destabilized regions of DNA following UV irradiation or chemical modification (10, 44, 99). Fluorimetric investigations of complexes of damaged DNA and KWK have shown that the peptide binds preferentially in the vicinity of pyrimidine dimers induced by UV irradiation (27), of guanine-aminofluorene derivative adducts (93), and of cis-diaminodichloroplatinum-modified bases (94). These damages are all known to induce a local opening of the DNA double helix. In contrast, chemical modification of double-stranded DNA by dimethylsulfate or by trans-diaminodichloroplatinum does not result in a significant destabilization of double-stranded DNA and KWK does not exhibit a higher affinity for these modified DNAs than for native DNA (94, 95). These observations must be related to the fact that the initial formation of a stacked complex requires a local distortion or unwinding of the double helix which probably involves an increase of the distance between two adjacent bases. Whether the stacked complex will form depends on the balance between the further energy required for structural change in the polynucleotide and that gained from stacking of the aromatic amino acid with nucleic acid bases. From this point of view, apurinic sites are expected to be the strongest peptide binding sites: the cavity left by the removal of a purine is exactly the correct size to accommodate an indole ring. Thus, no energy has to be provided for unstacking of bases. As a matter of fact, the equilibrium constant K_2 between the outside and the stacked complexes (see above) was found to be two orders of magnitude higher for association of KWK with apurinic sites than with native double-stranded sites (95, 96). Therefore, stacking of indole can be a very efficient way for a peptide to recognize single-stranded structures. For tyrosine, a similar mechanism is potentially even more discriminating: no stacking of the tyrosyl ring was detected when tyrosine-containing peptides bound to double-stranded polynucleotides, whereas, stacking was readily observed with single-stranded ones (20).

Stacking of aromatic amino acids could therefore be a very efficient way for repair enzymes to recognize their target. It has been reported that gp32 is able to bind to damaged regions in double-stranded DNA in so far as the damage will result in a local destabilization of nucleic acid (97). Stacking interactions involving tyrosine, phenylalanine, and possibly tryptophan could play a role in the binding of SSB proteins to single-stranded nucleic acids. The results presented above suggest that this is indeed the case for gp5 from fd and gp32 from T4 phages. Whether such stacking interactions are of general importance in interactions of other SSB proteins with DNA will await detailed investigations of those proteins and their interactions.

ACKNOWLEDGEMENTS

Thanks are due to Professor Claude Hélène (Paris) and Dr. Dennis Searcy (Amherst, MS) for critical reading of the manuscript. I am grateful to Drs. Alexander McPherson and Gary D. Brayer for permission to reprint their gp5-binding site model.

REFERENCES

1. HELENE, C. and LANCELOT, G. (1982). Interactions between functional groups in protein-nucleic acid associations. Prog. Biophys. Molec. Biol. 39, 1.

2. GOUREVITCH, M., PUIGDOMENECH, P., CAVE, A., ETIENNE, G., MERY, J., and PARELLO, J. (1974). Model studies in relation to the molecular structure of chromatin. Biochimie 56, 967.

3. PINKSTON, M.F. and LI, H.J. (1974). Studies on interactions between poly(L-lysine$_{40}$, L-alanine$_{60}$) and deoxyribonucleic acids. Biochemistry 13, 5227.

4. PORSCHKE, D. (1979). The binding of Arg- and Lys-peptides to single-stranded polyribonucleotides and its effect on the polymer conformation. Biophys. Chem. 10, 1.

5. LENG, M. and FELSENFELD, G. (1966). The preferential interactions of polylysine and polyarginine with specific base sequences in DNA. Proc. Natl. Acad. Sci. USA 56, 1325.

6. WEHLING, K., ARFMANN, H.A., STANDKE, K.H.C. and WAGNER, K.G. (1975). Specificity of DNA-basic polypeptide interactions. Influence of neutral residues incorporated into polylysine and polyarginine. Nucleic Acids Res. 2, 799.

7. RECORD, M.T., Jr., ANDERSON, C.F., and LOHMAN, T.M. (1978). Thermodynamic analysis of ion effects on the binding and conformational equilibria of proteins and nucleic acids: the roles of ion association or release, screening, and ion effects on water activity. Quarterly Rev. Biophys. 11, 103.

8. CARTER, C.W., and KRAUT, J. (1974). A proposed model for interaction of polypeptides and RNA. Proc. Natl. Acad. Sci. USA 71, 283.

9. GURSKY, G.V., ZASEDETELEV, A.S., TUMANYAN, V.G., ZHUZE, A.L., GROKHOVSKY, S.L., and GOTTIKH, B.P. (1979). Complementarity and recognition code between regulatory proteins and DNA. FEBS Proc. 51, 23.

10. HELENE, C. and MAURIZOT, J.C. (1981). Interactions of oligopeptides with nucleic acids. CRC Crit. Rev. Biochem. 10, 213.

11. HELENE, C. (1977). Specific recognition of guanine bases in protein-nucleic acid complexes. FEBS Lett. 74, 10.

12. McPHERSON, A. (1982). X-ray crystallographic studies of nucleic acid binding proteins. In: "Topics in Nucleic Acid Structure," (S. Neidle, ed.) part 2, p. 199. Mcmillan Press, New York.

13. MONTENAY-GARESTIER, T. and HELENE, C. (1968). Molecular interactions between tryptophan and nucleic acid components in frozen aqueous solutions. Nature 217, 844.

14. MONTENAY-GARESTIER, T. and HELENE, C. (1971). Reflectance and luminescence studies of molecular complex formation between tryptophan and nucleic acid components in frozen aqueous solutions. Biochemistry 10, 300.

15. DIMICOLI, J.L. and HELENE, C. (1973). Complex formation between purine and indole derivatives in aqueous solutions. Proton magnetic resonance studies. J. Amer. Chem. Soc. 95, 1036.

16. RASZKA, M. and MANDEL, M. (1971). Interaction of aromatic amino acids with neutral poly(adenylic) acid. Proc. Natl. Acad. Sci. USA 68, 1190.

17. BROWN, P.E. (1970). The interaction of basic dipeptide methyl ester with DNA. Biochim. Biophys. Acta 213, 282.

18. GABBAY, E.J., SANFORD, K., BAXTER, C.S., and KAPICAK, L. (1973). Specific interaction of peptides with nucleic acids. Evidence for a "selective bookmark" recognition hypothesis. Biochemistry 12, 4021.

19. DIMICOLI, J.L. and HELENE, C. (1974). Interactions of aromatic residues of proteins with nucleic acids. I. PMR studies of the binding of tryptophan-containing peptides to poly(adenylic acid) and deoxyribonucleic acid. Biochemistry 13, 714.

20. MAYER, R., TOULME, F., MONTENAY-GARESTIER, T., and HELENE, C. (1979). The role of tyrosine in the association of proteins and nucleic acids. J. Biol. Chem. 254, 75.

21. NOVAK, R.L. and DOHNAL, J. (1973). Tyrosyl peptide models for acidic protein-DNA interactions. Nature New Biol. 243, 155.

22. MONTENAY-GARESTIER, T. (1975). Singlet energy transfer between aromatic amino acids and nucleic acid bases. Theoretical calculations. Photochem. Photobiol. 22, 3.

23. ALEV-BEHMOARAS, T., TOULME, J.J., and HELENE, C. (1979). Effect of phosphate ions on the fluorescence of tryptophan derivatives. Implication in fluorescence investigation of protein-nucleic acid complexes. Biochimie 61, 957.

24. DURAND, M., MAURIZOT, J.C., BORAZAN, H.N., and HELENE, C. (1975). Interactions of aromatic residues of proteins with nucleic acids. Circular dichroism studies of the binding of oligopeptides to poly(adenylic acid). Biochemistry 14, 563.

25. MONTENAY-GARESTIER, T., BROCHON, J.C., and HELENE, C. (1981). Complex formation between tryptophan-containing peptides and nucleic acids: fluorescence decay studies using synchrotron radiation. Int. J. Quantum Chem. 20, 41.

26. BRUN, F., TOULME, J.J., and HELENE, C. (1975). Interactions of aromatic residues of proteins with nucleic acids. Fluorescence studies of the binding of oligopeptides containing tryptophan and tyrosine residues to polynucleotides. Biochemistry 14, 558.

27. TOULME, J.J. and HELENE, C. (1977). Specific recognition of single-stranded nucleic acids. Interaction of tryptophan-containing peptides with native, denatured, and ultraviolet-irradiated DNA. J. Biol. Chem. 252, 244.

28. PORSCHKE, D. (1980). Structure and dynamics of a tryptophan peptide-polynucleotide complex. Nucleic Acids Res. 8, 1591.

29. HELENE, C. (1973). Energy transfer between nucleic acid bases and tryptophan in aggregates and in oligopeptide-nucleic acid complexes. Photochem. Photobiol. 18, 255.

30. HELENE, C., TOULME, J.J., and LE DOAN, T. (1979). A spectroscopic probe of stacking interactions between nucleic acid bases and tryptophan residues of proteins. Nucleic Acids Res. 7, 1945.

31. CHA, T.A. and MAKI, A.H. (1982). Influence of the mercury blocking reagent 2-mercaptoethanol on the spectroscopic properties of complexes formed between lysyltryptophyl-lysine and mercurated poly(uridylic acid). Biochemistry 24, 6586.

32. LE DOAN, T., TOULME, J.J., SANTUS, R., and HELENE, C. (1981). The photophysical and photochemical processes of tryptophan in interaction with polynucleotides: laser-flash photolysis study. Photochem. Photobiol. 34, 309.

33. MAURIZOT, J.C., BOUBAULT, G., and HELENE, C. (1978). Interactions of aromatic residues of proteins with nucleic acids. Binding of oligopeptides to copolynucleotides of adenine and cytosine. Biochemistry 17, 2096.

34. GABBAY, E.J., ADAWADKAR, P.D., and WILSON, W.D. (1976). Stereospecific binding of diastereomeric peptides to salmon sperm DNA. Biochemistry 15, 146.

35. SHARDY, R.D. and GABBAY, E.J. (1983). Stereospecific binding of diastereomeric peptides to salmon sperm DNA. Further evidence for partial intercalation. Biochemistry 22, 2061.

36. COLOT, V., TOULME, J.J., and HELENE, C. (1984). Interaction of a tryptophan-containing peptide with chromatin core particles. A fluorescence study. FEBS Lett. 169, 205.

37. ALBERTS, B.M. and FREY, L. (1970). T4 bacteriophage gene 32: a structural protein in the replication and recombination of DNA. Nature 227, 1313.

38. ALBERTS, B.M., FREY, L., and DELIUS, H. (1972). Isolation and characterization of gene proteins of filamentous bacterial viruses. J. Mol. Biol. 68, 139.

39. MOLINEUX, I.J., FRIEDMAN, S., and GEFTER, M.L. (1974). Purification and properties of the E. coli DNA-unwinding protein. J. Biol. Chem. 249, 6090.

40. HERRICK, G. and ALBERTS, B.M. (1976). Purification and physical characterization of nucleic acid helix-unwinding proteins from calf-thymus. J. Biol. Chem. 251, 2124.

41. DUGUET, M. and DE RECONDO, A.M. (1978). A DNA unwinding protein isolated from regenerating rat liver. J. Biol. Chem. 253, 1660.

42. COLEMAN, J.E. and OAKLEY, J.L. (1980). Physical chemical studies of the structure and function of DNA binding (helix destabilizing) proteins. CRC Crit. Rev. Biochem. 7, 247.

43. KOWALCZYKOWSKI, S.C., BEAR, D.G., and VON HIPPEL, P.H. (1982). Single-stranded DNA binding proteins. In: "The Enzymes," Vol. XIV (P.D. Boyer, ed.) p. 373. Academic Press, New York.

44. HELENE, C., TOULME, J.J., and MONTENAY-GARESTIER, T. (1982). Recognition of natural and chemically-damaged nucleic acids by peptides and proteins. In: "Topics in Nucleic Acid Structure," part 2, (S. Neidle, ed.) p. 229. Mcmillan Press, New York.

45. NAKASHIMA, Y., DUNDER, A.K., MARVIN, D.A., and KONIGSBERG, W. (1974). The amino acid sequence of a DNA-binding protein, the gene 5 product of fd filamentous bacteriophage. FEBS Lett. 40, 290.

46. ANDERSON, R.A., NAKASHIMA, Y., and COLEMAN, J.E. (1975). Chemical modifications of functional residues of fd gene 5 DNA-binding protein. Biochemistry 14, 907.

47. ALMA, N.C.M., HARMSEN, B.J.M., DEJONG, E.A.M., VEN, J.V.D., and HILBERS, C.W. (1983). Fluorescence studies of the complex formation between the gene 5 protein of bacteriophage M13 and polynucleotides. J. Mol. Biol. 163, 47.

48. PORSCHKE, D. and RAUH, H. (1983). Cooperative, excluded-site binding and its dynamics for the interaction of gene 5 protein with polynucleotides. Biochemistry 22, 4737.

49. ALMA, N.C.M., HARMSEN, B.J.M., VAN BOOM, J.H., VAN DER MAREL, G., and HILBERS, G.W. (1983). A 500-MHz proton nuclear magnetic resonance study of the structure and structural alterations of gene 5 protein-oligo(deoxy-adenylic acid) complexes. Biochemistry 22, 2104.

50. SHIMAMOTO, N. and UTIYAMA, H. (1983). Mechanism and role of cooperative binding of bacteriophage fd gene 5 protein to single-stranded deoxyribonucleic acid. Biochemistry 22, 5869.

51. BRAYER, G.D. and McPHERSON, A. (1983). Refined structure of the gene 5 DNA binding protein from bacteriophage fd. J. Mol. Biol. 169, 565.

52. BRAYER, G.D. and McPHERSON, A. (1984). Mechanism of DNA binding to the gene 5 protein of bacteriophage fd. Biochemistry 23, 340.

53. DAY, L.A. (1973). Circular dichroism and ultraviolet absorption of a deoxyribonucleic acid binding protein of a filamentous phage. Biochemistry 12, 5329.

54. COLEMAN, J.E., ANDERSON, R.A., RATCLIFFE, R.G., and ARMITAGE, I.M. (1976). Structure of gene 5 protein-oligo-deoxynucleotide complexes as determined by ^{1}H, ^{19}F, and ^{31}P nuclear magnetic resonance. Biochemistry 15, 5419.

55. GARSSEN, G.J., KAPTEIN, R., SCHOENMAKERS, J.G.G., and HILBERS, C.W. (1978). A photo-CIDNP study of the interactions of oligonucleotides with gene 5 protein of bacteriophage M13. Proc. Natl. Acad. Sci. USA 75, 5281.

56. O'CONNOR, T.P. and COLEMAN, J.E. (1983). Proton nuclear magnetic resonance (500 MHz) of mono-, di-, tri-, and tetradeoxynucleotide complexes of gene 5 protein. Biochemistry 22, 3375.

57. GRAY, D.M., GRAY, C.W., and CARLSON, R.D. (1982). Neutron scattering data on reconstituted complexes of fd DNA and gene 5 protein show that the DNA is near the center. Biochemistry 21, 2702.

58. EPSTEIN, R.A., BOLLE, A., STEINBERG, C.M., KELLENBERGER, E., BOY DE TO LA TOUR, E., CHEVALLEY, R., EDGAR, R.S., SUSMAN, M., DENHARDT, G.H., and LIELAUSIS, A. (1963). Physiological studies of conditional lethal mutants of bacteriophage T4. Cold Spring Harbor Symp. Quant. Biol. 28, 375.

59. HIBNER, U. and ALBERTS, B.M. (1980). Fidelity of replication catalyzed in vitro on a natural DNA template by the T4 bacteriophage multi-enzyme complex. Nature 285, 300.

60. HUBERMAN, J.A., KORNBERG, A., and ALBERTS, B.M. (1971). Stimulation of T4 bacteriophage DNA polymerase by the protein product of T4 gene 32. J. Mol. Biol. 62, 39.

61. TOPAL, M.D. and SINHA, N.K. (1983). Products of bacteriophage T4 genes 32 and 45 improve the accuracy of DNA replication in vitro. J. Biol. Chem. 258, 12274.

62. MOSIG, G. and BRESCHKIN, A.M. (1975). Genetic evidence for an additional function of phage T4 gene 32 protein: interaction with ligase. Proc. Natl. Acad. Sci. USA 72, 1226.

63. CURTIS, M.J. and ALBERTS, B.M. (1976). Studies on the structure of intracellular bacteriophage T4 DNA. J. Mol. Biol. 102, 793.

64. KRISCH, H.M. and VAN HOUWE, G. (1976). Stimulation of the synthesis of bacteriophage T4 gene 32 protein by ultraviolet light irradiation. J. Mol. Biol. 108, 67.

65. KRISCH, H.M., BOLLE, A., and EPSTEIN, R.H. (1974). Regulation of the synthesis of bacteriophage T4 gene 32 protein. J. Mol. Biol. 88, 89.

66. LEMAIRE, G., GOLD, L., and YARUS, M. (1978). Autogenous translational repression of bacteriophage T4 gene 32 expression in vitro. J. Mol. Biol. 126, 73.

67. KRISCH, H.M. and ALLET, B. (1982). Nucleotide sequences involved in bacteriophage T4 gene 32 translational self-regulation. Proc. Natl. Acad. Sci. USA 79, 4937.

68. WILLIAMS, K.R., LO PRESTI, M.B., and SETOGUCHI, M. (1981). Primary structure of the bacteriophage T4 DNA helix-destabilizing protein. J. Biol. Chem. 256, 1754.

69. HOSODA, J. and MOISE, H. (1978). Purification and physicochemical properties of limited proteolysis products of T4 helix destabilizing protein (gene 32 protein). J. Biol. Chem. 253, 7547.

70. WILLIAMS, K.R. and KONIGSBERG, W. (1978). Structural changes in the T4 gene 32 protein induced by DNA and polynucleotides. J. Biol. Chem. 253, 2463.

71. KOWALCZYKOWSKI, S.C., LONBERG, N., NEWPORT, J.W., and VON HIPPEL, P.H. (1981). Interactions of bacteriophage T4-coded gene 32 protein with nucleic acids. I. Characterization of the binding interactions. J. Mol. Biol. 145, 75.

72. HELENE, C., TOULME, F., CHARLIER, M., and YANIV, M. (1976). Photosensitized splitting of thymine dimers in DNA by gene 32 protein from phage T4. Biochem. Biophys. Res. Commun. 71, 91.

73. JENSEN, D.E., KELLY, R.C., and VON HIPPEL, P.H. (1976). DNA "melting" proteins. II. Effects of bacteriophage T4 gene 32-protein binding on the conformation and stability of nucleic acid structures. J. Biol. Chem. 251, 7215.

74. DELIUS, H., MANTELL, N.J., and ALBERTS, B.M. (1972). Characterization by electron microscopy of the complexes formed between T4 bacteriophage gene 32-protein and DNA. J. Mol. Biol. 67, 341.

75. TOULME, J.J. and HELENE, C. (1980). Fluorescence study of the association between gene 32 protein of bacteriophage T4 and poly(1,N^6-ethenoadenylic acid). Evidence for energy transfer. Biochim. Biophys. Acta. 606, 95.

76. NEWPORT, J.W., LONBERG, N., KOWALCZYKOWSKI, S.C., and VON HIPPEL, P.H. (1981). Interactions of bacteriophage T4-coded gene 32 protein with nucleic acids. II. Specificity of binding to DNA and RNA. J. Mol. Biol. 145, 105.

77. VON HIPPEL, P.H., KOWALCZYKOWSKI, S.C., LONBERG, N., NEWPORT, J.W., LELAND, S.P., STORMO, G.D., and GOLD, L. (1982). Autoregulation of gene expression. Quantitative evaluation of the expression and function of the bacteriophage T4 gene 32 (single-stranded DNA binding) protein system. J. Mol. Biol. 162, 795.

78. FULFORD, W. and MODEL, P. (1984). Specificity of translational regulation by two DNA-binding proteins. J. Mol. Biol. 173, 211.

79. GREVE, J., MAESTRE, M.F., MOISE, H., and HOSODA, J. (1978). Circular dichroism studies of the interaction of a limited hydrolysate of T4 gene 32 protein with T4 DNA and poly[d(A-T)]-poly[d(A-T)]. Biochemistry 17, 893.

80. LONBERG, N., KOWALCZYKOWSKI, S.C., LELAND, S.P., and VON HIPPEL, P.H. (1981). Interactions of bacteriophage T4-coded gene 32 protein with nucleic acids. III. Binding properties of two specific proteolytic digestion products of the protein (G32 P*I and G32 P*III). J. Mol. Biol. 145, 123.

81. LOHMAN, T.M. and KOWALCZYKOWSKI, S.C. (1981). Kinetics and mechanisms of the association of the bacteriophage T4 gene 32 (helix destabilizing) protein with single-stranded nucleic acids. Evidence for protein translocation. J. Mol. Biol. 152, 67.

82. LOHMAN, T.M. (1984). The kinetics and mechanism of dissociation of cooperatively bound T4 gene 32 protein-single stranded nucleic acid complexes. I. Irreversible dissociation induced by [NaCl] jumps. Biochemistry 23, 4656.

83. McKAY, D.B. and WILLIAMS, K.R. (1982). Crystallization of a tryptic core of the single-stranded DNA binding protein of bacteriophage T4. J. Mol. Biol. 160, 659.

84. COHEN, H.A., CHIU, W., and HOSODA, J. (1983). Structural analysis of T4 helix destabilizing protein (gp 32*I) crystal by electron microscopy. J. Mol. Biol. 169, 235.

85. KELLY, R.C. and VON HIPPEL, P.H. (1976). DNA "melting" proteins. III. Fluorescence "mapping" of the nucleic acid binding site of bacteriophage T4 gene 32 protein. J. Biol. Chem. 251, 7229.

86. ANDERSON, R.A. and COLEMAN, J.E. (1975). Physicochemical properties of DNA binding proteins: gene 32 protein and E. coli unwinding protein. Biochemistry 14, 5485.

87. TOULME, J.J., LE DOAN, T., and HELENE, C. (1984). Role of tryptophyl residues in the binding of gene 32 protein from phage T4 to single-stranded DNA. Photochemical modification of tryptophan by trichloroethanol. Biochemistry 23, 1195.

88. CASAS-FINET, J.R., TOULME, J.J., CAZENAVE, C., and SANTUS, R. (1984). Role of tryptophan and cysteine in the binding of gene 32 protein from phage T4 to single-stranded DNA. Modification of crucial residues with selective free-radical anions. Biochemistry 23, 1208.

89. LE DOAN, T., TOULME, J.J., and HELENE, C. (1984). Involvement of tryptophyl residues in the binding of model peptides and gene 32 protein from phage T4 to single-stranded polynucleotides. A spectroscopic method for detection of tryptophan in the vicinity of nucleic acid bases. Biochemistry 23, 1202.

90. SAITO, I., ITO, S., MATSUURA, T., and HELENE, C. (1981). Specific photocoupling of 5-bromouridine to tryptophan in aqueous frozen solution. Photochem. Photobiol. 33, 15.

91. CASAS-FINET, J.R., TOULME, J.J., SANTUS, R., BUTLER, J., LAND, E.J., and SWALLOW, A.J. (1984). Influence of DNA binding of the formation and reactions of tryptophan and tyrosine radicals in peptides and proteins. Int. J. Radiat. Biol. 45, 119.

92. PRIGODICH, R.V., CASAS-FINET, J., WILLIAMS, K.R., KONIGSBERG, W., and COLEMAN, J.E. (1984). ^{1}H NMR (500 MHz) of gene 32 protein-oligonucleotide complexes. Biochemistry 23, 522.

93. TOULME, F., HELENE, C., FUCHS, R., and DAUNE, M. (1980). Binding of a tryptophan-containing peptide (lysyltryptophanyllysine) to deoxyribonucleic acid modified by 2-(N-acetoxy acetyl amino) fluorene. Biochemistry 19, 870.

94. HELENE, C., TOULME, J.J., BEHMOARAS, T., and CAZENAVE, C. (1982). Mechanisms for the recognition of chemically-modified DNA by peptides and proteins. Biochimie 64, 697.

95. BEHMOARAS, T., TOULME, J.J., and HELENE, C. (1981). Specific recognition of apurinic sites in DNA by a tryptophan-containing peptide. Proc. Natl. Acad. Sci. USA 78, 926.

96. BEHMOARAS, T., TOULME, J.J., and HELENE, C. (1981). A tryptophan-containing peptide recognizes and cleaves DNA at apurinic sites. Nature 292, 858.

97. TOULME, J.J., BEHMOARAS, T., GUIGUES, M., and HELENE, C. (1983). Recognition of chemically-damaged DNA by the gene 32 protein from bacteriophage T4. EMBO J. 2, 505.

98. CHA, T.A., and MAKI, A.H. (1984). Close range interactions between nucleotide bases and tryptophan residues in an Escherichia coli single-stranded DNA binding protein-mercurated poly(uridylic acid) complex. J. Biol. Chem. 259, 1105.

99. TOULME, J.J. and SAISON-BEHMOARAS, T. (1985). Recognition of damaged regions in DNA by oligopeptides and proteins. Biochimie 67, 301.

ORGANIZATION AND EVOLUTION OF THE PROTAMINE GENES OF SALMONID FISHES

Gordon H. Dixon, Judd M. Aiken, Jacek M.
Jankowski, Deborah I. McKenzie, Robert Moir,
and J. Christopher States

Department of Medical Biochemistry
Faculty of Medicine
Health Sciences Centre
University of Calgary
3330 Hospital Drive N.W.
Calgary, Alberta T2N 4N1 Canada

INTRODUCTION

In the final stages of development of the male gametes of almost all animals and some plants, new sperm-specific basic proteins are synthesized and form complexes with the DNA. This process, which is a characteristic feature of male gametogenesis, is frequently, but not always, accompanied by a replacement of the pre-existing, somatic histones by the new sperm proteins and very often correlates with a profound re-organization of the chromatin involving the formation of a highly-condensed and transcriptionally-inactive sperm nucleus. The variety of sperm basic proteins is extremely wide and the distribution of different types through phylogeny, apparently quite irregular. This is in strong contrast with the somatic histones which are amongst the proteins most strongly conserved in evolution. In 1969, Bloch (1), compiled a catalogue of sperm histones and in Figure 1, a classification based on the one suggested by Bloch is presented. There are four broad categories of sperm basic proteins classified in order of their increasing degree of difference from the somatic histones. They are as follows: (a) Histone-like proteins, whose amino-acid composition and electrophoretic behavior are very similar to the somatic histones, and which are found in some Echinoderms (2), some

287

Anuran amphibia and some fishes, e.g. Carpio; (b) <u>Intermediate proteins</u>, which are clearly different in composition (often being more arginine-rich) and electrophoretic mobility, and which are seen in some Echinoderms, some mollusks, some amphibia, and some fishes; (c) <u>True</u> <u>protamines</u>, which are very arginine-rich, often quite small, and which are seen in some mollusks, amphibia, fishes (particularly Salmonids and Clupeids), reptiles, birds, and metatherian mammals; (d) <u>Stable protamines</u> or basic keratins, which are somewhat longer polypeptides that, in addition to being arginine-rich, also contain ∿6-9 residues of ½-cystine; they occur in some insects, an amphibian, and all eutherian mammals. The presence of the high content of ½-cystine confers great mechanical and chemical stability to the sperm nuclei since a three-dimensional network of disulfide bridges is formed which can only be disrupted <u>in vitro</u> by thiols in the presence of strong denaturing agents (3-5). There is, in addition, another class of sperm proteins found in the very unusual ameboid sperm of crabs and crayfish (Arthropoda), which have nuclei in which the chromatin is less condensed than in somatic cells. These proteins of ameboid sperm are not basic but acidic in nature (5, 6) and have not been extensively characterized.

In Figure 2 is depicted the distribution of the four major classes of sperm basic proteins in a variety of organisms on the phylogenetic tree. The major conclusion is that there appears to be little regularity of occurrence of a particular class of sperm proteins in any group of organisms. For example, in

(a) **HISTONE-LIKE PROTEINS**
some Echinoderms, some Anurans, some Fishes

(b) **INTERMEDIATE PROTEINS**
between "True" Histones and "True" Protamines
some Echinoderms, Molluscs, Amphibians and Fishes

(c) **TRUE PROTAMINES**   V.Arginine–Rich
Molluscs, Amphibians, Fishes, Reptiles,
Birds and Metatherian Mammals

(d) **"STABLE" PROTAMINES or "BASIC KERATINS"**
rich in Arg. <u>and</u> ½ Cystine
some Insects, an Amphibian, all Eutherian Mammals

Figure 1. Classification of spermatozoan basic proteins after Bloch (1).

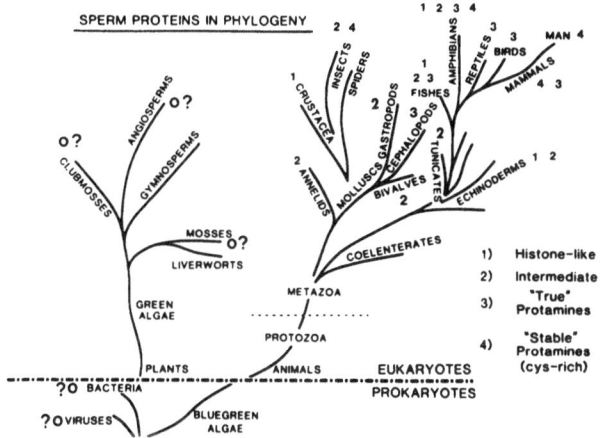

Figure 2. Distribution of the four classes of sperm proteins in the phylogenetic tree. The symbol 0? indicates that a basic male gamete protein has been observed.

vertebrates, among the amphibia, all four classes occur and among fishes, three different classes are represented. This apparently sporadic distribution suggests two possibilities. First, the genes that specify sperm basic proteins might diverge rapidly since they might be under less stringent selective forces than, for example, those for the somatic histones whose structures are under strong evolutionary constraints so as to maintain the histone-histone interactions essential for nucleosome core formation (7). The second possibility is that several alternative sperm protein gene sets might have evolved, any one of which would provide a functional sperm protein so that even in fairly closely related forms, e.g., among amphibia, different members of the gene set might have been expressed. It is difficult to choose between these two hypotheses, but evidence to be described below for the protamine genes of Salmonid and related fishes (8) tends to support the first possibility that these genes diverge in evolution much more rapidly than the histone genes in the same fishes. (See chapter by Kasinsky et al., this volume, for an extended analysis of the diversity of sperm basic proteins in vertebrates.)

We will be concerned in the remainder of this article with the true protamines, the third class of sperm basic proteins, since a substantial amount of data at the molecular level has accumulated for them. In Figure 3, the amino acid sequences for protamines from three species of teleost fishes, Pacific herring

(9), rainbow trout (9), and tuna (10) -- all of which are 30-33 residues long and contain approximately 2 out of every 3 residues as arginines -- are compared with chicken protamine (galline) (11, 12a), which is equally arginine-rich but twice as long (65 residues). It seems evident that the increased length of galline is due to a partial duplication event during bird evolution since there is evidence of an internal homology between residues 21-39 and residues 40-65. The amino acid sequences of true protamines of the fishes are characterized by the presence of four blocks of arginine residues. The blocks are most often four residues long. This pattern is expanded to 7 blocks in galline as a result of its doubling in length. There is a very limited range of non-arginine amino acids, usually, in order of frequency: proline, serine, alanine, glycine, valine, and isoleucine and, more rarely, threonine, glutamine, and tyrosine.

In Figure 4, the amino acid sequence of chicken protamine (galline) is compared with the known sequences, as compiled by Coelingh and Rozijn (4), of the fourth class of sperm basic proteins that are characteristic of eutherian mammalian sperm, the stable protamines or basic keratins. The proteins in this group are ~50 residues long and, as indicated in Figure 4, can be divided into a central 24 residue domain (residues 15-38) which is extremely arginine-rich but with 2-3 ½-cystines in the cases of boar and bull. The N-terminal 12-14 residues are

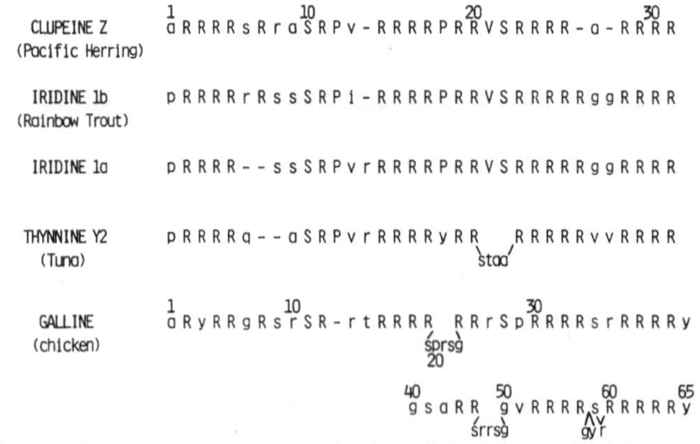

Figure 3. Amino acid sequences of various fish "true protamines" compared with that of chicken (galline). Sequences collected from references 9-12. Capitals indicates conserved residues.

quite well conserved with an identical tetrapeptide sequence at the N-terminus in chicken, rat, human I, boar, and bull, but again the mammalian protamines (with the exception of human I) differ from chicken in possessing a pair of ½-cystine residues at positions 5 and 6. There is less similarity in the C-terminal region amongst the mammalian protamines but there are additional ½-cystine residues -- 4 in boar for a total in the molecule of 9, and 2 in bull for a total of 6. From a comparison of sequence similarities in Figure 4, it seems most likely that the appearance of ½-cystine residues in the mammalian protamines may have been the result of single base mutations in the Arg codons CGU or CGC or the Ser codons AGU or AGC to the Cys codon UGU and UGC. Direct evidence of this must await isolation and sequencing of mammalian protamine cDNAs or genes.

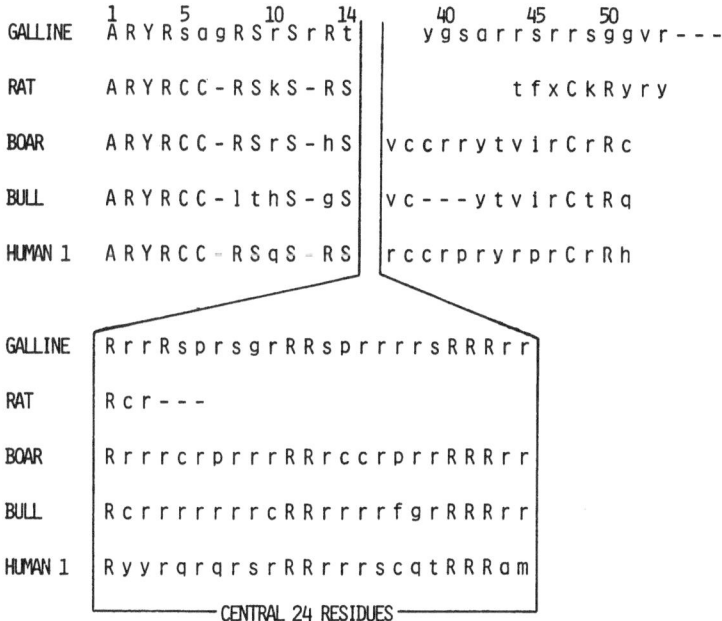

COMPARISON OF GALLINE WITH MAMMALIAN PROTAMINES

Figure 4. Amino acid sequences of the "stable protamines" of eutherian mammals compared with that of chicken protamine (galline) after Coelingh and Rozijn (4). Capitals indicate conserved residues. The data for Human 1 protamine are from a recently completed amino acid sequence analysis (12b).

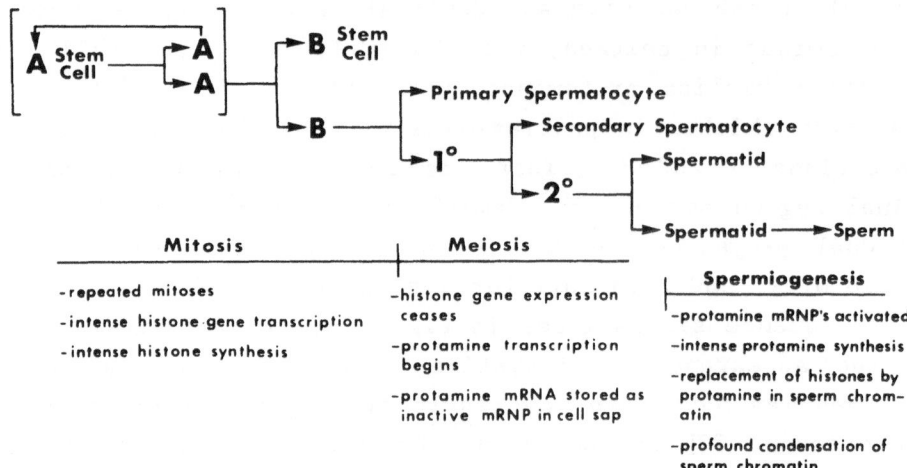

Figure 5. Scheme of spermatogenesis and spermiogenesis in rainbow trout.

Presumably, cysteine residues first appeared in the ancestors to eutherian mammals, and their ability to form covalent disulfide links, which resulted in increased mechanical stability of the DNA- protamine complex, had selective advantage and was conserved.

THE TRUE PROTAMINES

Our laboratory has been mainly concerned with the true protamines and their genes in the Salmonid fishes, and in our investigations of the regulation of protamine gene expression we have studied primarily the developing trout testis. In Figure 5, the processes of spermatogenesis and spermiogenesis in the rainbow trout are outlined in simplified form. The rainbow trout, like other Salmonids, spawns seasonally. The process is initiated by shortening of the day length which, in turn, stimulates pituitary gonadotrophin production. The immature testes are thin threads of tissue containing a relatively small number of spermatogonial stem cells. Under the influence of increased levels of gonadotrophins, these spermatogonia undergo repeated mitotic divisions (13), usually 11-12 in number, thus increasing their population by a factor of 2000-4000 (2^{11}-2^{12}). These type A spermatogonial stem cells then become committed to meiosis as

Type B spermatogonia and then primary spermatocytes. Two meiotic divisions follow, the first giving rise to secondary spermatocytes and the second to spermatids. The haploid spermatids no longer synthesize DNA and undergo terminal differentiation to mature sperm. Many interesting gene regulatory events occur in this complex process and in Figure 5 only those concerning the histone and protamine genes are depicted. During the early mitotic phase there is intense DNA synthesis accompanied by rapid histone gene transcription. DNA and histone synthesis probably continue in primary spermatocytes but at about this time there is a profound down-regulation of histone gene transcription and histone mRNA levels fall rapidly as judged by Northern hybridization with histone H2B and H4 gene probes (Winkfein and Dixon, in preparation). During this same time period, protamine mRNA sequences can be detected for the first time in the cell sap of primary spermatocytes (14), so that there appear to be reciprocal regulatory events, a "turn-off" of histone genes and a "turn-on" of protamine genes.

Gedamu et al. (15) and Sinclair and Dixon (16) were able to show that the protamine mRNAs are initially sequestered in translationally-inactive mRNP particles confined to the cell supernatant. Only much later, at the spermatid stage of development, can these mRNPs be bound to polysomes (15) in a translationally-active state (16) so that protamine polypeptide synthesis can begin. Protamine polypeptides do not accumulate in the cytoplasm but are mono-phosphorylated (17) and rapidly transported into the spermatid nucleus (18). Further phosphorylation to the di- or tri-phospho state follows in the nucleus (17). (This is dependent on the amino acid sequence: some protamines have only two serine residues while others have 3 or 4 (19).) The somatic histones, particularly H3 and H4, in early and middle spermatid nuclei become hyperacetylated (20, 21) in domains of up to 50 nucleosomes in length. These regions appear as smooth fibers in the electron microscope (21) (as if they contain unfolded nucleosomes). Hyperacetylation thus seems to "prime" spermatid chromatin for the displacement of the histones by phospho-protamine, starting in these domains. The displacement occurs as a highly ordered process in which the histones are removed in the order H3, H4 > H2A, H2B > H1 (22). After the displacement, the phospho-protamines are dephosphorylated and

the nucleoprotamine complex becomes progressively more condensed
and highly refractile. It is possible that the dephosphorylation reactions provide an energy-source which may drive the
physical condensation and packaging of the sperm nucleoprotamine.

PROTAMINE cDNAs

A family of 6S mRNAs coding for rainbow trout protamines
(23) was isolated, purified in milligram quantities (24), and
characterized by partial RNA sequence analysis (25). Both
poly(A)$^+$ and poly(A)$^-$ forms were isolated (26) and shown to fall
into four separable size classes upon electrophoresis in
denaturing gels (27). However, cell-free translation of the

```
                 Met Pro Arg Arg Arg Arg|     Ala Ser Arg Arg Val|Arg Arg Arg Arg Arg Pro
        pRTP 43  ATG CCC AGA AGA CGC AGA|--- GCC AGC CGC CGT GTC|CGC AGG CGC CGT CGC CCC

                 Met Pro Arg Arg Arg Arg|     Ala Ser Arg Arg Ile|Arg Arg Arg Arg Arg Pro
        pRTP 178 ATG CCC AGA AGA CGC AGA|--- GCC AGC CGC CGT ATC|CGC AGG CGC CGT CGC CCC

                                        |     Ala Ser Arg Arg Ile|Arg Arg Arg Arg Arg Pro
        pRTP 94                         |    GCC AGC CGC CGG ATC|CGC AGG CGC CGT CGC CCC

                                        |Ser Ser Ser Arg Pro Val|Arg Arg Arg Arg Arg Pro
        pRTP 59                         |TCC TCC AGC CGA CCT GTC|CGC AGG CGC CGC CGC CCC

                                        |Ser Ser Arg Arg Pro Val|Arg Arg Arg Arg Arg Pro
        pRTP 242                        |TCC TCC AGA CGA CCT GTT|CGC AGG CGC CGC CGC CCC


                 Arg Val Ser Arg     Arg Arg Arg Arg Gly Gly Arg Arg Arg Term
        pRTP 43  AGG GTG TCC CGG --- CGT CGC AGG AGA GGA GGC CGC AGG AGG CGT TAG

                 Arg Val Ser Arg     Arg Arg Arg Arg Gly Gly Arg Arg Arg Term
        pRTP 178 AGG GTG TCC CGG --- CGT CGC AGG AGA GGA GGC CGC AGG AGG CGT TAG

                 Arg Val Ser Arg     Arg Arg Arg Arg Gly Gly Arg Arg Arg Term
        pRTP 94  AGG GTG TCC CGG --- CGT CGC AGG AGA GGA GGC CGC AGG AGG CGT TAG

                 Arg Val Ser Arg Arg Arg Arg Arg Arg Gly Gly Arg Arg Arg Arg Term
        pRTP 59  AGG GTG TCC CGA CGT CGT CGC AGG AGA GGA GGC CGC AGG AGG CGT TAG

                 Arg Val Ser Arg Arg Arg Arg Arg Arg Gly Gly Arg Arg Arg Arg Term
        pRTP 242 AGG GTG TCC CGA CGT CGT CGC AGG AGA GGA GGC CGC AGG AGG CGT TAG
```

Figure 6. Nucleotide sequences of the amino acid coding por-
          tions of five cloned protamine cDNA components
          together with the predicted amino acid sequence
          The area enclosed in the box is the "variable
          region". Since the 5' boundary of this region is
          uncertain, a dotted line has been placed at the left
          margin. Dashes indicate gaps introduced for optimal
          sequences alignment. The sequences are from reference
          30.

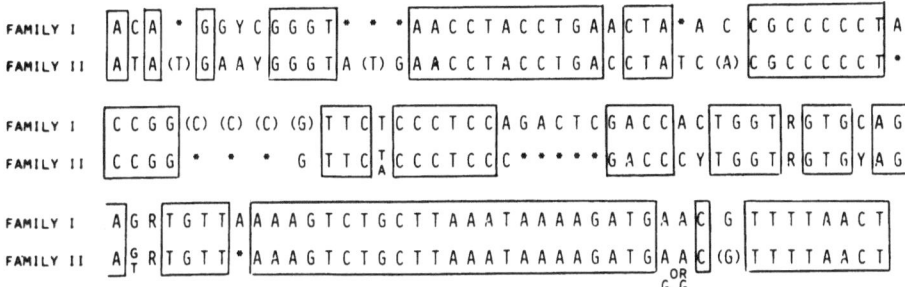

Figure 7. Family I and II 3'-untranslated sequences. Each sequence from Figure 6 has been placed into one of two distinct categories. Nucleotides in parentheses are present only in some sequences belonging to that family. Asterisks (*) indicate that no nucleotide occurs at that particular position, in that family, after alignment of sequences as in Figure 6. R = purine and Y = pyrimidine. Nucleotides enclosed in boxes are totally conserved in every sequence thus far examined. The sequences are from reference 30.

apparently physically separated protamine mRNAs (28) as well as hybridization studies of protamine cDNA to poly(A)$^+$ protamine mRNA (27, 29) showed that although enrichment of particular mRNA components was obtained, complete purification of the individual mRNA components from one another had clearly not been achieved. It was necessary, therefore, to prepare a cDNA library from the protamine mRNA fraction and clone and sequence individual clones. Five cDNA clones were sequenced by us (30) and several others by Jenkins (31). The sequences could be divided into two families, I and II, on the basis of systematic differences in both coding and 3'-untranslated sequences. The coding regions of Family I sequences (Figure 6) are 30 residues long and show an Ala residue at position 6, and Arg residue at position 9, and have a 5-membered Arg tract at positions 20-24. The systematic differences in Family II (Figure 6) are that the coding region is now 32 residues long with an additional Ser between positions 5 and 6 (of Family I), a second Ser replacing Ala at position 6 (now position 7 in Family II), a Pro relacing Arg at position 9, and a 6-membered Arg tract at positions 21-26. Position 10 in both families can be either Ile or Val. The sequences in the 3'-untranslated region are highly conserved within each of the two families but there are clear differences between Family I and II. For example, in Figure 7, in which the identities are boxed, it can be seen that tri- and penta-nucleotide sequences are present in Family I but deleted (marked by asterisks) in Family II. Similarly, an ATG sequence in Family II is deleted in Family I. These clear differences in a generally conserved 3'-untranslated sequence imply that the two present-day "families" of protamine sequences have diverged from a common ancestral gene sequence by a series of duplications.

The argument is further strengthened when the variable portion of the coding region (see Figure 6, box) is examined closely. Because a disproportionate number of nucleotide and amino acid replacements have occurred here, this region is quite valuable as an indication of the evolutionary history of the entire sequence and provides a convenient way to sub-classify individual members of each family. As depicted in Figure 8, it seems probable that the sequences represented by pRTP43 and pRTP59 diverged from one another first and have since formed the two families of sequences, I and II. Other variable region sequences can all be related to either pRTP43 or pRTP59 by one or two additional nucleotide replacements. These data strengthen the proposition that the present-day protamine genes have arisen from a common ancestral sequence through a series of gene duplication events. The two sequence families thus produced seem to

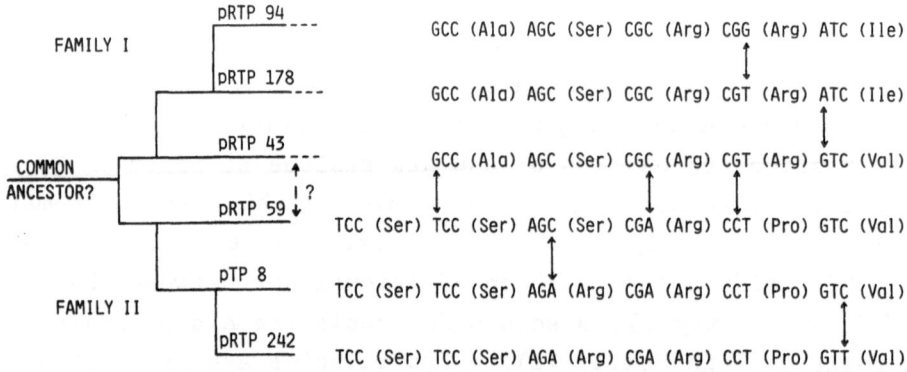

Figure 8. A possible scheme for the evolutionary divergence of known protamine sequences based upon nucleotide changes in the variable regions of the cDNA sequences. Branching order was determined by the principle of maximum economy; components having the fewest nucleotide changes (indicated by arrows) are considered to have diverged more recently. The sequence of pTP8 is from reference 31 and the remainder from reference 30.

be evolving in parallel, since the level of positional identities between sequences within the same family is high (greater than 90%), whereas inter-family sequence conservation is significantly lower. Nevertheless, these two families seem not to be diverging very rapidly from one another, since at least 70-80% identity still exists between any two sequences compared across family boundaries.

PROTAMINE GENES

The cDNA probes described above were used to isolate clones containing protamine genes from a partial Eco R1 genomic library of rainbow trout DNA prepared in the lambda vector, Charon 4A (32). From this library 1.2×10^6 clones equivalent to 1.8×10^{10} bp or 3.2 genomes of trout DNA were searched providing a 97% probability of isolating any given protamine gene (33). Forty-nine positive clones were isolated using the

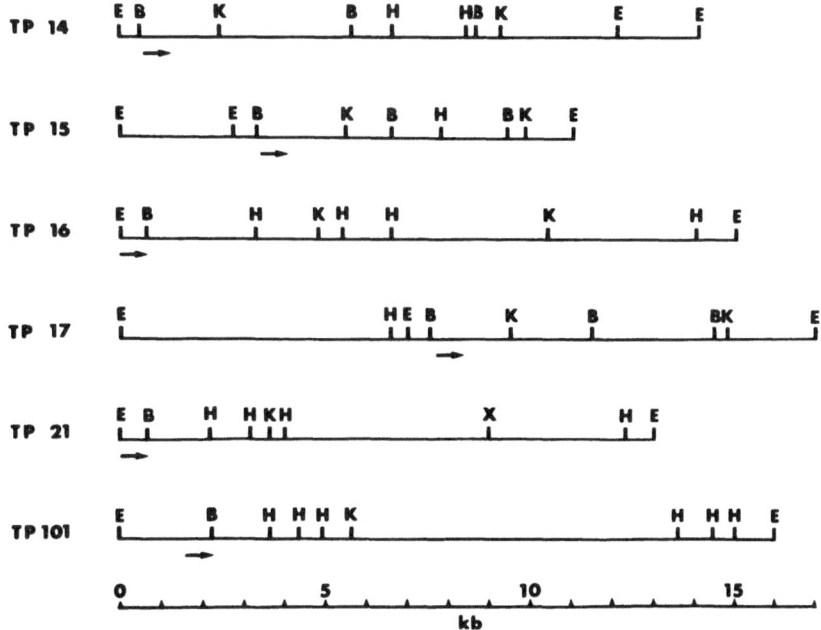

Figure 9. Restriction maps of six protamine gene-containing inserts isolated from an Eco R1 lambda charon 4A library. The arrows indicate the location, approximate size, and direction of transcription (5' to 3'). B = Bam H1, E = Eco R1, H = Hind III, K = KpnI, and X = Xba I. Taken from reference 35.

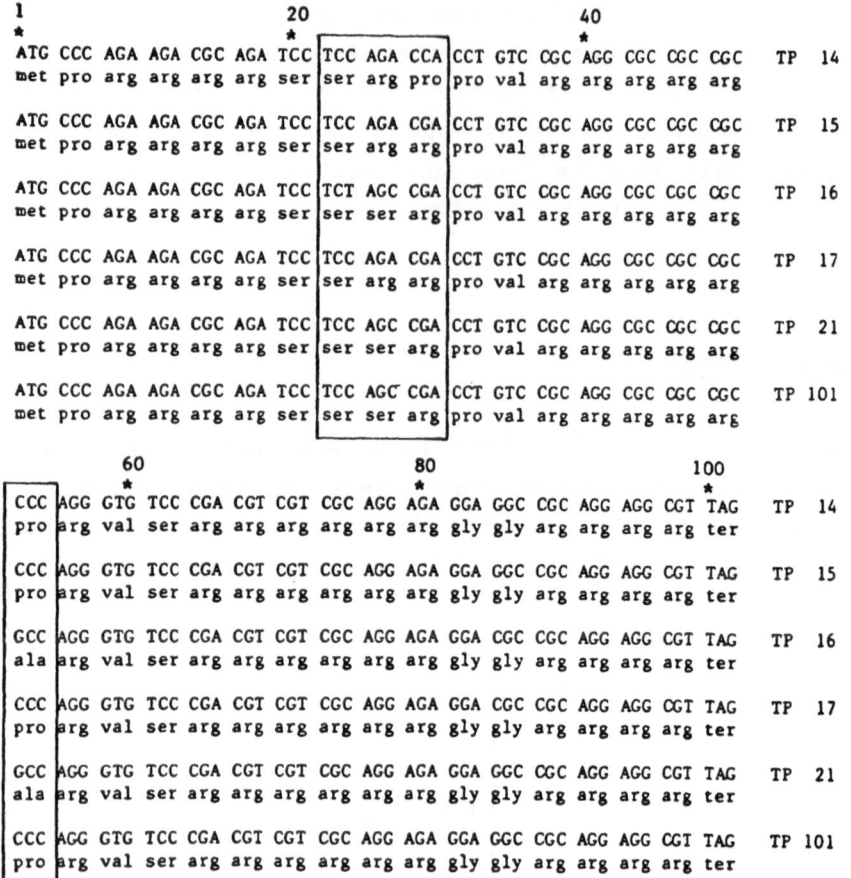

Figure 10. DNA sequences of the coding regions of six protamine genes. The potential amino acid residues are given beneath each codon. Regions of nucleotide heterogeneity are boxed. The TP101 sequence is from reference 32. Met, the initiating amino acid, is not found in mature protamine polypeptides. Taken from reference 35.

pRTP242 probe, and by restriction mapping these were classified into six different classes (Figure 9). The protamine gene-containing regions were located by Southern blotting of a series of restriction digests and hybridization with a pRTP242 probe, and approximately 1 kb of DNA containing the protamine gene was subcloned into pBR322 and sequenced. Sequences were determined 270 bp upstream and ~500 bp downstream of the 102 bp coding region.

The most important conclusions from the sequence data are as follows. The protamine genes are intron-less, a feature

shown by relatively few other genes including those for histones, human interferon and yeast glyceraldehyde-3- phosphate
dehydrogenase (34). The protamine genes are also precisely co-
linear with the expressed polypeptide so that no precursor to
the final peptide exists. However, unlike the organization of
most histone genes, including those of the rainbow trout (35),
the protamine genes occur singly and not clustered and, in TP17
at least, a seven kilobase region of each side of the gene does
not contain another protamine gene.

THE CODING REGION

The coding regions of the six different protamine genes
isolated are compared in Figure 10. The clones are either identical to or slightly changed from Family II cDNA clones, pRTP43
and 242 (36). The nucleotide sequence varies at only six positions and only three of these changes result in alterations of
the coding sequence. Position +27 can either contain a C or an
A coding for respectively Arg (AGA) or Ser (AGC) at amino acid
position 8. TP14 differs form the other genes in having a C
replacing a G at position 29, a change that leads to a Pro (CCA)
instead of an Arg (CGA). At position 52, either a G or C is
present, a shift that produces Ala (GCC) or a Pro (CCC) at the
17th amino acid residue.

The coding sequences in TP14, TP16, and TP21 provide the
first evidence of such genes in the trout. Protamine amino acid
sequences so far published (9) had not indicated a Pro-Pro
sequence at the 8th and 9th amino acid residues or that on Ala
could be present at residue 17, although more recently, following further fractionation of protamine polypeptides, first by
CM-52 and then by HPLC on a C-18 reverse phase column (McKay,
Renaux, and Dixon, in preparation), a protamine corresponding to
TP16 has been isolated and sequenced.

The 49 positive clones isolated from the Charon 4A library
and classified in the six classes in Figure 9, are 11 members
of protamine Family II. So far, no genomic clone representative
of the Family I cDNA sequences has been isolated. It is likely
that such sequences are not represented in this particular

library possibly because the distribution of Eco R1 sites flanking Family I genes is such that the partial Eco R1 digest yields fragments either too small or too large to be efficiently packaged in the λ vector. Other libraries in Charon 30 using a partial MboI digest of trout genomic DNA are presently under construction and should provide an alternative route to the missing Family I sequence.

THE 5' FLANKING REGION

In Figure 10, the 5' flanking regions of the group of cloned protamine genes are compared. The first 25 bp upstream of the coding region display a high degree of conservation between the six clones. Studies investigating the precise point of protamine mRNA initiation have produced conflicting results. States et al. (32) indicated initiation of protamine mRNA transcription of clone TP101 at position -19 (by S1 protection experiments). However, Gregory et al. (37), with a clone almost identical with TP101, found the transcription initiation point to be at -14, five base pairs downstream. Since it has been shown previously by Gedamu et al. (38) that the protamine mRNA population is heterogeneous at the 5' end, with both A and G as the 5' terminal nucleotide after removal of the m^7G cap, it is likely that both initiation sites are used in vivo. In either instance, the 5' untranslated leader sequence is exceptionally short, being either 14 or 19 nucleotides long. (For comparison, in the trout histone genes (35), it is 33-93 nucleotides long.)

The Goldberg-Hogness or TATA box is very well-defined and is located at -29 or -34 bp respectively upstream of the two alternative initiation points. An 11 bp sequence, TATAAAAGGGA, is totally conserved in all our clones and, as has been pointed out (37), this promoter element is identical to a sequence in the adenovirus major late promoter (39) and the chicken conalbumin promoter (40). In a series of experiments by Jankowski and Dixon (41), a 920 bp Bgl II-BAM H1 fragment containing the protamine gene from the lambda clone pTP101 was subcloned into pBR322 and transcribed in the _in vitro_ polymerase II HeLa lysate system (42) to yield a protamine mRNA transcript. It was found that restriction cutting at 15 bp upstream (Alu I) of the TATA gene's natural promoter. The effect is strongest in plasmid constructs in which the eukaryotic TATA box had been deleted.

There are three other regions upstream of the coding region that are conserved in all sequenced protamine genes (Figure 11). At approximately 90 bp upstream (positions -92 to -71) is a 22 bp region (CATCATTTATCCATAATGACA) about the same distance upstream as the CAAT or Chambon box found in many other gene systems (45). A second conserved region (ATTTAAACTGTCTTTAA) is A-T rich, 17 bp in length and located 121 bp upstream. The third conserved sequence (CTGCCATTGCTACTATGACGTCACA) in the 5'-flanking region extends for 25 bp (-165 to -141). Although we have not yet performed any functional assays for these regions (e.g., by injecting a series of deletion mutant plasmids into Xenopus oocytes), it might be expected that specific sequences involved in the co-ordinate expression of the apparently widely scattered protamine genes might be conserved among the protamine gene set. Such sequences might provide binding sites either for protamine-specific protein transcription factors or a set of factors that might control transcription of sperm-specific genes.

THE 3' FLANKING REGION

The sequence immediately downstream of the coding region (positions 103-232), corresponding to 3' untranslated region of the mRNA, is very highly conserved as summarized in Figure 12. Contained within this region is the canonical polyadenylation signal, AATAA, and the point at which the mRNA terminates AACTAAAA (positions 211-219). Thus, the first 4 adenines of the poly(A) tail are coded in the genome, whereas the coding region and most of the 3' untranslated region of the protamine mRNA is exceedingly G-C rich. The extreme 3' end of the mRNA (positions 179-215) and the adjacent 76-80 nucleotides (depending on the particular clone) is equally highly enriched in AT sequences, and there are 7 tandemly arranged motifs very similar to the polyadenylation signal AATAA in this region. Whether such regions are functional, perhaps in termination of transcription, remains to be determined.

box or at points further upstream at 117 (Fok I), 222 (RSa I), or 238 (Hpa II) had no effect on the efficiency of <u>in vitro</u> transcription. However, cutting 6 bp 3' to the TATA box into Ava II completely abolished transcription. Replacement of the native protamine TATA box with that of herpes virus thymidine kinase (43) led to a down regulation of protamine mRNA transcription by a factor of 10, indicating that the natural protamine promoter is a strong one. Some recent studies by Jankowski, Walczyk, and Dixon (44) have shown that in addition to the strong eukaryotic promoter in the 5' region of the protamine gene, there is also a strong prokaryotic promoter sequence upstream of the eucaryotic TATA box. This prokaryotic promoter can control the transcription of a prokaryotic gene, chloramphenicol acetyltransferase, almost as efficiently as does that

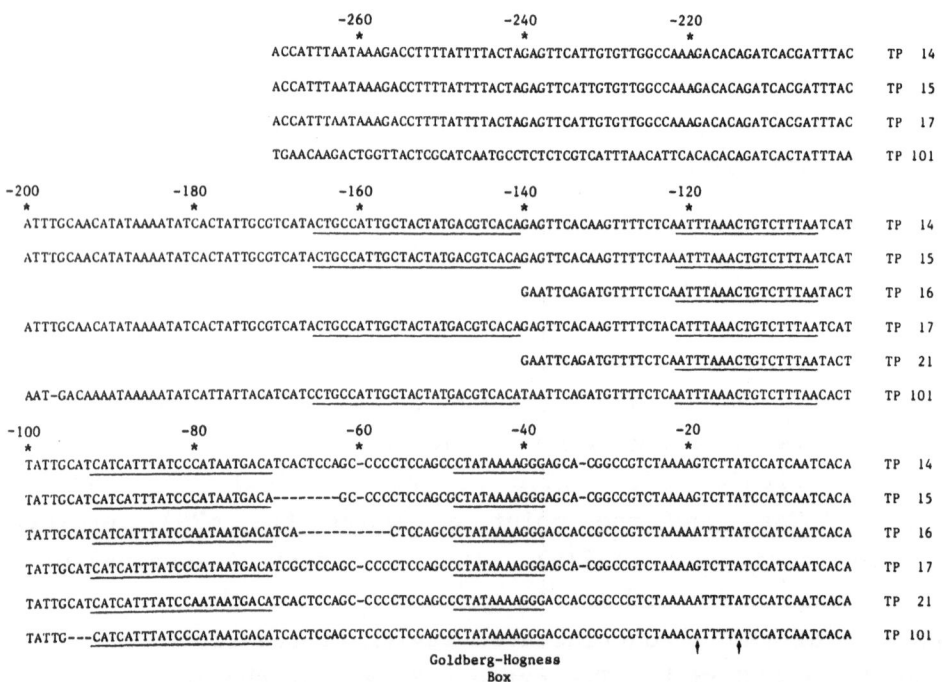

Figure 11. 5'-untranslated and flanking regions of the Family II protamine genes. The two vertical arrows below the lowest sequence indicate the two initiation (CAP) sites discussed in the text. The Goldberg-Hogness (TATA) box is underlined as are other extensive, totally conserved sequences. Hyphens have been inserted to indicate gaps which have been introduced for optimal sequence alignment. Sequence data beyond -140 are not available for TP16 and TP21. Taken from reference 35.

```
              120            140            160            180            200
               *              *              *              *              *
ATAGAATGGGTAGAACCTACCTGACCTATCCGCCCCCTCCGGGTTCTCCCTCCCGACCCTTGGTGGTGTAGAGGTGTTAAAGTCTGCTTAAATAAAAGAT     TP  14
ATAGAATGGGTAGAACCTACCTGACCTATCCGCCCCCTCCGGGTTCTCCCTCCCGACCCT-GGTGGTGTAGAGGTGTTAAAGTCTGCTTAAATAAAAGAT     TP  15
ATAGAACGGGTAGAACCTACCTGACCTATCCGCCCCCTCCGGGTTCTCCCTCCCGACCCTTGGTAGTGTAGATGTGTTAAAGTCTGCTTAAATAAAAGAT     TP  16
ATAGAATGGGTAGAACCTACCTGACCTATCCGCCCCCTCCGGGTTCTCCCTCCCGACCCTTGGTGGTGTAGAGGTGTTAAAGTCTGCTTAAATAAAAGAT     TP  17
ATAGAACGGGTAGAACCTACCTGACCTATCCGCCCCCTCCGGGTTCTCCCTCCCGACCCTTGGTAGTGTAGATGTGTTAAAGTCTGCTTAAATAAAAGAT     TP  21
ATAGAACGGGTAGAACCTACCTGACCTATCCGCCCCCTCCGGGTTCTCCCTCCCGACCCTTGGTAGTGTAGAGGTGTTAAAGTCTGCTTAAATAAAAGAT     TP 101
                                                                                          Poly A
                                                                                          Signal

              220            240            260            280            300
               *              *              *              *              *
GGGCTTTTAA-CTAAAACTGTTACGACTTTCTTTATTTTA-TAAGAT-GGTGTTC-TTTTT-GGC-ATAA--GTTTTAGGCAATAGAGTTAATCATAGAT     TP  14
GGACTTT-AAACTAAAACTGTTACGACTTTCTTTATTTTA-TAAGAT-GGTGATC-TTTTT-GGCTATAA--GTTTTAGGCAATAGAGTTAATCATAGAT     TP  15
GGGCTTTTAA-CTAAAACTGTTACGACTTT----ATATTA-GTAGATAGG-----TTTTTTAGGCTGTAAGAGTTTTTGGCGGTAGAGTTAATAAT--AT     TP  16
GGGCTTTTAA-CTAAAACTGTTACGACTTTCTTTATTTTA-TAAGAT-GGTGTTC-TTTTT-GGCTATAA--GTTTTAGGCAATAGAGTTAATCATAGAT     TP  17
GGGCTTTTAAACTAAAACTGTTACGACTTT----ATATTAGATAGATAG-T----TTTTTT-GGCTGTA-GAGTTTTTGGCGGTAGAGTTAATAAT--AT     TP  21
GGGCTTTTAA-CTAAAACTGTTACGACTTT----ATTTTA-GTAGATAGG-----TTTTTTAGGCTGTAAGAGTTTTTAGGCGGTAGAGTTAATAAT--AT     TP 101
           ↑
         mRNA
          end 320            340            360            380            400
               *              *              *              *              *
ATTTGA-ATAACTGTGTC--T-GTCCCTAACAAA-TGAATAAATAA--------------------------ATT-CAAACAATGTTTTT-------     TP  14
ATTTGA-ATAACTGTGTC--T-GTCCCTAACAAA-TGAATAAAT-AAATTCAAACAATGTTTTTATTTAAAACATATTTCAAACAATGTTTTT-------     TP  15
ATTTGAGATAATACAAATAATAG--CCTA-CTTACTG--TTAGTAATATATATATAATTAAAACGTTTTAATAATTGT---ATC--TGTCCCTAATAAAT     TP  16
ATTTGA-ATAACTGTGTC--T-GTCCCTAACAAA-TGAATAAAT-AAATTCAAACAATGTTTTTATTTAAAACATATTTCAAACAATGTTTTT-------     TP  17
ATTTGAGATAATACAAATAATAG--CCTA-CTTACTG--TTAGTAATATATATATAATTAAAACGTTTTAATAATTGT---ATC--TGTCCCTAATAAAT     TP  21
ATTTGAGATAATATAAATAATAG--CCTACT-A-TG--TTAGTAATATATATATAATTAAAACGTTTTAATA-T-----------TGTCCCTAATAAAT     TP 101

              420            440            460            480            500
               *              *              *              *              *
------ATTTAAAACA-----GTATTGAGAAA---TG-CAGTCTT--------AACCG-TCAAGTCAGATGATGCATTGCACTATTATGGTTCAAGT-C     TP  14
------ATTTAAAACA------GTATTGAGAAA---TG-CAGTCTT-----ATCAACCG-TCAAGTCAGATGATGCATTGCACTATTATGGTTCAAGT-C     TP  15
AAAATACA-TTAAACAATATATTTATTGA-AAAC--TGACACA-TTCAATCATCAACCGTTCAAGTCAGATAATGCTT-GTACCATTATGGTTTA-GTTC     TP  16
----A-ATTTAAAACA------GTATTGAGAAA--TG-CAGTCTT-----ATCAACCG-TCAAGTCAGATGATGCATTGCACTATTATGGTTCAAGT-C     TP  17
AAAATAC-ATTAAAACAATATATGTATTGA-AAAC--TGACACA-TTCAATCATCAACCG-TCAAGTCAGATAATGCTTTGTACCATTATGTTTAA---C     TP  21
AAAATACA-TTAAA-------------------CGGTGACACA-TTCAATCATC-ACCG-TCAAGTCAGATAATGCTTTGTACCATTATGGTTTA-GTTC     TP 101
```

Figure 12. 3'-untranslated and flanking regions of the protamine genes. The poly(A) signal is underlined and the mRNA termination is indicated. Hyphenated regions represent gaps introduced to optimize sequence similarity. Taken from reference 35.

A-C RICH REGION

Extensive A-C rich regions have been found in three different forms of organization approximately 1 kb downstream of several protamine genomic clones (36, 46). In Figure 13, it can be seen that these regions are organized as single, double, or triple tandem duplications of a perfect 46 bp repeat which contains a predominantly alternating purine-pyrimidine sequence. The 46 bp repeats are separated by highly conserved 20 bp spacers, also predominantly of alternating purine-pyrimidines. In Figure 13c, a catalogue of the dinucleotide repeats indicates that C-A is by far the most frequent dinucleotide and 86-88% of the sequences are composed of alternating purine-pyrimidines. Since, in studies of model synthetic repeating oligonucleotides,

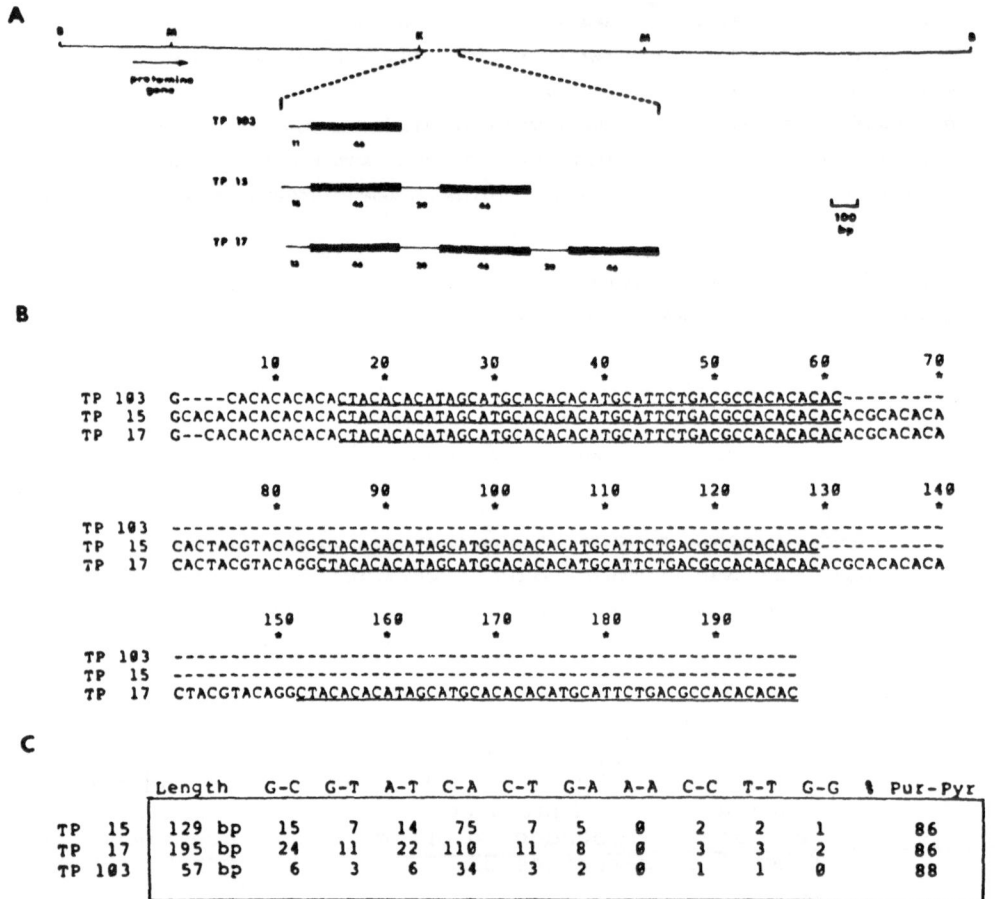

Figure 13.   A. Location and organization of the $(A-C)_n$ rich regions downstream of three protamine genomic clones. The $(A-C)_n$ rich sequences are schematically represented by the solid bars indicating the 46 bp repeat common to all three clones. B = Bam H1, M = MspI, K = KpnI.
     B. Comparison of the sequences of the $(A-C)_n$ rich regions of the three clones TP103, TP15, and TP17 determined by Maxam-Gilbert sequencing. The 46 bp repeat region is underlined. Base number one is 10 nucleotides downstream of the KpnI site in Figure 13A. Dashes have been inserted to indicate gaps introduced to optimize sequence alignment.
     C. Catalogue of dinucleotide frequencies in clones TP103, TP15, and TP17.

it has been shown that poly(dG-dC) (47) and poly(dA-dC)/(dT-dG) (48, 49) can assume the left-handed Z-DNA conformation under appropriate conditions of ionic strength, dielectric constant or negative supercoiling in closed covalent circular DNAs, we have been interested in the possibility that the A-C/T-G regions in the protamine clones could also become left-handed in the Z-conformation. Using three criteria -- competition with an authentic labelled Z-form DNA (poly(dG-m^{5}dC)) for binding to an anti-Z DNA antibody (50), immuno-electron microscopy to locate the sites of binding of anti-Z antibodies on plasmid TP17 (containing the triple tandem (A-C) repeat), and the mapping of B-Z junctions by S1 nuclease digestion -- it has been clearly established that these A-C rich regions could flip to the Z-conformation under conditions of physiological ionic strength if incorporated into a negatively supercoiled plasmid (46). The in vivo significance of this structural traansition can presently only be speculated upon, but in the Paulson-Laemmli model (51) of chromosome structure, which involves a series of DNA loops covalently closed by insertion upon a chromosome scaffold, it is possible to envisage that potential Z-DNA-forming regions could profoundly modify the topology of such a chromosomal loop and hence the susceptibility to transcription of sets of genes within the transcriptional unit defined by the loop, if mechanisms were available to achieve B to Z conformational "flipping." It is possible, for example, that developmental regulation of such gene sets could be controlled by the inter-action of Z-DNA binding proteins with the potential Z-region thus displacing the B to Z equilibrium in favor of Z-DNA.

PROTAMINE GENE EVOLUTION IN THE SALMONID AND CLUPEID FISHES

With the availability of rainbow trout probes for protamine genes, it was possible to pose the question of how rapidly have the protamine genes diverged among a series of fishes of varying degrees of taxonomic relatedness.

As noted in Figure 1, true protamines of very similar amino acid sequence were found in both Salmonoidea (salmons, trouts, chars, graylings, and whitefish) and Clupeomorpha (herrings and sprats). We therefore compared the hybridization of a rainbow

CLUPEINE Z $\overset{1}{a}$ R R R R s R r $\overset{10}{a}$ S R P v - R R R R P R $\overset{20}{R}$ V S R R R R - a - R $\overset{30}{R}$ $\overset{}{R}$ R
(Pacific Herring)

IRIDINE Ib   p R R R R r R s s S R P i - R R R R P R R V S R R R R g g R R R R
(Rainbow Trout)

IRIDINE Ia   p R R R R - - s s S R P v r R R R R P R R V S R R R R g g R R R R

RAINBOW   TROUT   ●

CUTTHROAT TROUT   ●

CHUM   SALMON   ●

PACIFIC   HERRING

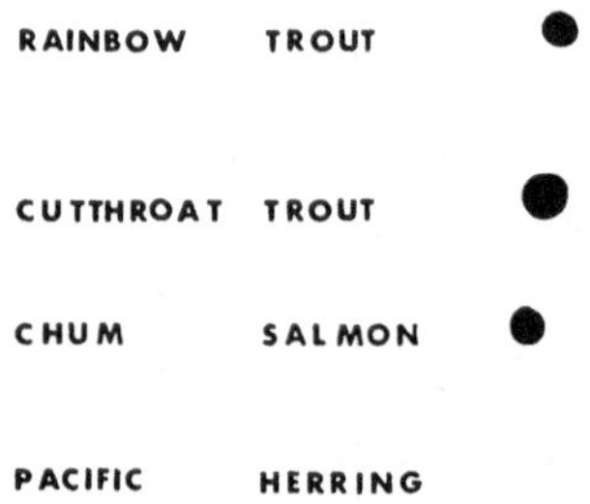

Figure 14. A comparison of the amino acid sequences of "true" protamines from Pacific herring (Clupeine Z) and two rainbow trout protamines, iridine Ia and Ib (9, 10) and a comparison by "dot-blot" of the hybridization of a ^{32}P-labelled rainbow trout protamine probe with total genomic DNA from the indicated species. Capitals indicates variable residues.

trout protamine probe with total genomic DNA from rainbow trout, a closely related member of the Salmo genus, S. clarkii, the cut-throat trout, the Pacific chum salmon, Oncorhynchus keta and the Pacific herring (Clupea pallasii). As indicated in Figure 14, the amino acid sequences of Pacific herring protamine (clupeine Z) and rainbow trout protamines (iridine Ia and Ib) are very similar, but at the DNA level the nucleotide sequences have diverged so much that no hybridization is possible (52). This suggests that divergence between the protamine genes of herring and trout has been far more rapid than, for example, among histone genes. However, the more closely related Salmonids, S. clarkii and O. keta, had protamine genes apparently homologous with those of S. gairdnerii under these conditions of hybridization stringency. In order to explore the

relationships of the protamine genes among the Salmonoidea in
more detail, DNA was collected from a series of fishes whose
relationships are given in Figure 15. These DNAs were then
spotted onto nitrocellulose filters and probed with the rainbow
trout probe under conditions of increasing stringency of hybridization by increasing progressively the temperature of washing.
The resultant dot-blot experiment (52) is illustrated in Figure
16, in which it may be seen at the highest stringency (80°C),
the protamine genes of the rainbow trout (genus Salmo) appear to
be most closely related to those of the lake trout (genus Salve-
linus) and less so to the mountain whitefish (genus Coregonus),
which is in a different family, Coregonidae. Interestingly, in
another member of the Salmonidae family, the Pacific chum salmon
(O. keta), the protamine genes seem to be more diverged and
there is no hybridization at 80°C.

The preliminary conclusion from this sort of experiment is
that, as suggested as a possibility at the beginning of this

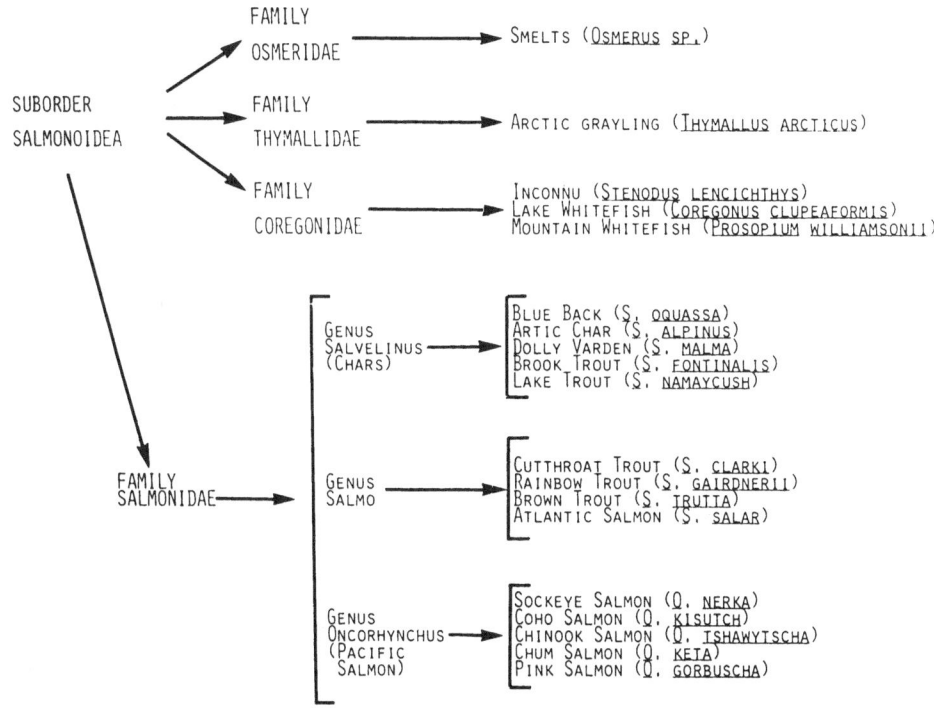

Figure 15. Taxonomic relationships in the Sub-order Salmonoidea
of the members of the Families Osmeridae, Thymal-
lidae, Coregonidae, and Salmonidae.

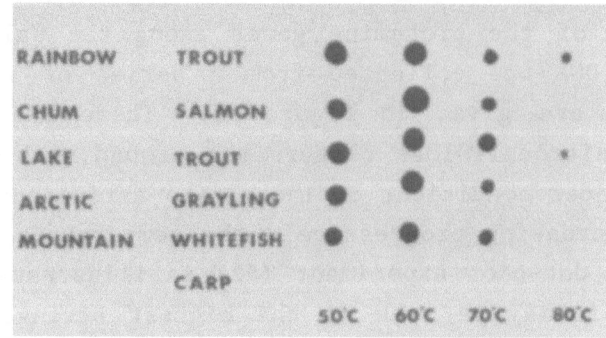

Figure 16. A "dot-blot" hybridization of a ^{32}P-labelled rainbow trout protamine probe with total genomic DNA blots of the indicated species. The increasing temperatures represent increasing stringencies of hybridization.

paper, protamine genes are inherently quite variable and, compared with the histones at least, perhaps hyper-variable.

PROTAMINE GENE NUMBER

It has proved quite difficult to determine the copy number of protamine genes in the rainbow trout. Initial studies using cDNA probes (53, 54) suggested that there were 1-2 copies of each of perhaps 4-6 protamine genes per sperm DNA equivalent for a total of 4-12 copies. Isolation of six different protamine cDNA sequences (30), three in Family I and three in Family II, was reasonably consistent with this copy number, as was the subsequent isolation of six classes of genomic Family II genes. However, the inability to isolate any Family I genes suggested that the gene number might be higher, perhaps by a factor of 2. When Southern blots of restricted DNA from individual rainbow trout were probed with the 920 bp Bgl II-Bam Hl probe from p101, complex band patterns were seen. Qualitative scanning of these gels indicated that up to 150 protamine genes might be present per sperm DNA equivalent (Aiken, unpublished result). It could also be seen that there was little similarity in the complex band patterns in other Salmonoidea. This copy number paradox has recently been cleared up by the finding that protamine genes are flanked by moderately repetitive DNA sequences, some of which are included in the Bgl II-Bam Hl probe. Using a smaller probe that comprises only the coding region of the gene (Ava II-Hpa II), the complexity of the Southern blot decreases and the copy number decreases to a value of 15-20. This value cannot yet be regarded as definitive since there appears to be appreciable variability between individual trout DNA samples. The majority of the protamine genes must be expressed, however, since recently, by careful fractionation of the protamine polypeptides from a single trout testis by CM-52 followed by HPLC chromatography, six different protamines have been resolved and sequenced (McKay, Renaux, and Dixon, in preparation).

ORIGIN OF PROTAMINE GENES

The sporadic occurrence of the "true protamines" primarily among certain fishes, reptiles, and birds suggests that the genes may have arisen by horizontal transmission from, for example, a virus rather than by vertical transmission as in normal evolutionary processes. There is some circumstantial evidence supporting this view as follows:

1. The lack of introns in the protamine genes is more characteristic of prokaryotic than eukaryotic genes.

2. The protamine promoter is identical to the viral adenovirus major late promoter over an 11 bp region.

3. There is an extremely arginine-rich sequence in the DNA-binding core protein of hepatitis B virus residues 149 181 (53) which is very similar to chicken protamine (galline) residues 9-36.

4. A small open reading frame exists 3' to the $tRNA_1^{Tyr}$ gene of E. coli (56) which encodes a putative polypeptide 33 amino acids in length and which bears a striking resemblance to fish protamines. This gene is apparently expressed and gives rise to the translation of two small basic proteins in E. coli.

5. A strong prokaryotic promoter exists 5' to the TATA box of the protamine gene, p101, which can control the expression of a prokaryotic gene, chloramphenicol acetyltransferase (57), and could represent a vestigial prokaryotic signal.

6. There are sequences flanking the protamine gene which are quite similar to the long terminal repeats (LTRs) characteristic of retrovirus genes inserted into genomic DNA (44).

The structural domain encoded by the fish and chicken protamine genes is extremely arginine-rich and has as its function the ability to complex with and condense DNA. The arginine-rich region in the hepatitis B core protein (55) has precisely the same structure and apparently the same function, i.e., to package and condense the viral DNA. Thus, a case can be made that if a primordial arginine-rich functional gene were to be introduced into the germ line of primitive vertebrates, its ability to condense DNA efficiently may have had sufficient advantage in the maturation of the sperm nucleus to have been conserved in evolution.

ACKNOWLEDGMENTS

The experimental work described above has been generously supported by the Medical Research Council of Canada, the Alberta Heritage Foundation for Medical Research, and the Alberta Cancer Board.

REFERENCES

1. BLOCH, D.P. (1969). A catalog of sperm histones. *Genetics* (suppl.) 61, I:93-111.

2. COELINGH, J.P., MONFOORT, C.H., ROZIJN, T.H., LEUVEN, J.A.G., SCHIPHOF, R., STEYN-PARVE, E.P., BRAUNITZER, G., SCHRANK, B., and RUHFUS, A. (1972). The complete amino-acid sequence of the basic nuclear protein of bull spermatozoa. *Biochim. Biophys. Acta* 285, 1-14.

3. VON HOLT, C., STRICKLAND, N., BRANDT, W.F., and STRICKLAND, M.S. (1979). More histone structures. *FEBS Lett.* 100, 201-218.

4. COELINGH, J.P. and ROZIJN, T.H. (1975). Comparative studies on the basic nuclear proteins of mammalian and other spermatozoa. *In*: "The Biology of the Male Gamete," (J.G. Duckett and P.A. Racey, eds.), pp. 245-246. Academic Press, New York.

5. MARUSHIGE, Y. and MARUSHIGE, K. (1978). Methods of isolation of nuclei from spermatozoa. *In*: "Methods in Cell Biology, Chromosomal Protein Research," Vol. XVII, pp. 59-73.

6. VAUGHAN, J.C. and HINSCH, G.W. (1972). Isolation and characterization of chromatin and DNA from the sperm of the spider crab, Libinia emarginata. J. Cell Sci. 11, 131-152.

7. ISENBERG, I. (1979). Histones. Ann. Rev. Biochem. 48, 159-191.

8. MOIR, R. and DIXON, G.H. (1985). Protamine gene evolution in Salmonoidea (in preparation).

9. ANDO, T, and WATANABE, S. (1969). A new method for fractionation of protamines and the amino acid sequences of salmine and three components of iridine. Int. J. Protein Peptide Res. 1, 221-224.

10. ANDO, T., YAMASAKI, M. and SUZUKI, K. (1973). "Protamines -- Isolation, Characterization, Structure, and Function." Springer-Verlag, Berlin and New York.

11. NAKANO, M., TOBITA, T., and ANDO, T. (1975). Studies on a protamine (galline) from fowl sperm. 2. The amino acid sequences of two components of galline. Int. J. Pep. Prot. Res. 7, 31-46.

12. NAKANO, M., TOBITA, T., and ANDO, T. (1976). Studies on a protamine (galline) from fowl sperm. 3. The total amino acid sequence of intact galline molecule. Int. J. Pep. Prot. Res. 8, 565-578.

12a. McKAY, D.J., RENAUX, B.S., and DIXON, G.H. (1985). The amino acid sequence of human protamine 1. Bioscience Rep. 5, 383-391.

13. LOUIE, A.J. and DIXON, G.H. (1972). The biosynthesis of protamine in trout testis. IV. Kinetics of enzymatic modifications of newly synthesized protamine. J. Biol. Chem. 247, 5490-5497.

14. IATROU, K., SPIRA, A.W., and DIXON, G.H. (1978). Protamine messenger RNA: evidence for early synthesis and accumulation during spermatogenesis in rainbow trout. Dev. Biol. 64, 82-98.

15. GEDAMU, L., DAVIES, P.L., and DIXON, G.H. (1977). Identification and isolation of protamine messenger ribonucleoprotein particles from rainbow trout testis. Biochemistry 16, 1383-1391.

16. SINCLAIR, G.D. and DIXON, G.H. (1982). Purification and characterization of cytoplasmic protamine messenger ribonucleoprotein particles from rainbow trout testis. Biochemistry 21, 1869-1877.

17. LOUIE, A.J. and DIXON, G.H. (1972). Trout testis cells. II. Synthesis and phosphorylation of histones and protamines in different cell types. J. Biol. Chem. 247, 7962-7968.

18. TREVITHICK, J.R., INGLES, C.J., and DIXON, G.H. (1969). The biosynthesis of protamine in trout testis. I. Intracellular site of synthesis. Can. J. Biochem. 47, 51-60.

19. SANDERS, M.M. and DIXON, G.H. (1972). The biosynthesis of protamine in trout testis. IV. Sites of phosphorylation. J. Biol. Chem. 249, 851-855.

20. CHRISTENSEN, M. and DIXON, G.H. (1982). Hyperacetylation of histone H4 correlates with the terminal, transcriptionally inactive stages of spermatogenesis in rainbow trout. Develop. Biol. 93, 404-415.

21. CHRISTENSEN, M., RATTNER, J.B., and DIXON, G.H. (1984). Hyperacetylation of histone H4 promotes chromatin decondensation prior to histone replacement by protamines during spermatogenesis in rainbow trout. Nucleic Acids Res. 12, 4575-4592.

22. MARUSHIGE, K. and DIXON, G.H. (1969). Developmental changes in chromosomal composition and template activity during spermatogenesis in trout testis. Develop. Biol. 19, 397-414.

23. GEDAMU, L. and DIXON, G.H. (1976). Purification and properties of biologically active trout testis protamine mRNA. J. Biol. Chem. 251, 1455-1463.

24. GEDAMU, L., IATROU, K., and DIXON, G.H. (1977). A simple procedure for the large-scale isolation and purification of protamine mRNA from trout testis. Biochem. J. 171, 589-599.

25. DAVIES, P.L., DIXON, G.H., SIMONCSITS, A., and BROWNLEE, G.G. (1979). Sequences of large T1 ribonuclease-resistant oligoribonucleotides from protamine mRNA: the overall architecture of protamine mRNA. Nucleic Acids Res. 7, 2323-2345.

26. IATROU, K. and DIXON, G.H. (1977). The distribution of poly(A)$^+$ and poly(A)$^-$ protamine messenger RNA sequences in the developing trout testis. Cell 10, 433-441.

27. GEDAMU, L., IATROU, K., and DIXON, G.H. (1977). Isolation and characterization of trout testis protamine mRNAs lacking poly(A). Cell 10, 443-451.

28. GEDAMU, L., and DIXON, G.H. (1979). Heterogeneity of biologically active deadenylated protamine mRNA components isolated from rainbow trout testes. Nucleic Acids Res. 6, 3661-3672.

29. IATROU, K., GEDAMU, L., and DIXON, G.H. (1979). Protamine messenger RNA: partial purification and characterization of a heterogeneous family of polyadenylated messenger components. Can. J. Biochem. 57, 945-956.

30. GEDAMU, L., WOSNICK, M.A., CONNOR, W., WATSON, D.C., and DIXON, G.H. (1981). Molecular analysis of the protamine multi-gene family in rainbow trout testis. Nucleic Acids Res. 9, 1463-1482.

31. JENKINS, J. (1979). Sequence divergence of rainbow trout protamine mRNAs: comparison of coding and non-coding nucleotide sequences in three protamine cDNA plasmids. Nature 279, 809-811.

32. STATES, J.C., CONNOR, W., WOSNICK, M.A., AIKEN, J.M., GEDAMU, L., and DIXON, G.H. (1982). Nucleotide sequence of a protamine component C_{II} gene of Salmo gairdnerii. Nucleic Acids Res. 10, 4551-4563.

33. CLARK, L. and CARBON, J. (1978). A colony bank containing synthetic col El hybrid plasmids representative of the entire E. coli genome. Cell 9, 91-99.

34. LIPMAN, P.J. and MAIZEL, J. (1982). Comparative analysis of nucleic acid sequences by their general constraints. Nucleic Acids Res. 10, 2723-2739.

35. CONNOR, W., MEZQUITA, J., WINKFEIN, R.J., STATES, J.C., and DIXON, G.H. (1984). Tandem repeats of a specific alternating purine-pyrimidine DNA sequence adjacent to protamine genes in the rainbow trout that can exist in the Z-conformation. J. Molec. Evol. 20, 227-235.

36. AIKEN, J.M., McKENZIE, D., ZHAO, H-Z., STATES, J.C., and DIXON, G.H. (1983). Sequence homologies in the protamine gene family of rainbow trout. Nucleic Acids Res. 11, 4907-4922.

37. GREGORY, S.P., DILLON, N.D., and BUTTERWORTH, P.H.W. (1982). The localisation of the 5'-termini of in vivo and in vitro transcripts of a cloned rainbow trout protamine gene. Nucleic Acids Res. 10, 7581-7592.

38. GEDAMU, L., IATROU, K., and DIXON, G.H. (1979). Translation of partially-purified poly(A)$^+$ protamine mRNA components in wheat germ and rabbit reticulocyte cell-free systems. Biochim. Biophys. Acta 562, 481-494.

39. ZIFF, E.B. and EVANS, R.M. (1978). Coincidence of the promoter and capped 5' terminus of RNA from the adenovirus 2 major late transcription unit. Cell 15, 1463-1475.

40. CORDEN, J., WASYLYK, B., BUCHWALDER, A., SASSONE-CORSI, P., KEDINGER, C., and CHAMBON, P. (1980). Promoter sequences of eukaryotic protein-coding genes. Science 209, 1406-1414.

41. JANKOWSKI, J.M. and DIXON, G.H. (1984). Transcription of the trout protamine gene in vitro: the effects of alterations in promoters. Can. J. Biochem. Cell. Biol. 62, 291-300.

42. MANLEY, J.L., FIRE, A., CANO, A., SHARP, P.A., and GEFTER, M.L. (1980). DNA-dependent transcription of adenovirus genes in a soluble whole-cell extract. Proc. Natl. Acad. Sci. USA 77, 3855-3859.

43. WAGNER, M.J., SHARP, P.A., and SUMMERS, W.C. (1981). Nucleotide sequence of the thymidine kinase gene of herpes simplex virus type 1. Proc. Natl. Acad. Sci. USA 78, 1441-1445.

44. JANKOWSKI, J.M. and DIXON, G.H. (1985). The in vitro transcription of a rainbow trout (Salmo gairdnerii) protamine gene. II. Controlled mutation of the cap site region. Bioscience Rep. 5, 113-120.

45. BENOIST, C., O'HARE, K., BREATHNACH, A., and CHAMBON, P. (1980). The ovalbumin gene -- sequence of putative control regions. Nucleic Acids Res. 8, 127-142.

46. AIKEN, J.M., MILLER, F.D., HAGEN, F., McKENZIE, D.I., KRAWETZ, S., VAN DE SANDE, J.H., RATTNER, J.B., and DIXON, G.H. (1985). Biochemistry (in press).

47. PECK, L.J., NORDHEIM, A., RICH, A., and WANG, J.C. (1982). Flipping of cloned d(pCpG)$_n$-d(pCpG)$_n$ DNA sequences from right- to left-handed helical structure by salt, Co(III), or negative supercoiling. Proc. Natl. Acad. Sci. USA 79, 4560-4569.

48. HANIFORD, D.B. and PULLEYBLANK, D.E. (1983). Facile transition of [d(TG)-d(CA)] into a left-handed helix in physiological conditions. Nature 302, 632-634.

49. HANIFORD, D.B. and PULLEYBLANK, D.E. (1983). The in vivo occurrence of A-DNA. J. Biomolec. Structure Dynamics 1, 593-609.

50. ZARLING, D.A., ARNDT-JOVIN, D.J., ROBERT-NICOUD, M., McINTOSH, L.P., THOMAE, R., and JOVIN, T.M. (1984). Immunoglobulin recognition of synthetic and natural left-handed Z-DNA conformations and sequences. J. Mol. Biol. (submitted).

51. PAULSON, J.R. and LAEMMLI, U.K. (1977). The structure of histone-depleted metaphase chromosomes. Cell 12, 817-828.

52. MOIR, R.M. and DIXON, G.H. (1985). Evolution of histone genes in Salmonoidea (in preparation).

53. LEVY-WILSON, B. and DIXON, G.H. (1977). Reiteration frequency of the protamine genes in rainbow trout (Salmo gairdnerii). J. Biol. Chem. 252, 8062-8065.

54. SAKAI, M., KURIYAMA, Y.F., and MURAMATSU, M. (1978). Number and frequency of protamine genes in rainbow trout testis. Biochemistry 17, 5510-5515.

55. PASEK, M., GOTO, T., GILBERT, W., ZINLE, B., SCHALLER, H., MacKAY, P., LEADBETTER, G., and MURRAY, K. (1979). Hepatitis B virus genes and their expression in E. coli. Nature 282, 575-579.

56. ALTMAN, S., MODEL, P., DIXON, G.H., and WOSNICK, W.A. (1981). An E. coli gene coding for a protamine-like protein. Cell 26, 299-304.

57. JANKOWSKI, J.M., WALCZYK, E., and DIXON, G.H. (1985). Functional prokaryotic gene control signals within a eukaryotic rainbow trout protamine promoter. Bioscience Rep. 5, 453-461.

CHROMATIN PROTEINS AND CHROMATIN STRUCTURE IN SPERMATOGENESIS

Cristóbal Mezquita

Department of Physiology
Faculty of Medicine
University of Barcelona
Casanova, 143
08036 Barcelona, Spain

INTRODUCTION

During spermatogenesis, the population of stem cells (diploid spermatogonia) divides and differentiates into tetraploid spermatocytes. Spermatocytes undergo meiosis, in which genetic recombination occurs, producing haploid spermatids. Spermatids, through an extraordinary process of metamorphosis called spermiogenesis, develop into a highly specialized motile vector for transportation of genetic information: the spermatozoa (Figure 1).

Spermatogenesis offers an excellent model to investigate the relationship between changes in nuclear protein composition and the structural and functional transitions that chromatin undergoes during the differentiation of the germinal cell line. Although the replacement of histones by protamine is not an obligatory step in the spermiogenesis of every organism studied, this chapter is addressed especially to the phenomena that take place during the nucleohistone-nucleoprotamine transition. In this transition, the nucleosomes are disassembled and the DNA is organized in a new, highly condensed structure. This chapter summarizes the evidence obtained so far on the correlation between changes in chromatin composition and chromatin structure during spermatogenesis.

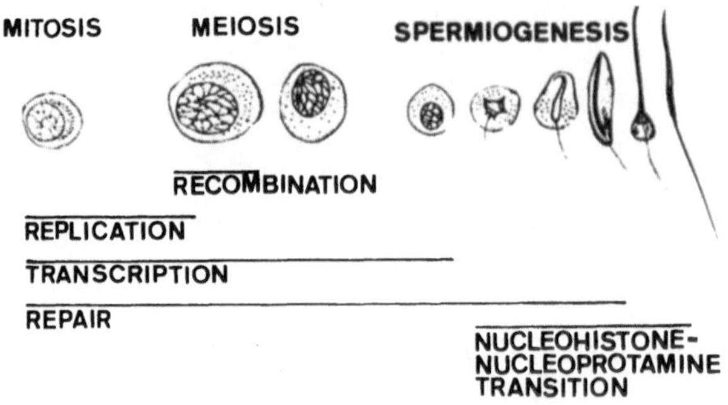

Figure 1. Diagram illustrating changes in gene activity and chromatin structure during spermatogenesis. Spermatogenesis -- the process by which spermatogonia are transformed into spermatozoa -- can be divided into three major phases: mitosis, meiosis and spermiogenesis. During the mitotic phase, chicken spermatogonial cells (left) replace themselves and generate spermatocytes (center). Spermatocytes undergo meiosis and give rise to haploid spermatids (right). During spermiogenesis, spermatids, through an extraordinary process of metamorphosis, develop into spermatozoa. DNA replication occurs in spermatogonia and in spermatocytes before meiosis (1, 2). Spermatogonia, spermatocytes, and early spermatids are active in nuclear transcription, whereas spermatids undergoing differentiation and spermatozoa are totally inactive (3-6). In meiotic cells at pachytene, occurs a period of regulated DNA nicking probably related with genetic recombination (7). Late spermatids are genetically inactive in DNA replication and transcription, however, these cells still are able to repair their genetic damage before the final nuclear condensation of chromatin occurs at the end of spermiogenesis (8, 9). The most dramatic changes in chromatin structure observed in eukaryotes take place during the nucleohistone-nucleoprotamine transition when nucleosomes are disassembled in late spermatids and replaced by a highly condensed nucleoprotamine complex (10, 11).

QUANTITATIVE AND QUALITATIVE CHANGES OF NUCLEAR PROTEINS DURING SPERMATOGENESIS

Basic Proteins

In premeiotic, meiotic, and early spermatid chicken cell nuclei, obtained by sedimentation at unit gravity, the ratio of basic proteins to DNA is 1.0 (w/w). This ratio decreases to 0.5 in nuclei of late spermatids and spermatozoa (3). Similar losses of nuclear basic proteins have been reported to occur in different species during the nucleohistone-nucleoprotamine transition (12, 13).

Chicken spermatids, at successive stages of differentiation, show decreasing levels of nucleosomal histones and increasing amounts of protamine galline (Figure 2). Galline is a very basic protein (63% arginine residues) with a higher molecular weight, 9,829, than other protamines (14). Transitory proteins present in the spermatid nuclei and disappearing in spermatozoa could play a role during the nucleohistone-nucleoprotamine transition of certain species (12, 15-17). Such proteins have not been detected in trout or chicken spermiogenesis (3, 18, 19). In addition to nuclear spermatidal proteins, several testis-specific histone variants synthesized in the pachytene stage of mammalian spermatogenesis have been

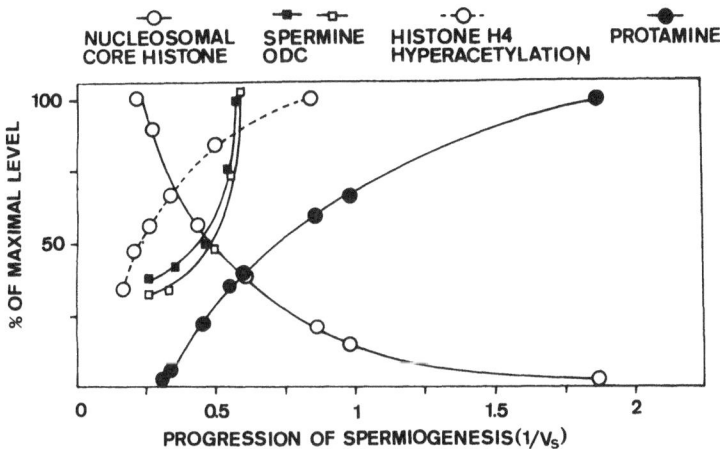

Figure 2. The nucleohistone-nucleoprotamine transition during chicken spermiogenesis. Chicken spermatids at successive stages of differentiation, obtained by sedimentation at unit gravity (66), show decreasing levels of nucleosomal core histones (o——o) and increasing amounts of protamine (•——•). A significant increase in the steady-state level of hyperacetylation of histone H4, i.e., the ratio three-acetylated H4/non-acetylated H4 (o---o), occurs during spermiogenesis (R. Oliva, unpublished results). A concomitant rise in the cellular concentration of spermine (ratio spermine/sedimentation velocity) (■——■) and ornithine decarboxylase activity (□——□) also has been detected during the nucleohistone-nucleoprotamine transition (66). V_s, sedimentation velocity, mm/h.

detected. Their biological function is as yet unknown (15, 20-22).

Acetylation of Histones

During spermatogenesis, a significant increase in the steady-state level of hyperacetylation of histones -- especially of histone H4 -- has been reported in trout, rat, and chicken spermatids (Figures 2 and 3) (23-26). Long chromatin domains enriched in hyperacetylated H4, have been isolated from rainbow trout spermatids (24). A rapid turnover of histone H4 acetyl groups takes place in chicken elongated spermatids (26).

Through hyperacetylation of histone H4 and rapid turnover of its acetyl groups, sites on DNA can be rapidly and reversibly exposed, facilitating the binding of chromosomal proteins to DNA (Figure 3). In genetically active cells, such proteins could be enzymes and regulatory proteins involved in DNA replication, transcription, recombination, or repair. DNA repair is the only genetic activity that remains in late spermatids before the condensation of the chromatin occurs (8, 9). In these cells, extensive histone hyperacetylation could be involved in exposing DNA binding sites to protamine and to other nuclear proteins with a putative role in the nucleohistone-nucleoprotamine transition.

Although it is not clear at present to what extent the hyperacetylation of histones changes the stability of the nucleosome or polynucleosomal structures (27, 28), it has been suggested that the level of acetylation of lysine residues per nucleosome during spermiogenesis (8-10 residues) would open up the nucleosome (24, 29, 30). Using core particles with hyper-acetylated histones, a nearly stoichiometric complex of all core histones can be liberated by the protamine (29). Acetylation of histones is not detectable at the end of spermiogenesis in species which retain histones in the sperm nucleus (31, 32). In addition to hyperacetylation of histones, other changes in nuc-leosomal proteins could be necessary for the disassembling of nucleosomes during spermiogenesis.

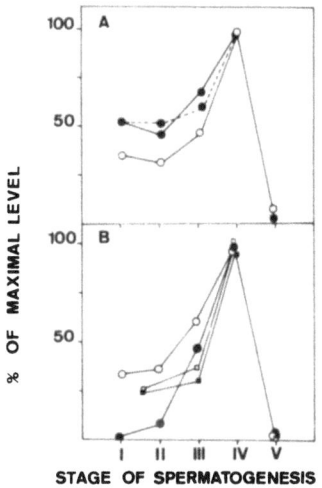

Figure 3. Changes in chromatin structure and composition during chicken spermatogenesis. Nuclei at different stages of chicken spermatogenesis were separated by sedimentation at unit gravity (3). Stage I, tetraploid primary spermatocytes; Stage II, small primary spermatocytes, secondary spermatocytes, and spermatogonia; Stage III, early spermatids; Stage IV, late spermatids and testicular spermatozoa; Stage V, spermatozoa from the vas deferens.
A. Unmasking of DNA during spermiogenesis: initiation sites for E. coli RNA polymerase in vitro (●—●); rate of propagation of growing RNA chains in vitro (●---●); actinomycin D binding (o—o).
B. Changes in chromatin proteins during spermatogenesis: ubiquitin-H2A (●—●); HMG-1 (■—■); HMG-2 (□—□); acetylation of histone H4 (o—o).

Ubiquitin-Histone H2A Conjugates

The covalent conjugate, ubiquitin-histone H2A (uH2A or A24), resulting from the post-translational addition of ubiquitin to histone H2A at Lys 119, increases markedly in chicken testis cell nuclei during spermiogenesis, reaching a maximum (3.5% and 11% of the total amount of nucleosomal core histones) in late spermatids (Figure 3) (33). This modification has little influence at the level of individual nucleosomes (34). However, ubiquitin attachment to the core histone H2A may prevent formation of higher order chromosomal structures by modifying nucleosome-nucleosome interactions (35, 36), and may contribute to the relaxation of chromatin that occurs during the nucleohistone-nucleoprotamine transition in chicken spermiogenesis.

Protein-ubiquitin conjugates in the cytoplasm are substrates for an ATP-dependent proteolytic system (37-39).
Although proteolysis has been proposed as a mechanism for
histone removal during the nucleohistone-nucleoprotamine transition (40, 41), it is not yet known if an ubiquitin-ATP proteolytic system could be involved in the disassembly of nucleosomes
during chicken spermiogenesis. Ubiquitin is an abundant protein
in testicular cells (42, 43). This protein, in addition to
forming protein conjugates, might be involved in modulation of
certain enzymatic activities as has been observed in the case of
histone deacetylase activity assayed in vitro (44).

ADP-Ribosylation of Nuclear Proteins

The nuclear enzymatic activity ADP-ribosyltransferase
(ADPRT) attaches ADP-ribose moieties from NAD to chromatin proteins to form mono- or poly-ADP-ribosyl derivatives (45). The
structural and functional changes of chromatin during spermatogenesis could be potentially regulated by ADP-ribosylation of
nuclear proteins.

ADPRT activity and the turnover of ADP-ribosyl residues
decrease drastically during the transition from premeiotic,
meiotic cells, and early spermatids to late spermatids and
spermatozoa. ADPRT activity can be induced in late spermatids
but not in mature spermatozoa by agents that damage DNA, such as
dimethyl sulfate (46).

ADPRT activity is induced by DNA strand breaks (47). ADP-
ribosylation could be involved in DNA ligation during the interval of regulated DNA nick-repair activity at pachytene and in
other gene rearrangements occurring in genetically active meiotic and premeiotic cells. Late spermatids, where no physiological breaking and rejoining of DNA presumably occur, show low
ARPRT activity and no turnover of ADP-ribosyl residues. The
ADPRT activity can be stimulated in these cells by dimethyl
sulfate treatment in a process probably related with DNA repair
activity. Late spermatids still are able to repair their genetic damage before the final condensation of chromatin occurs at
the end of spermiogenesis (8).

ADP-ribosylation of nuclear proteins might be involved in a shuttle mechanism for removing proteins from DNA (48). In the absence of removal of ADP-ribosyl residues from nuclear proteins in late spermatids, ADP-ribosylation could contribute to protein displacement at the end of spermiogenesis. Determination of the steady-state level of ADP-ribosylation of nuclear proteins in elongated spermatids in vivo is necessary to test this possibility.

Nonhistone Proteins

The nuclear nonhistone protein content decreases from 0.8 (w/w ratio of nonhistone protein to DNA) in chicken meiotic and premeiotic cells to 0.2-0.3 in elongated spermatids and spermatozoa (3). Similar losses of nonhistone nuclear proteins have been reported to occur during spermiogenesis in many species. The electrophoretic analysis of nonhistone proteins at different stages of chicken spermatogenesis reveals a drastic decrease in the heterogeneity of the nonhistone proteins at the end of spermiogenesis (3). Most of the nonhistone nuclear proteins have not yet been identified. Changes in the components of the nuclear lamina and other nuclear matrix proteins occur during spermatogenesis (49, 50).

The quantitative changes of a well characterized group of nonhistone chromosomal proteins -- the high mobility group HMG-1 and HMG-2 -- have been determined in chicken spermatogenesis (51). The ratios of HMG-1/nucleosomal histones and HMG-2/nucleosomal histones increase at the end of spermiogenesis during the transition from nucleohistone to nucleoprotamine, when nucleosomes are being disassembled in elongated spermatids. The HMG-1 and HMG-2 are not detectable in the nuclei of chicken spermatozoa (Figure 3).

The enrichment of nucleohistone in HMG-1 and HMG-2 in chicken late spermatids could play some role in the nucleohistone-nucleoprotamine transition. A remarkable feature of the sequence of HMG-1 and HMG-2 is the presence of an unbroken run of glutamic acid and aspartic residues in the carboxyl half of the molecules (52). Polyglutamic acid is capable of facilitating the assembly of core histones and DNA into nucleosomes

(53). Rat liver HMG-1 has been considered as a physiological nucleosome assembly factor (54). Through interaction of the highly acidic C-terminal domain of HMG-1 and HMG-2 with the very basic regions of histones, the high mobility group proteins might contribute to the nucleosome disassembly during chicken spermiogenesis.

CHANGES IN METHYLATION OF DNA DURING SPERMATOGENESIS

Although many genes are hypermethylated in spermatozoa, DNA isolated from sperm of different species is undermethylated in relation to the 5-methyl cytosine content of DNA isolated from the corresponding somatic tissues (55, 56). To study whether changes in methylation of DNA were related to the structural and functional changes that chromatin undergoes throughout spermatogenesis, the 5-methyl cytosine content of DNA, purified from chicken testis cells at successive stages of differentiation, has been determined. The DNA of meiotic and postmeiotic cells appears hypomethylated, containing 30% less methyl cytosine than the DNA obtained from premeiotic and somatic cells (57).

An important question to be solved is which DNA sequences are undermethylated in meiotic and postmeiotic cells. Certain chicken genes are undermethylated in expressing tissues and heavily methylated in sperm, while other genes are not differentially methylated (58). The major cause of undermethylation in bovine sperm DNA (2.5% of cytosine is methylated in sperm compared with 5.4% in calf thymus DNA) is the presence of methyl deficient satellite DNA (55). Undermethylation of satellite DNA has also been detected in the germinal cell line of the mouse (59, 60).

Hypomethylation stimulates genetic recombination of bacteriophage lambda (61). The methyl cytosine content of certain DNA sequences might control the extent of the recombination reaction that occurs in meiotic and somatic cells. Highly repetitive DNA sequences are found in constitutive heterochromatin; undermethylation of satellite DNA might cause structural changes in this particular domain during spermatogenesis. A drastic reduction in the number of constitutive heterochromatin blocks per cell nucleus has been detected during the transition from diploid chicken spermatogonia to haploid late spermatids (62).

Thus, two classes of DNA sequences operate differently with respect to methylation during spermatogenesis (63). One group of genes, active in somatic cells, are highly methylated in sperm DNA. These genes are activated in the zygote through replication and cell division. Another group of DNA sequences, including some of the highly repeated DNA and certain integrated viral genomes, arrive in the zygote in an unmethylated state and are methylated during development (64). Hypomethylated sequences in the germinal cell line could play some role during spermatogenesis and are inactive in somatic cells. The integrated viral genomes could be inactivated by the same mechanism acting in the other hypomethylated sequences if their methylation signal came under the same control system (63).

CHANGES IN CELLULAR CONTENT AND BIOSYNTHESIS OF POLYAMINES DURING SPERMATOGENESIS

In addition to the changes observed in nuclear proteins and DNA at different stages of spermatogenesis, the ionic ambiance in which the structural and functional transitions of chromatin occur should be considered. The organic polycations spermine, spermidine, and putrescine, are particularly interesting, because these polyamines are involved in genetic activities and structural changes of DNA (65, 66).

During chicken spermiogenesis, the ratio of polyamines/cell volume undergoes a marked increase (Figure 2) (67). This fact, together with the high enzymatic activities of polyamine biosynthesis detected in late spermatids (Figure 2) (67), suggests that the dramatic changes in chromatin composition and structure that take place during spermiogenesis could occur in an ambiance of high polyamine concentration. The transition from nucleohistone to nucleoprotamine could be facilitated by polyamines in a similar way to that proposed for DNA compaction in several bacteriophages. In this system, during DNA condensation, the entropic force, the work required to bring the negatively charged segments of DNA into close proximity, and the energy necessary to bend the stiff DNA chains would be strongly diminished by the presence of polyamines (66).

In addition to a possible structural role of polyamines, facilitating the condensation of DNA at the end of spermiogenesis, other possible functions of polyamines can be postulated: 1) competition with histones for DNA binding during histone removal; 2) protection of DNA against nucleases and mutagens; and 3) regulation of enzymatic activities involved in the nucleohistone-nucleoprotamine transition. The polyamine spermine stimulates histone acetyl transferase in chicken testis chromatin in vitro (68).

CHANGES IN CHROMATIN STRUCTURE DURING SPERMATOGENESIS

Most of the information on the structural changes of chromatin during spermatogenesis has been obtained with the electron microscope and by cytochemical methods. Nucleases and RNA polymerases have also been used as a tool for probing the major changes in chromatin structure that take place during the nucleohistone-nucleoprotamine transition.

Primary Spermatocytes

Chromosomal structures in primary spermatocytes have been identified as lampbrush loop-like differentiations. RNA synthesis and processes other than transcription have been demonstrated in association with these chromatin loops by electron microscopy and autoradiography of pachytene chromosomes (2, 69).

In species that are as phylogenetically distant as lily and mouse, a common pattern for the organization of meiotic recombination has been reported (7). Four components of the pattern have been identified: transient appearance of a protein that facilitates DNA reannealing; the programmed introduction of single strand nicks, many of them extended to gaps; the repair of endogenous formed nicks and gaps; and the preferential localization of nicks and gaps in specific DNA sequences.

During meiotic prophase, the sites of DNA that undergo nick-repair activity and could be initiation sites for genetic recombination, show a unique organization (70, 71). A group of small nuclear RNA molecules (PsnRNA) transcribed by RNA polymerase III during the zygoten-pachytene interval, are associated with these sites by sequence complementarity. The altered chromatin domains containing PsnRNA become accessible to meiotic endonuclease and DNase II. These domains probably contain a nonhistone protein rather than the usual histones.

Spermiogenesis

The initiation pattern of RNA synthesis in vitro, obtained on chromatin of chicken spermatids by using E. coli RNA polymerase differs from the pattern observed on chromatin of previous stages of spermatogenesis (10) in the following ways: 1) increased number of binding sites; 2) increased rate of propagation of growing RNA chains; 3) presence of strong and weak polymerase binding sites; 4) shorter half time of formation of high-affinity enzyme-chromatin complexes (RS complexes) and higher temperature dependence of the RS-complex formation. The eukaryotic RNA polymerase II enzyme is far less efficient for initiation of RNA synthesis and shows higher temperature dependence on chromatin obtained from elongated spermatids than on chromatin of meiotic and premeiotic cells.

The characteristics of the initiation pattern of RNA synthesis in vitro on chromatin of genetically inactive spermatids undergoing the nucleohistone-nucleoprotamine transition, together with the high capacity for binding of actinomycin D in vivo and in vitro of this chromatin, reflects unmasking of DNA during spermiogenesis (Figure 3) (3, 10). Electron microscopic visualization of the nucleohistone-nucleoprotamine transition also shows marked changes in chromatin structure. The beaded chromatin fibers of mouse testis cells are replaced in late spermatids by smooth fibers (4). Domains of chromatin obtained from late stage trout testis also show a highly relaxed structure with reduced nucleosome density or an entirely nucleosome-free appearance (24).

Bulk cell chromatin consists of static nucleosomes with DNA supercoils constrained and packaged by the nucleosomal core histones and histone H1 (72, 73). As a consequence of histone modification or by the presence of additional proteins, together with a possible induction of DNA supercoiling by an active energy-driven process, static nucleosomes undergo transition into dynamic nucleosomes (72, 73). The generated supercoils of dynamic nucleosomes are able to interact with proteins. The term "dynamic chromatin" is used instead of "active chromatin" because gene expression requires the presence of enzymes and protein factors in addition to supercoiled DNA.

Bulk chromatin of spermatids, before the nucleohistone-nucleoprotamine transition commences, might consist of static nucleosomes. Histones cannot be removed from static nucleosomes by protamine (29). The canonical histone octamer in static nucleosomes should be altered during spermiogenesis by histone modifications (hyperacetylation, ubiquitination, or ADP-ribosylation) or by the presence of additional components (HMG proteins) to become dynamic nucleosomes. The DNA of these dynamic nucleosomes is genetically inactive, but it is able to interact with the protamine. Through interaction with the protamine in an ambiance of high polyamine concentration, the disassembly of nucleosomes would become possible. We do not yet know if an active ATP-driven reaction is necessary for the transition from static to dynamic nucleosomes during spermiogenesis.

Besides the possibility of an active induction, supercoiled DNA in spermatids must be present as a consequence of histone removal. The superhelical density induced by histone removal (74) is in the range sufficient to drive alternating dG-dC sequences towards the Z-DNA conformation. Polyamines and proteins containing large numbers of basic residues stabilize Z-DNA (75). Supercoiled DNA and Z-DNA conformation could play some role during the nucleohistone-nucleoprotamine transition.

The binding of the protamine to DNA occurs when the serine residues of the protamine are phosphorylated (76). Phosphorylated protamines displace histones and avoid a premature condensation of DNA, which probably could interfere with the process. Once histones have been displaced, the dephosphorylation of protamine produces extensive DNA condensation (77). Granules of about 400 Å in diameter appear scattered throughout the nucleus of chicken late spermatids, and then coalesce to form the compact nucleoprotamine complex of the head of spermatozoa (78). The volume of chicken early spermatid nuclei (25 μm^3) decreases to 2 μm^3 during spermiogenesis. The DNA becomes tightly packed (0.6 g of DNA/cm^3), reaching a degree of condensation similar to that found in the bacteriophage head. The condensed DNA of spermatozoa possesses a low capacity for binding of actinomycin D and shows a drastic decrease in the number of initiation sites for RNA polymerase (Figure 3) (3, 10). The process of condensation of DNA during spermiogenesis is similar to that observed when viral DNA is packaged into virions. Arginine-rich peptides, responsible for the condensation of the nucleoprotein cores released from virions, are remarkably similar to protamines (79). Extensive sequence similarity is found in nucleotides of the core protein gene of hepatitis B virus compared with nucleotides of the trout protamine gene (80).

The end product of differentiation of the male germinal cell line, the spermatozoan, shows an efficient packaging of DNA. The DNA of this cell is highly protected against nucleases and mutagens in a nucleus that possesses a hydrodynamic shape that facilitates the delivery of the genetic information with a minimum energetic cost.

REFERENCES

1. MONESI, V. (1971). Chromosome activities during meiosis and spermiogenesis. J. Reprod. Fert. (suppl.) 13, 1-14.

2. KIERSZENBAUM, A.L. and TRES, L.L. (1978). RNA transcription and chromatin structure during meiotic and postmeiotic stages of spermatogenesis. Federation Proc. 37, 2512-2516.

3. MEZQUITA, C. and TENG, C.S. (1977). Changes in nuclear and chromatin composition and genomic activity during spermatogenesis in the maturing rooster testis. Biochem. J. 164, 99-111.

4. KIERSZENBAUM, A.L. and TRES, L.L. (1975). Structural and transcriptional features of the mouse spermatid genome. J. Cell Biol. 65, 258-270.

5. IATROU, K. and DIXON, G.H. (1978). Protamine messenger

RNA: its life history during spermatogenesis in rainbow trout. <u>Federation Proc</u>. <u>37</u>, 2526-2533.

6. DISTEL, R.J., KLEENE, K.C., and HECHT, N.B. (1984). Haploid expression of a mouse testis α-tubulin gene. <u>Science</u> <u>224</u>, 68-70.

7. STERN, H. and HOTTA, I. (1977). Biochemistry of meiosis. <u>Phil</u>. <u>Trans</u>. <u>Roy</u>. <u>Soc</u>. <u>B</u> <u>277</u>, 277-293.

8. LAHDETIE, J., KAUKOPURO, S., and PARVINEN, M. (1983). Genotoxic effects of ethyl methanesulfonate and X-rays at different stages of rat spermatogenesis, studied by inhibition of DNA synthesis and induction of DNA repair in vitro. <u>Hereditas</u> <u>99</u>, 269-278.

9. SEGA, G.A. (1974). Unscheduled DNA synthesis in germ cells of male mice exposed in vivo to the chemical mutagen ethyl methanesulfonate. <u>Proc</u>. <u>Natl</u>. <u>Acad</u>. <u>Sci</u>. <u>USA</u> <u>71</u>, 4955-4959.

10. MEZQUITA, C. and TENG, C.S. (1977). Changes in chromatin structure during spermatogenesis in maturing rooster testis as demonstrated by the initiation pattern of ribonucleic acid synthesis in vitro. <u>Biochem</u>. <u>J</u>. <u>170</u>, 203-210.

11. LOIR, M. and COURTENS, J.L. (1979). Nuclear reorganization in ram spermatids. <u>J</u>. <u>Ultrastruct</u>. <u>Res</u>. <u>67</u>, 309-324.

12. LOIR, M. and LANNEAU, M. (1978). Transformation of ram spermatid chromatin. <u>Exp</u>. <u>Cell Res</u>. <u>155</u>, 231-243.

13. VAUGHN, J.C. and THOMSON, L.A. (1972). A kinetic study of DNA and basic protein metabolism during spermatogenesis in the sand crab, Emerita analoga. <u>J</u>. <u>Cell Biol</u>. <u>52</u>, 322-337.

14. NAKANO, M., TOBITA, T., and ANDO, T. (1976). Studies on a protamine (galline) from fowl sperm. <u>Int</u>. <u>J</u>. <u>Peptide Protein Res</u>. <u>8</u>, 565-578.

15. MEISTRICH, M.L., BROCK, W.A., GRIMES, S.R., PLATZ, R.D., and HNILICA, L.S. (1978). Nuclear protein transitions during spermatogenesis. <u>Federation Proc</u>. <u>37</u>, 2522-2525.

16. KAYE, J.S. and McMASTER-KAYE, R. (1982). Characterization of the ususual basic proteins of cricket spermatid nuclei on the basis of their molecular weights and amino acid compositions. <u>Biochim</u>. <u>Biophys</u>. <u>Acta</u> <u>696</u>, 44-51.

17. CHAUVIERE, M., LAINE, B., SAUTIERE, P., and CHEVAILLIER, P. (1983). Purification and characterization of two basic spermatid-specific proteins isolated from the dog-fish Scylliorhinus caniculus. <u>FEBS Lett</u>. <u>152</u>, 231-235.

18. MARUSHIGE, K. and DIXON, G.H. (1969). Developmental changes in chromosomal composition and template activity during spermatogenesis in trout testis. <u>Dev</u>. <u>Biol</u>. <u>19</u>, 397-414.

19. MARUSHIGE, K. and DIXON, G.H. (1971). Transformation of trout testis chromatin. <u>J</u>. <u>Biol</u>. <u>Chem</u>. <u>246</u>, 5799-5805.

20. BUCCI, L.R., BROCK, W.A., and MEISTRICH, M.L. (1982). Distribution and synthesis of histone 1 subfractions during spermatogenesis in the rat. <u>Exp</u>. <u>Cell Res</u>. <u>140</u>, 111-118.

21. SEYEDIN, S.M. and KISTLER, W.S. (1983). H1 histones from mammalian testes. <u>Exp</u>. <u>Cell Res</u>. <u>143</u>, 451-454.

22. TROSTLE-WEIGE, P.K., MEISTRICH, M.L., BROCK, W.A.,

NISHIOKA, K., and BREMER, J.W. (1982). Isolation and characterization of TH2A, a germ cell-specific variant of histone 2A in rat testis. J. Biol. Chem. 257, 55560-55567.

23. CHRISTENSEN, M.E. and DIXON, G.H. (1982). Hyperacetylation of histone H4 correlates with the terminal transcriptionally inactive stages of spermatogenesis in rainbow trout. Dev. Biol. 93, 404-415.

24. CHRISTENSEN, M.E., RATTNER, J.B., and DIXON, G.H. (1984). Hyperacetylation of histone H4 promotes chromatin decondensation prior to histone replacement by protamines during spermatogenesis in rainbow trout. Nucleic Acids Res. 12, 4575-4592.

25. GRIMES, S.R. and HENDERSON, N. (1983). Acetylation of histones during spermatogenesis in the rat. Arch. Biochem. Biophys. 221, 108-116.

26. OLIVA, R. and MEZQUITA, C. (1982). Histone H4 hyperacetylation and rapid turnover of its acetyl groups in transcriptionally inactive rooster testis spermatids. Nucleic Acids Res. 10, 8049-8059.

27. McGHEE, J.D., NICKOL, J.M., FELSENFELD, G., and RAU, D.C. (1983). Histone hyperacetylation has little effect on the higher order folding on chromatin. Nucleic Acids Res. 11, 4065-4074.

28. CARY, P.D., CRANE-ROBINSON, C., BRADBURY, E.M., and DIXON, G.H. (1982). Effect of acetylation on the binding of N-terminal peptides of histone H4 to DNA. Eur. J. Biochem. 127, 137-143.

29. BODE, J., HENCO, K., and WINGENDER, E. (1980). Modulation of the nucleosome particles open as the histone core becomes hyperacetylated. Eur. J. Biochem. 110, 143-152.

30. BODE, J., GOMEZ-LIRA, M.M., and SCHROTER, H. (1983). Nucleosomal particles open as the histone core becomes hyperacetylated. Eur. J. Biochem. 130, 437-445.

31. RUIZ-CARRILLO, A. and PALAU, J. (1973). Histones from embryos of the sea urchin Arbacia lixula. Dev. Biol. 35, 115-123.

32. WANGH, L., RUIZ-CARRILLO, A., and ALLFREY, V.G. (1972). Separation and analysis of histone subfractions differing in their degree of acetylation: some correlation with genetic activity in development. Arch. Biochem. Biophys. 150 44-54.

33. AGELL, N., CHIVA, M., and MEZQUITA, C. (1983). Changes in nuclear content of protein conjugate histone H2A-ubiquitin during rooster spermatogenesis. FEBS Lett. 155, 209-212.

34. KLEINSCHMIDT, A.M. and MARTINSON, H.G. (1981). Structure of nucleosome core particles containing uH2A (A24). Nucleic Acids Res. 9, 2423-2331.

35. MATSUI, S., SEON, B.K., and SANDBERG, A.A. (1979). Disappearance of a structural chromosomal protein A24 in mitosis: implications for molecular basis of chromatin condensation. Proc. Natl. Acad. Sci. USA 76, 6386-6390.

36. LEVINGER, L. and VARSHAVSKY, A. (1982). Selective arrangement of ubiquitinated and D1 protein-containing nucleosomes within the Drosophila genome. Cell 28, 375-385.

37. HERSHKO, A. (1983). Ubiquitin: roles in protein modification and breakdown. Cell 34, 11-12.

38. FINLEY, D., CIECHANOVER, A., and VARSHAVSKY, A. (1984). Thermolability of ubiquitin-activating enzyme from the mammalian cell cycle mutant ts85. Cell 37, 43-55.

39. CHIECHANOVER, A., FINLEY, D., and VARSHAVSKY, A. (1984). Ubiquitin dependence of selective protein degradation demonstrated in the mammalian cell cycle mutant ts85. Cell 37, 57-66.

40. MARUSHIGE, U. and MARUSHIGE, K. (1983). Proteolysis of somatic type histones in transforming rat spermatid chromatin. Biochim. Biophys. Acta 761, 48-57.

41. KUMAROO, K.K. and IRVING, J.L. (1984). Diisopropyl fluorophophate-interacting proteinases of nuclei of rat testis cells. Biochim. Biophys. Acta 782, 320-327.

42. WATSON, D.C., LEVY, B., and DIXON, G.H. (1978). Free ubiquitin is a non-histone protein of trout testis chromatin. Nature 276, 196-198.

43. LOIR, M., KARATY, A., LANNEAU, M., MENEZO, Y., MUH, J.P., and SAUTIERE, P. (1984). Purification and characterization of ubiquitin from mammalian testis. FEBS Lett. 169, 199-204.

44. MEZQUITA, J., CHIVA, M., VIDAL, S., and MEZQUITA, C. (1982). Effect of high nobility group non-histone proteins HMG-20 (ubiquitin) and HMG-17 on histone deacetylase activity in vitro. Nucleic Acids Res. 10, 1781-1797.

45. HAYAISHI, O. and UEDA, K. (1982). Poly- and mono(ADP-ribosyl)ation reactions: their significance in molecular biology. In: "ADP-ribosylation Reactions," (O. Hayaishi, and K. Ueda, eds.) pp. 3-16. Academic Press, New York.

46. MEZQUITA, C. and COROMINAS, M. (1983). ADP-ribosyl transferase activity during rooster spermatogenesis. J. Cell Biol. 97, 138a.

47. SHALL, S. (1982). ADP-ribose in DNA repair. In: "ADP-ribosylation Reactions," (O. Hayaishi and K. Ueda, eds.) pp. 477-520. Academic Press, New York.

48. ZAHRADKA, P. and EBISUZAKI, K. (1982). A shuttle mechanism for DNA-protein interactions. The regulation of poly(ADP-ribose)polymerase. Eur. J. Biochem. 127, 579-585.

49. STICK, R. and SCHWARZ, H. (1982). The disappearance of the nuclear lamina during spermatogenesis: an electron microscopic and immunofluorescence study. Cell Differentiation 11, 235-243.

50. PRUSLIN, F.H. and ROSMAN, T.C. (1983). Proteins of demembraned protamine-depleted mouse sperm. Homology with proteins of somatic cell nuclear envelope/matrix. Exp. Cell Res. 144, 115-126.

51. CHIVA, M. and MEZQUITA, C. (1983). Quantitative changes of high mobility group non-histone chromosomal proteins HMG-1 and HMG-2 during rooster spermatogenesis. FEBS Lett. 162, 324-328.

52. REECK, G.R., ISACKSON, P.J., TELLER, D.C. (1982). Domain structure in high molecular weight high mobility group nonhistone chromatin proteins. Nature 300, 76-78.

53. STEIN, A., WHITLOCK, J.P., and BINA, M. (1979). Acidic polypeptides can assemble both histones and chromatin in vitro at physiological ionic strength. Proc. Acad. Sci. USA 76, 5000-5004.

54. BONNE-ANDREA, C., HARPER, F., SOBCZAK, J., and DE RECONDO, A.M. (1984). Rat liver HMG-1: a physiological nucleosome assembly factor. EMBO J. 3, 1193-1199.

55. ADAMS, R.L.P., BURDON, R.H., and FULTON, J. (1983). Methylation of satellite DNA. Biochem. Biophys. Res. Commun. 113, 695-702.

56. GAMA-SOSA, M.A., WANG, R.Y.H., KUO, K.C., GEHRKA, C.X., and EHRLICH, M. (1983). The 5-methylcytosine content of highly repeated sequences in human DNA. Nucleic Acids Res. 11, 3087-3095.

57. ROCAMORA, N. and MEZQUITA, C. (1984). Hypomethylation of DNA in meiotic and postmeiotic rooster testis cells. FEBS Lett. 177, 81-84.

58. RAHE, B., ERICKSON, R.P., and QUINTO, M. (1983). Methylation of unique sequence DNA during spermatogenesis in mice. Nucleic Acids Res. 11, 7947-7959.

59. PONZETTO-ZIMMERMAN, C. and WOLGEMUTH, D.J. (1984). Methylation of satellite sequences in mouse spermatogenic and somatic DNA's. Nucleic Acids Res. 12, 2807-2821.

60. SANFORD, J., FORRESTER, L., and CHAPMAN, V. (1984). Methylation patterns of repetitive DNA sequences in germ cells of Mus musculus. Nucleic Acids Res. 12, 2823-2836.

61. KORBA, B.E. and HAYS, J.B. (1982). Partially deficient methylation of cytosine in DNA at CC_T^AGG sites stimulates genetic recombination of bacteriophage lambda. Cell 28, 531-541.

62. DRESSLER, B. and SCHMID, M. (1976). Specific arrangement of chromosome in the spermiogenesis of Gallus domesticus. Chromosoma (Berl.) 58, 387-391.

63. TAYLOR, J.H. (1984). DNA methylation and cellular differentiation. In: "Cell Biology Monographs," Vol. 11. Springer-Verlag, Wien.

64. JAHNER, D., STUHLMANN, H., STEWART, C.L., HARBERS, K., LOHLER, J., SIMON, I., and JAENISCH, R. (1982). De novo methylation and expression of retroviral genomes during mouse embryogenesis. Nature 298, 623-628.

65. MORRIS, D.R. (1978). Polyamine function in rapidly proliferating cells. In: "Advances in Polyamine Research," Vol. 1 (R.A. Campbell, R.D. Morris, D. Bartos, G.D. Daves, and F. Bartos, eds.) pp. 105-115. Raven Press, New York.

66. GOSULE, L.C., CHATTORAJ, D.K., and SCHELLMAN, J.A. (1978). Condensation of phage DNA by polyamines. In: "Advances in Polyamine Research," Vol. 1 (R.A. Campbell, R.D. Morris, D. Bartos, G.D. Daves, and F. Bartos, eds.) pp. 201-215. Raven Press, New York.

67. OLIVA, R., VIDAL, S., and MEZQUITA, C. (1982). Cellular content and biosynthesis of polyamines during rooster spermatogenesis. Biochem. J. 208, 269-273.

68. MEZQUITA, C., MEZQUITA, J., VIDAL-SIVILLA, S. (1980). The polyamine spermine stimulates enzymatic acetylation of his-

tones in rooster testis chromatin in vitro. <u>Eur</u>. <u>J</u>. <u>Cell</u> <u>Biol</u>. <u>22</u>, 82.

69. GROND, C.J., RUTTEN, R.G.J., and HENNIG, W. (1984). Ultra-structure of the Y chromosomal lampbrush loops in primary spermatocytes of Drosophila hydei. <u>Chromosoma</u> (Berl.) <u>89</u>, 85-95.

70. HOTTA, Y. and STERN, H. (1981). Small nuclear RNA mole-cules that regulate nuclease accessibility in specific chromatin regions of meiotic cells. <u>Cell</u> <u>27</u>, 309-319.

71. HOTTA, Y. and STERN, H. (1984). The organization of DNA segments undergoing repair synthesis during pachytene. <u>Chromosoma</u> (Berl.) <u>89</u>, 127-137.

72. RYOJI, M. and WORCEL, A. (1984). Chromatin assembly in Xenopus oocytes: in vivo studies. <u>Cell</u> <u>37</u>, 21-32.

73. GLIKIN, G.C., RUBERTI, I., WORCEL, A. (1984). Chromatin assembly in Xenopus oocytes: in vitro studies. <u>Cell</u> <u>37</u>, 33-41.

74. SINGLETON, C.K., KLYSIK, J., STIRDIVANT, S.N., and WELLS, R.D. (1982). Left-handed Z-DNA is induced by supercoiling in physiological ionic conditions. <u>Nature</u> <u>299</u>, 312-316.

75. WANG, A.H.J., QUIGLEY, G.J., KOLPAK, F.J., CRAWFORD, J.L., BOOM, J.H., MAREL, G., and RICH, A. (1979). Molecular structure of a left-handed double helical DNA fragment at atomic resolution. <u>Nature</u> <u>282</u>, 680-686.

76. LOUIE, A.J. and DIXON, G.H. (1972). Trout testis cells. II. Synthesis and phosphorylation of histones and prota-mines in different cell types. <u>J</u>. <u>Biol</u>. <u>Chem</u>. <u>247</u>, 5498-5505.

77. WARRANT, R.W. and KIM, S.H. (1978). α-Helix-double helix interaction shown in the structure of a protamine-transfer RNA comples and a nucleoprotamine model. <u>Nature</u> <u>271</u>, 130-135.

78. McINTOSH, J.R. and PORTER, K.R. (1967). Microtubules in the spermatids of domestic fowl. <u>J</u>. <u>Cell</u> <u>Biol</u>. <u>35</u>, 153-173.

79. WAYDA, M.E., ROGERS, A.E., and FLINT, S.J. (1983). The structure of nucleoprotein cores released from adeno-virions. <u>Nucleic</u> <u>Acids</u> <u>Res</u>. <u>11</u>, 441-460.

80. ALTMAN, S., MODEL, P., DIXON, G.H., and WOSNICK, M.A. (1981). An E. coli gene coding for a protamine-like protein. <u>Cell</u> <u>26</u>, 299-304.

DIVERSITY OF SPERM BASIC CHROMOSOMAL PROTEINS IN THE
VERTEBRATES: A PHYLOGENETIC POINT OF VIEW

Harold E. Kasinsky[1,2], Mairi Mann[1], Michael Lemke[1],
and Sue-Ying Huang[1]

[1]Department of Zoology
University of British Columbia
Vancouver, B.C. V6T 2A9

[2]Unitat de Quimica Macromolecular
CSIC
Escuela T.S. d'Enginyers Industrials
Diagonal, 647
08028 Barcelona, Spain

INTRODUCTION

In this paper we try to explain the variability of sperm basic proteins in nature by taking the subphylum Vertebrata as a starting point. The data presently available indicate that the appearance of unique sperm basic proteins has not been a sporadic phenomenon during vertebrate evolution. Rather there is a general macroevolutionary trend; namely, extreme variability of sperm basic proteins in bony fish and frogs gives way to a relative constancy of sperm protein types within urodeles, snakes, lizards, turtles, birds, metatherian and eutherian mammals. Cartilaginous fish also have similar sperm basic proteins. Furthermore, within particular orders of frogs and bony fish, certain families of sperm basic proteins are characteristic for particular genera and even individual species can be distinguished by their typical set of sperm proteins. This burst of sperm protein variability in the bony fish and frogs during vertebrate phylogeny coincides with the absence of internal fertilization in these orders, the appearance of sperm motility in the testis rather than the excurrent duct, the existence of polyploidy and the general absence of heteromorphic sex chromosomes. This seems to relieve selection pressure to maintain some relative constancy of sperm protein type in these orders. We speculate that perhaps the set of basic chromosomal proteins required to produce a functional sperm in a particular species of frog or bony fish is due to the time of onset of sexual maturity in that species. Thus, from a phylogenetic point of view, although sperm basic protein evolution in the vertebrates has been much less conservative than that of the nucleosomal histones, it has not been entirely a random affair.

DIVERSITY OF BASIC PROTEINS IN SPERM

The basic proteins associated with DNA in the mature sperm are highly variable amongst the numerous animals and plants that have been examined (1, 2), in contrast to the evolutionarily conservative nucleosomal histones. It is possible to cluster this diversity into five main groups (2): type 1, the salmon type of arginine-rich monoprotamines; type 2, the mouse-grasshopper type of stable protamine or keratinous protamine containing cystine; type 3, the Mytilus or mussel intermediate type of di- or tri-protamine with lysine and/or histidine as well as arginine; type 4 or the Rana somatic-like histone type; type 5, the crab type, in which no basic protein is found in the mature sperm.

Subirana (3) has attempted to rationalize the classifi-cation of the first four types by distinguishing two broad groups: protamine-containing and histone-containing sperm. Proteins in the protamine class contain lysine and arginine to the extent of 45-80 mole percent and usually no histidine. Proteins in the histone class may be one of four types having: a) no detectable change in the histone H1 family, b) slight change in the histone H1 family, c) additional sperm-specific basic proteins, and d) considerable changes in histones H1 and H2B.

Whatever scheme of classification one adopts, the variety of sperm basic chromosomal proteins in nature appears to be overwhelming. There does not seem to be any logic to the seemingly sporadic appearance of different sperm proteins in different phylogenetic groups of animals and plants (2).

Is this indeed the case? Perhaps evolutionary trends in sperm basic protein evolution do not emerge because one is trying to compare too many taxa all at once. What if we limit our observations on the diversity of sperm proteins to a single subphylum: the vertebrates? If we do so, we believe that a macroevolutionary trend is apparent; namely, that the variability of sperm basic chromosomal proteins is limited to particular orders of bony fish and the frogs. Within other vertebrate orders these proteins are relatively constant (4). Furthermore, within the anurans and the teleosts, another evolutionary trend is apparent: variability may extend to the species level. Thus, different frog genera are characterized by different families of sperm basic proteins (4, 5, 6) and in the genus Xenopus, each species has a typical electrophoretic profile for its sperm proteins (7a,b). In the case of the stickleback, Gasterosteus aculeatus, variability extends even to the slightly different electrophoretic profiles between saltwater and freshwater forms of this species of teleost (Lemke and Kasinsky, in preparation).

In this paper we present a brief systematic survey of the sperm basic proteins found in a limited number of the approximately 40,000 vertebrate species (8). From these data we develop the phylogenetic point of view that sperm protein evolution has not been entirely sporadic in the vertebrates. We speculate that where sperm basic chromosomal proteins are most variable, in frogs and bony fish, consideration of the time of the onset of sperm protein synthesis in ontogeny might help to explain the phylogenetic patterns.

SUPERCLASS AGNATHA

The sperm proteins of jawless fishes have not yet been examined.

CLASS CHONDRICHTHYES

Amongst the cartilaginous fishes are found the sharks, skates, and rays in the subclass Elasmobranchi and the ratfish or chimaeras in the subclass Holocephali. Each animal has a

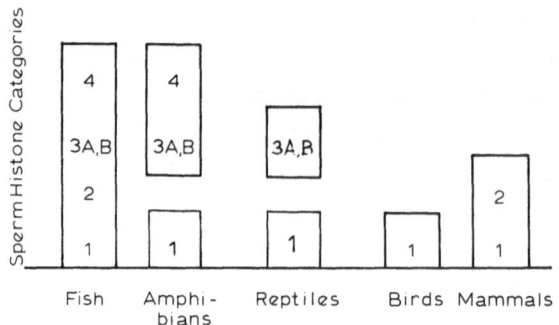

Figure 1. A composite cytochemical classification of sperm histones in the vertebrates. This is the classification scheme of Bloch (1976, 1969) as modified for type 3 sperm basic proteins by Kasinsky et al. (1985). The type 3 intermediate category can be subdivided into two subtypes depending on the extractability of sperm basic chromosomal proteins in 5% trichloroacetic acid at different temperatures. Type 3A requires higher temperatures of 95-100°C for extraction whereas type 3B can be extracted at the lower temperatures of 85-90°C.

zonate testis (9) in which spherical ampullae contain germ elements at the same stage of development. Cytochemically, the sperm basic chromosomal proteins of representatives of this class are type 2 keratinous protamines (Figure 1). Squalus acanthias, the spiny dogfish, Raja rhina, the longnose skate, and Hydrolagus colliei, the ratfish, all show a transition from somatic histones to the salmon type 1 protamine in the elongating spermatid to the mouse/grasshopper type 2 protamine, containing cystine, in the mature sperm (10-12). Biochemically, this transition is marked by the appearance of two basic proteins, S1 and S2, that replace the somatic histones at the beginning of nuclear elongation in Scyliorhinus caniculus, the lesser spotted dogfish (13, 14). These proteins are in turn replaced by four proteins in elongated spermatids, one of which (Z3) is arginine-rich, and the other three (Z1, Z2, S4) are arginine- and cysteine-rich (15). In mature sperm cells the latter proteins form disulfide crosslinks. Bloch (1, 2) had originally characterized the sperm protein type of another shark and a ray as intermediate or type 3 based on the earlier studies of Kossel (16). Recent electrophoretic analysis on acid/urea polyacrylamide gels of the testis-specific proteins of Hydrolagus (Kasinsky, Bols, Stanley, and Tsui, in preparation) and of Squalus and Rhina (Kasinsky and Tsui, in preparation) indicate that the basic proteins of these cartilaginous fish are similar to those of <u>Scyliorhinus</u>. In Figure 2 the <u>Squalus</u> electrophoretic profile for the rapidly migrating sperm proteins is shown as a representative for Chondrichthyes. The top four bands also represent the electrophoretic profile for <u>Hydrolagus</u>. We can conclude from these data that the basic sperm proteins of four representatives of the class Chondrichtyes are similar to each other and fall into the category of type 2 keratinous protamines.

One totally unexpected difference between spermatogenesis in Holocephali and Elasmobranchi deserves mention. Cytochemical and electron microscopic examination of ratfish testis has revealed an instance of the rare phenomenon of germline chromatin diminution during meiotic prophase in this holocephalan fish (17). This does not occur in the dogfish or skate. About 10% of the DNA is set aside as a chromatin diminution body in the ratfish secondary spermatocyte which then is phagocytized in a sertoli cell in mid-spermiogenesis.

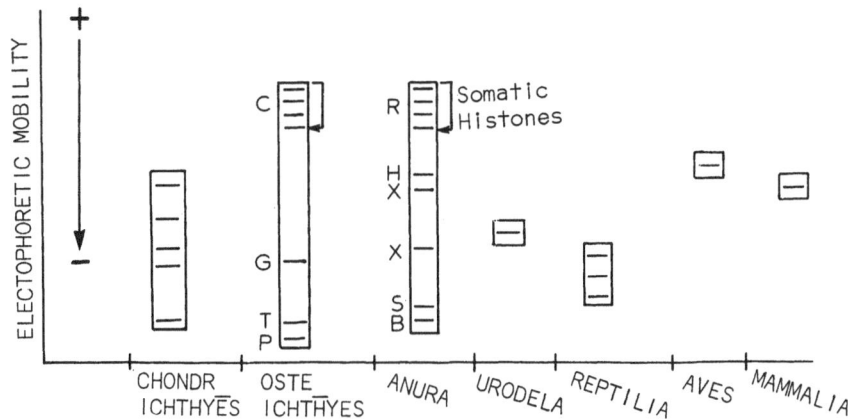

Figure 2.   A   composite,   schematic   electrophoretogram   of vertebrate   sperm   basic   proteins   on   acid/urea polyacrylamide   gels.   Abbreviations are: C = carp; G = Gasterosteus; T = Tilapia; P = herring protamine; R = Rana; H = Hyla; X = Xenopus; S = Scaphiopus; B = Bufo.

CLASS OSTEICHTHYES

The bony fish contain a great variety of sperm basic chromosomal proteins falling into the cytochemical categories (Figure 1) of type 1 protamines, type 3 intermediate proteins and type 4 somatic-like histones (1, 2). Great variability is the rule. Salmonid fishes (infraclass Teleostei, suborder Salmonoidei) display the protamine type, as Gordon Dixon and colleagues have elegantly demonstrated at both the protein (18) and DNA levels (see chapter by Dixon et al., this volume). This is also true of herring (infraclass Teleostei, order Clupeiformes) and sturgeon (infraclass Chondrostei, order Acipenseriformes). (See the reviews by Subirana (3) and Chevaillier (15) for pertinent references.) Perciformes, with nearly 7000 species, is the largest order of vertebrates (19) with sperm basic proteins falling into cytochemical types 1, 3, and 4 (1, 2).

The goldfish Carassius auratus (infraclass Teleostei, suborder Cyprinoidei) is the only organism studied thus far in which the sperm basic proteins appear to be identical with the somatic histones (20). In Figure 2 a variety of electrophoretic profiles is displayed for Osteichthyes ranging from the type 4 somatic-like histones of the carp to the type 3 intermediate proteins of Gasterosteus and Tilapia to the type 1 protamines of trout and herring. In the case of Gasterosteus, the threespine stickleback, Lemke and Kasinsky (in preparation) can detect slight electrophoretic differences between saltwater and freshwater forms and the electrophoretic profiles of fish related to sticklebacks all appear to be species- ecific. Thus, the variability of sperm proteins amongst the bony fishes extends to the species level.

CLASS AMPHIBIA

The phenomenon of marked sperm protein diversity is also apparent in anurans but not in urodeles. Thus, cytochemical (Figure 1), electrophoretic (Figure 2), and amino acid analysis of sperm basic proteins in anurans distinguishes type 4 somatic-like histones in Rana with its testis-specific histone H1 (21) from the type 3 intermediate proteins of <u>Bufo</u>, <u>Xenopus</u>, <u>Hyla</u>, and <u>Schaphiopus</u> (4-6, 22). In fact extractability in 5% trichloroacetic acid at different temperatures can split the type 3 intermediate histone category into two cytochemical groups: type 3B intermediate sperm proteins of <u>Bufo</u> (as well as the "typical" representative of this category, the mussel <u>Mytilus</u>) are extractable at 85-90°C while <u>Xenopus</u> intermediate type 3A sperm proteins require temperatures of 95-100°C for extraction (Figure 1). This may be due to the higher content of arginine in the more rapidly migrating (Figure 2) <u>Bufo</u> proteins (5). Both <u>Xenopus</u> and <u>Bufo</u> sperm proteins contain histidine and lysine as well as arginine. In the case of the genus <u>Xenopus</u>, the spermatid/sperm basic chromosomal protein profile is specific for each of 12 species in the genus and shows subtle variations within subspecies of <u>laevis</u> (7a, 7b). This is a polyploid genus with two lines. One line, including <u>Xenopus laevis</u> (diploid chromosome number 2n = 36) has representatives with 2n = 36, 72, and 108 (23, 24). The other line includes <u>Xenopus tropicalis</u> (2n = 20) and <u>Xenopus epitropicalis</u> (2n = 40) (25). Both cytochemically and electrophoretically, each of these polyploid lines can be distinguished from the other. Thus, representatives of species with 2n = 36, 72, and 108 have intermediate type basic proteins migrating more rapidly than histone H4 (7). On the other hand, <u>Xenopus tropicalis</u> (2n = 20) has type 4 somatic-like histones and does not reveal any intermediate bands upon electrophoresis. We predicted a similar result for <u>Xenopus epitropicalis</u> (2n = 40) and this proved to be the case (7). To our knowledge, this is the first instance of a sperm protein profile being successfully predicted with regard to an evolutionary trend. In addition, the fact that sperm nuclei of different species in the genus <u>Xenopus</u> look similar but have different sperm protein profiles (26) rules out the possibility that these basic proteins determine the shape of the nucleus (27), at least amongst these anurans. It will be interesting to find out whether the electrophoretically distinguishable intermediate sperm proteins present in some species of <u>Xenopus</u> are multiples of a basic peptide unit, as appears to be the case in the protamines of certain molluscs (28).

In the newts and salamanders (order Urodela), the story is one of constancy of sperm basic protein type, rather than variability, as in the frogs. <u>Notophthalmus viridescens</u>, the eastern red spotted newt, has a type 1 protamine in the sperm according to cytochemical tests (Figure 1), but the fast moving testis-specific band moves more slowly than trout protamine upon electrophoresis (Figure 2) and closer to the type 3 intermediate protein of <u>Bufo</u> (29). <u>Pleurodeles waltl</u> shows similar cytochemical properties (30) as do <u>Taricha torosa</u>, <u>Plethodon vehiculum</u>, and <u>Cynops pyrrhogaster</u> (Kasinsky and Mann, in preparation). The electrophoretic profiles of <u>Pleurodeles waltl</u>, <u>Ambystoma</u>, <u>Salamandra</u>, and <u>Euproctes</u> (31) are similar to each other and to that of <u>Notophthalmus</u> (29). During spermiogenesis, a transition from somatic histones to type 2 keratinous protamines in spermatids to type 1 protamines in the sperm takes place in both <u>Notophthalmus</u> (29) and <u>Pleurodeles</u> (30). Bedford and Calvin (32) utilized disulfide reducing reagents and detergents to arrive at a similar conclusion. Bloch (1, 2) might have been looking at <u>Triturus viridescens</u> spermatids rather than mature sperm in classifying these as type 2 keratinous protamines.

In the class Amphibia, then, we see an evolutionary pattern similar to that in fish. Cartilaginous fish display little variability in sperm protein type; bony fish are very diverse in this respect. Frogs, like bony fish, have many different kinds of sperm basic chromosomal proteins. These proteins appear to differ between genera and are species-specific within genera in some instances. Urodeles, like cartilaginous fish, display little variability in their sperm protein type. In the concluding section of this paper we will attempt to indicate where an explanation for these phenomena might be sought.

CLASS REPTILIA

In the order Squamata, the sperm basic proteins of snakes and lizards appear to be very similar both electrophoretically (4), with several proteins migrating in the vicinity of <u>Bufo</u> type 3 intermediate protamine (Figure 2), and with respect to their amino acid compositions (7b), containing histidine and lysine as well as arginine (Kasinsky, Mann, Huang, Coyle, Fabre, and Byrd, in preparation). They, therefore, appear to be intermediate type 3 sperm proteins. Cytochemically, four represen-

340

tatives of snakes in the suborder Serpentes fall into the intermediate type 3B category (Figure 1) rather than the type 1 protamine category for the snake <u>Natrix</u> <u>natrix</u> as reported by Bloch (1, 2). Kharchenko et al. (33) also observe that sperm proteins of the grass snake <u>Natrix</u> migrate close to those of <u>Bufo</u>. However, three lizards in the suborder Lacertilia show type 1 protamine staining in the mature sperm after passing through a type 3A intermediate protein transition (Kasinsky et al., in preparation). Bloch (1, 2) also observes type 1 protamines in the lizard <u>Holbrookia</u> <u>texana</u>. Why the cytochemical and biochemical results are reconciled for snakes but not for lizards cannot be explained at this time. The overall impression, however, is one of similarity of sperm basic proteins amongst snakes and also amongst lizards.

Turtles (order Testudines) show cytochemical staining that places their sperm basic proteins in the type 3A category (Figure 1). Electrophoretically, they show a principal fast moving band in the middle of the triplet displayed for reptiles (Figure 2) (Kasinsky et al., in preparation). Sperm basic proteins extracted from the epididymis of the Mississippi alligator (order Crocodylia) show three principal bands migrating close to that of the turtle. The amino acid composition suggests that, like turtle sperm proteins, these are the intermediate type of proteins containing histidine, lysine, and arginine. Both alligator and turtle sperm proteins are somewhat higher in arginine composition than the proteins of Squamata. Thus, in reptiles as a class, the evolutionary trend is towards relative constancy of sperm basic protein type, rather than variability.

CLASS AVES

The type 1 protamine of rooster (<u>Gallus</u> <u>domesticus</u>, order Galliformes) has been sequenced by Nakano et al. (34a). An electrophoretically similar protein can also be extracted from the ductus deferens of the mallard, <u>Anas</u> <u>platyrhynchos</u> (order Anseriformes) (34b). Because of an active acrosomal protease, characterization of intact bird protamines is difficult (35). However, Mezquita and colleagues (see his chapter in this volume) have succeeded in characterizing the biochemical transitions associated with histone changes during spermiogenesis in the rooster. Examination of five species of birds, representing four avian orders (Kasinsky and Mann, in preparation), shows that, cytochemically, avian sperm proteins can be classified as type 1 protamines (Figure 1). Once again, it is the relative constancy of sperm basic proteins in this class that stands out, although the total number of bird species examined is too low to come to a firm conclusion as yet.

CLASS MAMMALIA

Here too the sperm proteins show relative constancy with one important exception: seven species of mammals in the infraclass eutheria display a type 2 keratinous protamine containing cystine and arginine and, more occasionally, lysine and histidine (36). However, two mammals of the infraclass Metatheria, the oppossum _Didelphis_ _virginiana_ and the marsupial rat _Smithopsis_ _crassacaudata_, cytochemically (1) show type 1 protamines without cystine, similar to the sperm histones of birds (Figure 1). Preliminary data (Kasinsky and Johnstone, unpublished) indicate that a single protamine can be extracted from both the testis and the epididymis of _Antechinus_ _stuarti_, a brown marsupial mouse representing Mammalia in Figure 2. This protein has somewhat slower mobility than that of mouse protamine (36).

CONCLUSION

Several evolutionary trends in the diversity of sperm basic chromosomal proteins in the vertebrates are apparent from this analysis. First, there is a macroevolutionary trend from relative constancy of sperm proteins in cartilaginous fish to great variability in bony fish and frogs to relative constancy within the reptiles, birds, and mammals. Thus, particular sperm basic proteins do not appear sporadically in vertebrate phylogeny. Most of the variability is present in particular orders of bony fish and frogs and here sperm proteins of the type 1 protamines, type 3 intermediate proteins, and type 4 somatic-like histones are to be found. Why might this be the case? The sperm basic

TABLE 1. Occurrence of internal fertilization, sex chromosomes, and sperm protein diversity in the vertebrates.

	Chondrichthyes	Osteichthyes	Anura	Urodela	Reptilia	Aves	Mammalia
Sperm Basic Protein Diversity	-	+	+	-	-	-	-
Internal Fertilization[1]	+	rare	very rare	+	+	+	+
Appearance of Optimum Sperm Motility[1]	excurrent duct	testis	testis	duct	duct	duct	duct
Polyploidy[2]	?	+[3]	+[4]	rare	rare	-	-
Heteromorphic Sex Chromosomes[2]	?	rare	rare	rare	+	+	+

[1] from reference 38
[2] from references 44, 45
[3] from reference 39
[4] from reference 46

proteins in these taxa are probably under less selective pressure than they are in cartilaginous fish, reptiles, birds, and mammals. Bloch (1, 2) and Subirana (37) have suggested that the significance of these sperm-specific proteins may lie simply in their ability to condense the sperm chromatin and provide protection for the male genome, roles which may be rather unspecific. The only requirement of the sperm proteins appears to be their basicity. The trend away from sperm protein diversity towards a more narrow range of arginine-rich or arginine- and cystine-rich proteins in the higher vertebrates appears to correlate with the occurrence of internal fertilization and the appearance of heteromorphic sex chromosomes in vertebrate evolution (4).

A study of post-testicular sperm maturation has shown that, in teleosts and anurans, optimal sperm motility is attained in the testis. The sperm of elasmobranchs, urodeles, reptiles, birds, and mammals require a period of maturation in the excurrent ducts before becoming optimally motile (38). The necessity for post-testicular maturation appears to coincide with the situation in which internal fertilization has been adopted by these taxa, precisely those vertebrate groups which show relative constancy of sperm protein type (Table 1). Thus, in contrast to Bloch's (2) view, there does appear to be a correlation between the relative constancy of sperm basic proteins and animal groups where internal fertilization is the general rule. In teleosts and anurans, where fertilization is generally external, sperm proteins may have diverged in response to environmental factors and perhaps partly in response to the effects of genetic drift on proteins with an unspecific function. By contrast, the internal environment of the female tract may have placed certain selective constraints on the evolution of the sperm basic proteins in animals with internal fertilization.

In order to explain the variability of highly arginine-rich sperm basic proteins, Bloch (2) proposed that, in organisms with chromosomally-based sex-determination, some of the histone genes might have been freed from the selective restraints which ordinarily cause histones to be evolutionarily conservative. By becoming localized on the heteromorphic regions of the sex chromosomes, these genes would be protected from crossing over and point-mutations might accumulate. Since the only requirement of sperm proteins appears to be their basicity and since there are more codons for arginine than for lysine or histidine, the genetic code might provide a "statistical trap." Thus, genetic drift would be the motivating force behind the accumulation of arginine in these proteins.

Kasinsky et al. (4) proposed that the trend toward relative constancy of sperm basic proteins might be explained by a modification of Bloch's (1, 2) original hypothesis. We suggested that the amount of sperm protein variability in a given class might be related to the degree to which sex-determination is chromosomally based. As Ohno (39) has pointed out, it is with the reptiles that nature ceased its experimentation with polyploidization, as seen by the appearance of heteromorphic sex chromosomes that determine sex-differentiation. Heteromorphic sex chromosomes are known to be present in certain species of lizards and then appear more regularly in snakes, birds, and mammals (40). Rare cases of sex chromosome heteromorphism have also been noted among the urodeles (41a). With some exceptions (41b), fish and anurans do not generally display heteromorphic sex chromosomes and sex reversal is still possible by hormone treatment (42). Numerous cases of polyploidy are also observed in fish and frogs (43-46). It may not be simply a coincidence that in organisms where sex-determination is less chromosomally based (fish and anurans), we see diversity of sperm protein types, whereas in reptiles, birds, and mammals, with their chromosomal sex-determination, we see only the arginine-rich types of sperm proteins (Table 1). According to Bloch's (1, 2) hypothesis, one would expect to see an accumulation of arginine if the genes coding for sperm basic proteins had become located on the sex chromosomes. The appearance of cysteine in the mammalian protamines could be attributed to a simple point mutation from an arginine codon and its significance might be related to the fact that the mammalian sperm head must be highly stabilized in order to get through the very thick zona pellucida of the ovum in eutherian mammals (32). As we noted earlier (4), this hypothesis might be testable by isolation of protamine mRNA from reptiles, birds, or mammals and localization of the sperm protein genes on the sex chromosomes by in situ hybridization. The recent isolation of a mammalian protamine mRNA by Krawetz and Dixon (47) makes this experiment a distinct possibility. Nevertheless, the overall conclusion is that sperm protein evolution has not been entirely sporadic. Sperm basic proteins might be more variable precisely in those species of bony fish and frogs that are not constrained by internal fertilization and chromosomally-based sex-determination typified by heteromorphic sex chromosomes (Table 1).

While these considerations may help to explain the overall macroevolutionary trend in vertebrate sperm protein diversity, they still do not account for the species-specific sperm basic proteins in particular anurans, such as the genus Xenopus, or in bony fish like the sticklebacks. To explain these findings it may be necessary to compare the timing of the onset of sperm protein synthesis in different species of a single genus. Gould (48) has developed a clock model to explain heterochrony, or "changes in the relative time of appearance and rate of development for characters already present in ancestors." Such a model emphasizes that changes in developmental timing produce parallels between the stages of ontogeny and phylogeny. Bloch (1) has suggested that we might examine such relationships between ontogeny and phylogeny to search out clues for sperm protein variability.

Perhaps we can apply a clock model similar to Gould's (48), but at the molecular level, to the genus Xenopus. In Figure 3, above such a clock is the pathway for sperm protein changes in development: somatic-like sperm histones to intermediate type sperm proteins to protamines to keratinous protamines. Both the 2n = 20, 40 polyploid line represented by Xenopus tropicalis and the 2n = 36, 72, 108 polyploid line starting with Xenopus laevis and Xenopus borealis have diverged from a common ancestor which may have had 2n = 24, judging by the diploid chromosome numbers of closely related primitive frog genera (44, 45). Let us assume that tropicalis has more closely maintained the ontogenic pathway of that ancestor (49) so that, on a relative basis, laevis and borealis have shifted their age of first maturity ("puberty") to a longer interval after fertilization. This means that tropicalis, closer to the ancestral form, synthesizes a set of somatic-like sperm histones to produce a functional sperm. Therefore, both the "genus hand" and the "species hand"

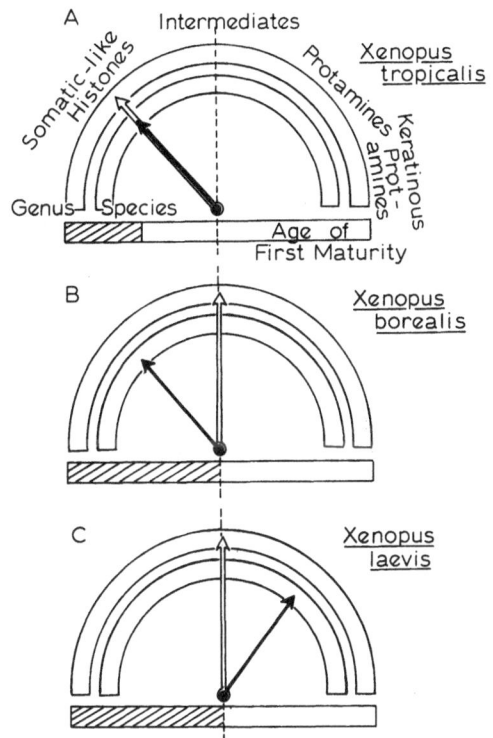

Figure 3. A speculative model based on the heterochrony clock of Gould (1977) to explain sperm basic chromosomal protein variability in the polyploid genus Xenopuls.

of the clock in the Figure 3 are set to the left. In <u>borealis</u> and <u>laevis</u>, less like the ancestral form, the longer time interval to sexual maturity means that these organisms utilize genes for intermediate type proteins to produce functional sperm. They both have their "genus hand" facing up in Figure 3. The "species hand" varies somewhat since, electrophoretically, <u>laevis</u> has a somewhat different sperm protein profile than <u>borealis</u> (7a, 7b).

The essence of this proposal is that individuals in a species select their particular set of sperm basic chromosomal proteins from the entire family of sperm protein genes phylogenetically available in that genus depending on the timing of sexual maturity. As the sperm basic proteins are relatively non-specific and any one of a variety of basic protein sets are suitable to incorporate into a functional sperm, the particular set of proteins chosen is due to other features of the organism's development. The protein set is species-specific not because it has been intensely selected but because it reflects the time after fertilization at which the organism terminates the possible ontogenetic pathway for sperm basic protein production and actually synthesizes a functional sperm. Such a proposal predicts that Xenopus tropicalis has a set of intermediate sperm protein genes similar to those of laevis, borealis, and other species in the genus Xenopus but does not express them. This possibility is open to experimental verification, as is the determination of relative rates for the onset of sexual maturity in the two polyploid lines of the genus. From a phylogenetic point of view, therefore, it may be possible to understand why sperm basic protein evolution in the vertebrates, although far from being as conservative as that of the nucleosomal histones, has not been a completely sporadic phenomenon.

ACKNOWLEDGMENTS

We are pleased to acknowledge the helpful comments offered by Professor Juan Subirana. We also thank NSERC of Canada for operating and equipment grants that supported the studies cited in this review as originating from our laboratory, as well as for a travel grant to H.E.K.

REFERENCES

1. BLOCH, D.P. (1976). Histones of sperm. In: "Handbook of Genetics," Vol. 5 (R.C. King, ed.) pp. 136-167. Plenum Press, New York.

2. BLOCH, D.P. (1969). A catalog of sperm histones. Genetics (suppl.) 61, 93-111.

3. SUBIRANA, J.A. (1983). Nuclear proteins in spermatozoa and their interactions with DNA. In: "The Sperm Cell," (J. Andre, ed.) pp. 197-213. Martinus Nijhoff, The Hague.

4. KASINSKY, H..E., HUANG, S.Y., KAWUK, S., MANN, M., SWEENEY, M.A.J., and YEE, B. (1978). On the diversity of sperm histones in the vertebrates. III. Electrophoretic variability of testis-specific histone patterns in Anura contrasts with relative constancy in Squamata. J. Exp. Zool. 203, 109-126.

5. KASINSKY, H.E., HUANG, S.Y., MANN, M., ROCA, J., and SUBIRANA, J.A. (1985). On the diversity of sperm histones in the vertebrates. IV. Cytochemical and amino acid analysis in Anura. J. Exp. Zool. 234, 33-46.

6. BOLS, N.C. and KASINSKY, H.E. (1973). An electrophoretic comparison of histones in anuran testes. Can. J. Zool. 51, 203-208.

7a. MANN, M., RISLEY, M.S., ECKHARDT, R.A., and KASINSKY, H.E. (1982). Characterization of spermatid/sperm basic chromosomal proteins in the genus Xenopus (Anura, Pipidae). J. Exp. Zool. 222, 173-186.

7b. KASINSKY, H.E. (1985). Why are sperm histones more variable in frogs than in salamanders, snakes, and lizards? Biologia Molecular: Segones Jornades, Societat Catalana de Biologia, 34-40.

8. VILLEE, C.A., WALKER, Jr. C.A., and BARNES, R.D. (1973). "General Zoology," fourth edition. W.B. Saunders, Philadelphia.

9. DODD, J.M. (1983). Reproduction in cartilaginous fishes (Chondrichthyes). In: "Fish Physiology," Vol. IX, Part A (W.S. Hoar, D.J. Randall, and E.M. Donaldson, eds.) pp. 31-95.

10. BOLS, N.C., BOLISKA, S.A., RAINVILLE, J.B., and KASINSKY, H.E. (1980). Spermiogenesis in the longnose skate and the spiny dogfish. J. Exp. Zool. 212, 423-433.

11. BOLS, N.C. and KASINSKY, H.E. (1976). On the diversity of sperm histones in the vertebrates. II. A cytochemical study of basic protein transitions during spermiogenesis in the cartilaginous fish Hydrolagus colliei. J. Exp. Zool. 198, 109-114.

12. BOLS, N.C. and KASINSKY, H.E. (1974). Cytochemistry of sperm histones in three cartilaginous fish. Can. J. Zool. 52, 437-439.

13. GUSSE, M. and CHEVAILLIER, P. (1981). Microelectrophoretic analysis of basic protein changes during spermiogenesis in the dogfish Scylliorhinus caniculus(L). Exp. Cell Res. 136, 391-397.

14. GUSSE, M. and CHEVAILLIER, P. (1978). Etude ultrastructurale et chimique de la chromatine au cours de la spermiogenese de la roussette Scyliorhinus caniculus(L). Cytobiologie 16, 421-443.

15. CHEVAILLIER, P. (1983). Some aspects of chromatin organization in sperm nuclei. In: "The Sperm Cell," (J. Andre, ed.) pp. 179-196. Martinus Nijhoff, The Hague.

16. KOSSEL, A. (1928). "The Protamines and Histones." Longmans Green, London.

17. STANLEY, H.P., KASINSKY, H.E., and BOLS, N.C. (1984). Meiotic chromatin diminution in a vertebrate, the holocephalon fish Hydrolagus colliei (Chondrichthyes, Holocephali). Tissue and Cell 16, 203-215.

18. DIXON, G.H. (1974). The basic proteins of trout testis chromatin: Aspects of their synthesis, post-synthetic modifications, and binding to DNA. Karolynska Symp. on Res. Meth. in Reprod. Biol. 5, 130-154.

19. MOYLE, P.B. and CECH, Jr. J.J. (1982). "Fishes: An Introduction to Ichthyology." Prentice Hall, New Jersey.

20. MUNOZ-GUERRA, S., AZORIN, F., CASAS, M.T., MARCET, X., MARISTANY, M.A., ROCA, J., and SUBIRANA, J.A. (1982). Structural organization of sperm chromatin from the fish Carassius auratus. Exp. Cell Res., 137, 47-53.

21. ALDER, D. and GOROVSKY, M.A. (1975). Electrophoretic analysis of liver and testis histones of the frog Rana pipiens. J. Cell Biol. 64, 389-397.

22. BOLS, N.C. and KASINSKY, H.E. (1972). Basic protein composition of anuran sperm: a cytochemical study. Can. J. Zool. 50, 171-177.

23. TYMOWSKA, J. (1977). A comparative study of the karyotypes of eight Xenopus species and subspecies possessing a 36-chromosome complement. Cytogen. Cell Genet. 18, 165-181.

24. TYMOWSKA, J. and FISCHBERG, M. (1973). Chromosome complements of the genus Xenopus. Chromosoma 44, 335-342.

25. FISCHBERG, M., COLOMBELLI, B., and PICARD, J.J. (1982). Diagnose preliminaire d'une espece naturelle de Xenopus du Zaire. Alytes 1, 53-55.

26. RISLEY, M.S., ECKHARDT, R.A., MANN, M., and KASINSKY, H.E. (1982). Determinants of nuclear shaping in the genus Xenopus. Chromosoma 84, 557-569.

27. FAWCETT, D.W., ANDERSON, W.A., and PHILLIPS, D.M. (1971). Morphogenetic factors influencing the shape of the sperm head. Dev. Biol. 26, 220-251.

28. COLOM, J. and SUBIRANA, J.A. (1979). Protamines and related proteins from spermatozoa of molluscs. Characterization and molecular weight determination by gel electrophoresis. Biochim. Biophys. Acta 581, 217-227.

29. BOLS, N.C., BYRD, E.W., Jr., and KASINSKY, H.E. (1976). On the diversity of the sperm histones in the vertebrates. I. Changes in basic proteins during spermiogenesis in the newt Notophthalmus viridescens. Differentiation 7, 31-38.

30. PICHERAL, B. (1970). Nature et evolution des proteines basiques au cours de la spermiogenese chez Pleurodeles waltl Michah, amphibien urodele. Histochemie 23, 189-206.

31. PICHERAL, B. (1979). Structural, comparative, and functional aspects of spermatozoa in urodeles. In: "The Spermatozoon," (D.W. Fawcett and J.M. Bedford, eds.) pp. 267-287. Urban and Schwarzenberg, Baltimore.

32. BEDFORD, J.M. and CALVIN, H.I. (1974). The occurrence and possible functional significance of -S-S- crosslinks in sperm heads with particular reference to eutherian mammals. J. Exp. Zool. 188, 137-156.

33. KHARCHENKO, E.P., NALIVAEVA, N.N., and SOKOLOVA, T.V. (1980). Heterogeneity of cationic proteins of the chromatin of various tissues. Biokhimiya 45, 1630-1638 (in Russian). English translation in 1981, Plenum Publishing Corp.

34a. NAKANO, M., TOBITA, T., and ANDO, T. (1976). Studies on a protamine (galline) from fowl sperm. 3. The total amino acid sequence of the intact galline molecule. Int. J. Pept. Prot. Res. 8, 565-578.

34b. CHIVA, M., KASINSKY, H.E., and SUBIRANA, J.A. (1985). Caracteritzacio de protamines d'aus. I. Proteins testiculars d'anec (Anas platyrhynchos). Biologia Molecular: Segones Jornades, Societat Catalana de Biologia, 64-65.

35. MEZQUITA, C. and TENG, C.S. (1977). Studies on sex organ development: changes in nuclear and chromatin composition and genomic activity during spermiogenesis in the maturing rooster testis. Biochem. J. 164, 99-111.

36. BELLVE, A.R. (1979). The molecular biology of mammalian spermatogenesis. In: "Oxford Reviews of Reproductive Biology," Vol. 1 (C.A. Finn, ed.) pp. 159-261. Clarendon Press, Oxford.

37. SUBIRANA, J.A. (1975). On the biological role of basic proteins in spermatozoa and during spermiogenesis. In: "The Biology of the Male Gamete," (J.G. Duckett and P.A. Racey, eds.) pp. 239-244. Academic Press, New York.

38. BEDFORD, J.M. (1979). Evolution of the sperm maturation and sperm storage functions of the epididymis. In: "The Spermatozoon," (D.W. Fawcett and J.M. Bedford, eds.) pp. 7-21. Urban and Schwarzenberg, Baltimore.

39. OHNO, S. (1969). The role of gene duplication in vertebrate evolution. In: "The Biological Basis of Medicine," Vol. 4 (E.E. Bittar, ed.) pp. 109-132. Academic Press, New York.

40. BECAK, M.L., BECAK, W., CHEN, T.R., and SCHOFFNER, R.M. (1975). "Chromosome Atlas: Fish, Amphibians, Reptiles and Birds," Vol. 3, Springer Verlag, New York.

41a. SCHMID, M., OLERT, J., and KLETT, C. (1979). Chromosome banding in amphibia. III. Sex chromosomes in Triturus. Chromosoma 71, 29-55.

41b. SCHMID, M., HAAF, T., GEILE, and SIMS, S. (1983). Unusual heteromorphic sex chromosomes in a marsupial frog. Experientia 39, 1153-1155.

42. REINBLOTH, R. (1975). "Intersexuality in the Animal Kingdom." Springer-Verlag, New York.

43. OHNO, S., WOLF, S., and ATKIN, N.B. (1968). Evolution from fish to mammals by gene duplication. Hereditas 59, 179-193.

44. MORESCALCHI, A. (1973). Amphibia. In: "Cytotaxonomy and Vertebrate Evolution," (A.B. Chiarelli and E. Capanna, eds.) pp. 233-348. Academic Press, New York.

45. MORESCALCHI, A. (1977). Phylogenetic aspects of karyological evidence. In: "Major Patterns in Vertebrate Evolution," (M.K. Hecht, P.C. Goody, and B.M. Hecht, eds.) pp. 149-167. Plenum Press, New York.

46. BOGART, J.P. (1980). Evolutionary implications of polyploidy in amphibians and reptiles. In: "Polyploidy: Biological Relevance," (W.H. Lewis, ed.) pp. 341-378. Plenum Press, New York.

47. KRAWETZ, S.A. and DIXON, G.H. (1984). Isolation and in vitro translation of a mammalian protamine messenger RNA. Bioscience Rep. 4, 593-604.

48. GOULD, S.J. (1977). "Ontogeny and Phylogeny." Belknap Press, Harvard. Cambridge, Mass.

49. MORESCALCHI, A. (1979). New developments in vertebrate cytotaxonomy. I. Cytotaxonomy of the amphibians. <u>Genetica</u> <u>50</u>, 179-193.

CONTRIBUTORS

AIKEN, J.M., Department of Medical Biochemistry, Health Sciences Centre, University of Calgary, 3330 Hospital Drive N.W., Calgary, Alberta T2N 4N1 CANADA

ARENDES, J., Institut für Physiologische Chemie, Johannes-Gutenberg-Universität, D-6500 Mainz, WEST GERMANY

AZORIN, F., Departament Quimica Macromolecular, E.T.S.I.I.B., Diagonal 999, 08028 Barcelona, SPAIN

BALDWIN, J., Department of Physics, Liverpool Polytechnic, Byrom Street, Liverpool L3 3AF UK

BEATO, M., Institut für Physiologische Chemie, der Phillipps-Universität, Deutschhausstrasse 1-2, D-3550 Marburg, FRG

CATO, A.C.B., Institut für Physiologische Chemie, der Phillipps-Universität, Deutschhausstrasse 1-2, D-3550 Marburg, FRG

CRANE-ROBINSON, C., Biophysics Laboratories, Portsmouth Poly-technic, St. Michael's Building, White Swan Road, Portsmouth PO1 2DT, Hants UK

DAVIES, S.G., Dyson Perrins Laboratory, South Parks Road, Oxford, OX1 3QU ENGLAND

DIXON, G.H., Department of Medical Biochemistry, Health Sciences Centre, University of Calgary, 3330 Hospital Drive N.W., Calgary, Alberta T2N 4N1 CANADA

EARNSHAW, W.C., Department of Cell Biology and Anatomy, Johns Hopkins School of Medicine, 725 North Wolfe Street, Baltimore, Maryland 21205 USA

ELGIN, S.C.R., Department of Biology, Washington University, St. Louis, Missouri 63130 USA

GOODWIN, G.H., Chester Beatty Laboratories, Institute of Cancer Research, Fulham Road, London, SW3 6JB UK

HEGUY, A., Department of Microbiology, University of Southern California, School of Medicine, 2011 Zonal Avenue, Los Angeles, California 90033 USA

HOLTLUND, J., Department of Biochemistry, University of Oslo, Oslo, NORWAY

HUANG, S.-Y., Department of Zoology, University of British Columbia, Vancouver, B.C. V6T 2A9 CANADA

JANICH, S., Institut für Physiologische Chemie, der Phillipps-Universität, Deutschhausstrasse 1-2, D-3550 Marburg, FRG

JANKOWSKI, J.M., Department of Medical Biochemistry, Health Sciences Centre, University of Calgary, 3330 Hospital Drive N.W., Calgary, Alberta T2N 4N1 CANADA

JOSE, M., Institut de Biologia de Barcelona del CSIC, Jordi Girona Salgado, 18, 08034 Barcelona, SPAIN

KARIN, M., Department of Microbiology, University of Southern California, School of Medicine, 2011 Zonal Avenue, Los Angeles, California 90033 USA

KASINSKY, H.E., Department of Zoology, University of British Columbia, Vancouver, B.C. V6T 2A9 CANADA

KORANT, B.D., Central Research & Development Department, Experimental Station, E.I. du Pont de Nemours & Company, Inc., Wilmington, Delaware 19898 USA

KRAUTER, P., Institut für Physiologische Chemie, der Phillipps-Universität, Deutschhausstrasse 1-2, D-3550 Marburg, FRG

LALAND, S.G., Department of Biochemistry, University of Oslo, Oslo, NORWAY

LASTERS, I., Labo Algemene Biologie, Vrije Universiteit Brussel, Paardenstraat 65, 1640 Sint Genesius Rode, Brussels, BELGIUM

LEMKE, M., Department of Zoology, University of British Columbia, Vancouver, B.C. V6T 2A9 CANADA

LUND, T., Department of Biochemistry, University of Oslo, Oslo, NORWAY

MANN, M., Department of Zoology, University of British Columbia, Vancouver, B.C. V6T 2A9 CANADA

McKENZIE, D.I., Department of Medical Biochemistry, Health Sciences Centre, University of Calgary, 3330 Hospital Drive N.W., Calgary, Alberta T2N 4N1 CANADA

MEZQUITA, C., Department of Physiology, Faculty of Medicine, University of Barcelona, Casanova, 143, 08036 Barcelona, SPAIN

MOIR, R., Department of Medical Biochemistry, Health Sciences Centre, University of Calgary, 3330 Hospital Drive N.W., Calgary, Alberta T2N 4N1 CANADA

MUYLDERMANS, S., Labo Algemene Biologie, Vrije Universiteit Brussel, Paardenstraat 65, 1640 Sint Genesius Rode, Brussels, BELGIUM

NAHON, J.-L., Laboratoire d'Enzymologie de C.N.R.S., 91190 Gif-sur-Yvette, FRANCE

NICOLAS, R.H., Chester Beatty Laboratories, Institute of Cancer Research, Fulham Road, London, SW3 6JB UK

PUIGDOMÈNECH, P., Institut de Biologia de Barcelona del CSIC, Jordi Girona Salgado, 18, 08034 Barcelona, SPAIN

REECK, G.R., Department of Biochemistry, Kansas State University, Manhattan, Kansas 66506 USA

RICH, A., Department of Biology, Massachusetts Institute of Technology, Cambridge, Massachusetts 02139 USA

SALA-TREPAT, J.M., Laboratoire d'Enzymologie de C.N.R.S., 91190 Gif-sur-Yvette, FRANCE

SCHEIDEREIT, C., Institut für Physiologische Chemie, der Phillipps-Universität, Deutschhausstrasse 1-2, D-3550 Marburg, FRG

SCHOFIELD, P.N., Department of Zoology, University of Oxford, South Parks Rd., Oxford, OX1 3PS, ENGLAND

SCHÜMPERLI, D., Institut für Molekularbiologie II, Universität Zürich, ETHZ-Hönggerberg, 8093 Zürich, SWITZERLAND

SIEGFRIED, E., Department of Biology, Washington University, St. Louis, Missouri 63130 USA

STATES, J.C., Department of Medical Biochemistry, Health Sciences Centre, University of Calgary, 3330 Hospital Drive N.W., Calgary, Alberta T2N 4N1 CANADA

SUSKE, G., Institut für Physiologische Chemie, der Phillipps-Universität, Deutschhausstrasse 1-2, D-3550 Marburg, FRG

THOMAS, G.H., Department of Biology, Washington University, St. Louis, Missouri 63130 USA

TOULMÉ, J.-J., Laboratoire de Biophysique, INSERM U.201, Muséum National d'Histoire Naturelle, 61, Rue Buffon, F-75005 Paris, FRANCE

VON DER AHE, D., Institut für Physiologische Chemie, der Phillipps-Universität, Deutschhausstrasse 1-2, D-3550 Marburg, FRG

WASYLYK, B., Laboratoire de Génétique Moléculaire des Eucaryotes de CNRS, Unité 184 de Biologie Moléculaire et de Génie Génétique de l'INSERM, 11, rue Humann, 67085 Strasbourg-Cedéx, FRANCE

WESTPHAL, H.M., Institut für Physiologische Chemie, der Phillipps-Universität, Deutschhausstrasse 1-2, D-3550 Marburg, FRG

WRIGHT, C.A., Chester Beatty Laboratories, Institute of Cancer Research, Fulham Road, London, SW3 6JB UK

WYNS, L., Labo Algemene Biologie, Vrije Universiteit Brussel, Paardenstraat 65, 1640 Sint Genesius Rode, Brussels, BELGIUM

ZAVOU, S., Chester Beatty Laboratories, Institute of Cancer Research, Fulham Road, London, SW3 6JB UK